Soda Science

∴

Soda Science

∵

MAKING THE WORLD SAFE
FOR COCA-COLA

Susan Greenhalgh

THE UNIVERSITY OF CHICAGO PRESS
CHICAGO AND LONDON

The University of Chicago Press, Chicago 60637
The University of Chicago Press, Ltd., London
© 2024 by The University of Chicago
Published 2024
Printed and bound by CPI Group (UK) Ltd, Croydon, CR0 4YY

33 32 31 30 29 28 27 26 25 24   1 2 3 4 5

ISBN-13: 978-0-226-82914-2 (cloth)
ISBN-13: 978-0-226-83473-3 (paper)
ISBN-13: 978-0-226-82915-9 (e-book)
DOI: https://doi.org/10.7208/chicago/9780226829159.001.0001

Library of Congress Cataloging-in-Publication Data

Names: Greenhalgh, Susan, author.
Title: Soda science : making the world safe for Coca-Cola / Susan Greenhalgh.
Description: Chicago : The University of Chicago Press, 2024. | Includes
    bibliographical references and index.
Identifiers: LCCN 2023058493 | ISBN 9780226829142 (cloth) |
    ISBN 9780226834733 (paperback) | ISBN 9780226829159 (ebook)
Subjects: LCSH: Cola drinks—Marketing—Moral and ethical aspects—United
    States. | Coca Cola (Trademark)—Marketing—Moral and ethical aspects. |
    Research, Industrial—Moral and ethical aspects—United States. | Science
    and industry—Moral and ethical aspects—United States. | Pseudoscience—
    United States. | Cola drinks—Health aspects. | Obesity—United
    States—Prevention. | Nutritionally induced diseases—United States. |
    Communication in public health—United States. | Cola drinks—China—
    Marketing. | Science and state—China.
Classification: LCC HD9349.S634 C63366 2024 | DDC 338.4/
    7663620973—dc23/eng/20240131
LC record available at https://lccn.loc.gov/2023058493

♾ This paper meets the requirements of ANSI/NISO Z39.48-1992
(Permanence of Paper).

Contents

Figures and Tables

Figures

Tables

Abbreviations

ACSM	American College of Sports Medicine
AND	Academy of Nutrition and Dietetics
AOM	America on the Move
ASN	American Society for Nutrition
BMI	body mass index
CAPM	Chinese Academy of Preventive Medicine (forerunner to the Chinese CDC)
CDC	Centers for Disease Control and Prevention (US); Center for Disease Control and Prevention (China)
CHP	Center for Health Promotion, unit within ILSI
CSPI	Center for Science in the Public Interest
EB	energy balance
EBAL	Committee on Energy Balance and Active Lifestyle, technical committee within ILSI-NA
EIM	Exercise Is Medicine
FAO	Food and Agriculture Organization (division of the United Nations)
GEBN	Global Energy Balance Network
ICN	International Congress of Nutrition
IFIC	International Food Information Council
ILSI	International Life Sciences Institute
ILSI-CHP	ILSI Center for Health Promotion
ILSI-NA	International Life Sciences Institute, North America Branch
JAMA	*Journal of the American Medical Association*
MOH	Ministry of Health of the People's Republic of China (1954–2013), superseded by the National Health and Family Planning Commission (2013–18) and the National Health Commission (2018–present)
NCD	noncommunicable disease (also known as chronic disease)
NGO	nongovernmental organization

NIH	National Institutes of Health (US)
NWCR	National Weight Control Registry
UCSFL	University of California San Francisco Library (home of Food Industry Documents collection)
UN	United Nations
UNICEF	United Nations Children's Fund (division of the United Nations)
WHO	World Health Organization (division of the United Nations)

What Is Soda Science, and Why Does It Matter?

The age of obesity has arrived; so declared the cover of *Newsweek* on July 3, 2000, over a photo of a big-bellied boy downing a generous helping of ice cream (figure 0.1).[1] "Six million kids are seriously overweight," the text said, warning they could be "fat for life." A year later, the US surgeon general made it official, warning starkly that "overweight and obesity have reached nationwide epidemic proportions." Displaying graphs of excess weight rising rapidly in every age group, in this first-ever Call to Action on obesity, he urged Americans to recognize overweight and obesity as major public health problems and to make the fight against excess body fat part of our daily routines.[2] For those of us who lived through them, few can forget those anxious, fearful years of national alarm about the obesity crisis, roughly 2000–2015, when an excess of pounds, long considered an aesthetic defect of individuals, was now defined by our government as a medical problem—an actual disease that leads to more serious diseases—and a public health emergency that was crippling the country.

The media were obsessed with fatness, and their gripping stories were hard to escape. The problem, we learned in article after article, was rooted in our lifestyle choices. "Our super-sized kids," *Time* wrote, showing a photo of a round-bodied boy, with an ice cream cone, on a skateboard that was bowed down in the middle because of his weight (figure 0.2). "An in-depth look at how our lifestyle is creating a juvenile obesity epidemic," it promised readers. How had we made our kids so heavy that they damaged their skateboards? The answer would become depressingly familiar. Starting in the 1970s, as the price of junk food and sugary drinks dropped, supersizing became the norm, and our intake of junk food soared, and as the lure of our televisions, video games, cell phones, and other toys kept us away from the gym, Americans of all ages began packing on the pounds.[3]

Our bulging bodies were harming the country and undermining its place in the world. Over a cover image of a jowly Statue of Liberty so big around she filled nearly the entire cover, the highbrow monthly *Atlantic* stamped

FIGURE 0.1. "Fat for Life?" (*Newsweek* cover)

this message: "Fat Nation: It's Worse Than You Think" (figure 0.3). With body sizes exaggerated to ridiculous effect and heads often left off in dehumanizing fashion, such images contained a grammar of morality that took no time to grasp. Lacking in willpower, unable to control their appetites, and too lazy to get to the gym, such images told us, fat people were morally flawed.[4] The message to everyone was unmistakable. If we wanted to avoid social shame, no matter how we did it, we had to avoid being fat.

Many of us needed to diet and exercise, that much was clear, but exactly *how* we didn't know. The media messages were confusing. Was dietary fat the enemy, as we had been told for years, or was it now our friend? Were carbs weight neutral, were they bad, or were they good? Every year, it seemed, the answers shifted. On June 5, 2017, after the intense anxiety and fear about weight had dulled a bit, the cover of *Time* read: "The Weight Loss Trap. Low Carb. Low Fat. Paleo. Vegan. Flexitarian. Why Your Diet Isn't Working." That was one message that rang true. The advice on exercise was

FIGURE 0.2. "Our Super-Sized Kids" (*Time* cover)

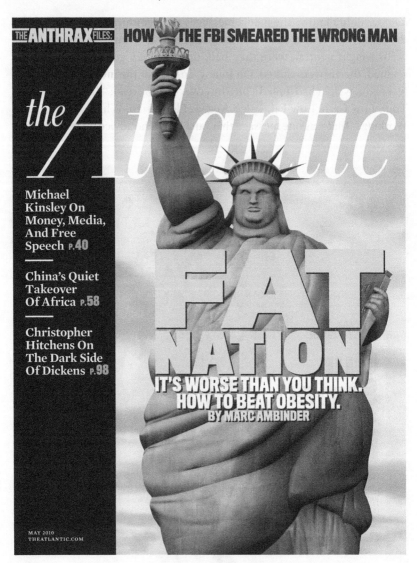

FIGURE 0.3. "Fat Nation" (*Atlantic* cover)

equally perplexing. In early 2010, First Lady Michelle Obama launched her celebrated "Let's Move" program to fight childhood obesity by encouraging kids to play more sports and get more exercise. That program seemed promising, but a few months earlier (on August 17, 2009), *Time* had unspooled "the myth about exercise." "Of course it's good for you," the magazine's cover lectured, "but it won't make you lose weight." So which is it? Does exercise take off the pounds, or is that just a myth? Even without clear and consistent messages, many Americans, worried about their health and feeling bad about themselves, did everything they could to lose unwanted pounds, putting all the weight-loss strategies available to the test.

Even as we tried this lifestyle tactic and that one, rarely with lasting success, the epidemic continued to worsen. By 2017–20, 73.5 percent of American adults were classified as overweight or obese. That hard-to-wrap-your-mind-around number was about 9 percentage points higher than the number in 1999–2000, when the US surgeon general had called us to action, and 17 percentage points higher than in the years 1988 to 1994, when the rise in excess weight was just coming into view.[5] Far from winning the battle, we as a country had lost significant ground.

Though the stinging stories about the obesity epidemic faded away after 2015, Americans continue to be obsessed with slimming down.[6] In any given year, half of American adults attempt to lose weight, largely because of worries about their health.[7] Over four-fifths of Americans see weight loss as their individual responsibility; not surprisingly, lifestyle modification is the primary approach. But which lifestyle behaviors should be modified? Confused as ever about what works, in a recent survey 63 percent of Americans tried exercising to shed pounds, while the same proportion tried eating less food. (Most used two or more methods.) Neither proved very helpful over the long run. Even the most extreme lifestyle interventions produced little lasting success. Contestants on the TV program *The Biggest Loser*, who ate a meager thirteen hundred kilocalories and engaged in 3.1 hours of vigorous exercise a day, lost substantial weight. Within six years, though, the majority had regained on average two-thirds of the pounds they had shed. The intense interventions needed to produce a large weight loss proved unsustainable.[8]

Over the last few decades, the US government has poured huge sums into building up a field of obesity research in hopes of finding ways to slow and then stop the rise in unhealthy weights. Despite the investment, a safe, effective, and affordable long-term solution that works for most people remains beyond reach, and obesity continues to climb.[9] What has gone wrong?

Big Food: Engineering Food Products and Obesity Science

While the media were publicizing the dangers of obesity, and many ordinary Americans were struggling to sort out changing messages about how to rid ourselves of unwanted pounds, the giant food companies were quietly enriching themselves selling us the addictive sugary, salty, and fat-laden junk foods and drinks that were making us fat. (From here on, I'll call all these just junk food.) The public health scholar Nicholas Freudenberg has described a dynamic—global in scope—in which the huge food and beverage companies have been actively promoting hyperconsumption by aggressively marketing these sickening foods and by making unhealthy food ubiquitous and inexpensive.[10] This unhealthy dynamic is part and parcel of the larger, corporate-dominated political and economic order that took shape in the 1970s and 1980s and prevailed until the late 2010s, when it began fracturing.[11] In this corporate order of things, generally called neoliberalism, large companies are among the most powerful forces in the world, markets are considered the best shapers of public policy, governments are expected to get out of the way of business by dismantling consumer-friendly regulations (among other things), and individuals—you and I and the kids on the covers of *Time* and *Newsweek*—are made responsible for our own health through our (or our parents') lifestyle choices and consumer practices, which we should optimize by becoming entrepreneurs of the self.[12] This is the larger context in which the modern obesity epidemic and humankind's failure to fix it became possible. These three figures—the all-powerful company, the weakened government, and the responsibilized consumer—will be with us throughout this book.

The latest thinking on why obesity rates have been steadily rising in the United States and globally points a finger directly at the world's largest food and beverage companies: Big Food. The high-level Lancet Commission on Obesity attributes soaring levels of obesity principally to an increased intake of energy-dense (high-fat, high-sugar), nutrient-poor foods and drinks marketed by the transnational ultra-processed food and beverage companies that dominate the global food system.[13] In his 2013 book *Salt, Sugar, Fat: How the Food Giants Hooked Us*, the investigative reporter Michael Moss documented how the food giants have been engaging in taste engineering, strategically manipulating the sugar, salt, and fat content of their processed foods and sugary drinks to create products so irresistible (the industry term is "alluring") that it's hard to stop consuming them.[14] *Hooked: How We Became Addicted to Processed Food*, his 2021 book, digs deeper, revealing how the industry is exploiting our biology and emotions to addict us to the processed foods that are slowly undermining our health.[15] What's become clear is that, in their relentless pursuit of shareholder profit, the giant food companies have been literally making us sick. The landmark clinical trials of Kevin Hall and his colleagues at the National Institutes of Health show that people who ate solely ultraprocessed foods consumed more than 508 extra calories per day, ate more carbohydrates and fats but not proteins, and rapidly gained weight.[16] Such foods and drinks make up more than half the total dietary energy consumed in high-income countries like the United States. (By ultraprocessed foods I mean "formulations of food substances often modified by chemical processes and then assembled into ready-to-consume hyper-palatable food and drink products using flavors, colors, emulsifiers, and a myriad of other cosmetic additives. Most are made and promoted by ... giant corporations. Their ultra-processing makes them highly profitable, intensely appealing, and intrinsically unhealthy."[17] The term was introduced in 2009, but I use it for earlier years as well. I use *ultraprocessed foods*, *highly processed foods*, and *junk foods* interchangeably.)

As the US surgeon general and many other authorities have warned, obesity poses serious threats to our health. A complex metabolic condition, obesity (understood as the excessive accumulation of body fat that may impair health) is not just an often-stigmatizing disease; it increases the risk of developing other serious chronic diseases such as cardiovascular disease, type 2 diabetes, musculoskeletal disorders, and some cancers.[18] It also increases the chances of getting severe COVID-19 and dying from it. While infectious disease outbreaks, and especially COVID-19, have produced unimaginable numbers of excess deaths in recent years, what the World Health Organization (WHO) calls the "invisible epidemic of chronic diseases" remains the leading cause of death and disability, responsible for seven in ten American deaths. The proportion is even higher on a global level.[19]

But corporate malfeasance does not end with the manufacturing and mass marketing of unhealthy foods. Beginning in the late 1990s, behind the scenes, far from the public's view, those same companies were at work building up a body of scientific knowledge that minimized the role of junk food and magnified the role of exercise in solving the obesity problem. For the companies, science was not a truth above interests and politics; it was a commodity like, say, caffeine or high-fructose corn syrup, to be mixed and manufactured in ways that maximize profits and minimize government regulation. No sooner was obesity recognized as an enormous public health problem than the companies, working with their nonprofit organization tasked with creating industry-friendly science, and their allies in the academy and government, began systematically building a body of corporate knowledge on obesity, turning it into a set of practical weight-loss tools, and promoting it as the basis for individual and public health understandings of the condition. At the same time that everyday Americans were struggling to parse the media messages on diet versus exercise, the companies were secretly massaging those very messages by sponsoring media-training workshops that taught science journalists that exercise was best and junk food was not a big problem because junk calories could be easily burned off.[20] This invisibly distorted science, which found its way to us through multiple mass media and social media channels, added to the confusion about what we should do to stay healthy. And because that science did not support measures that would hurt the soda industry (taxes, marketing restrictions), it had the potential to undermine efforts at community, state, and national levels to stop the obesity epidemic.

A prime mover in these twin efforts to grow the market for junk food while promoting a science to hide its ill effects was the Coca-Cola Company. Since that day in May 1886 when John S. Pemberton sold the first glass of fizzy Coke at Jacob's Pharmacy in Atlanta, Georgia, Coca-Cola has grown into the world's largest nonalcoholic beverage company, and the producer of the most recognizable symbol of American modernity on earth. (It is said that 94 percent of people around the world can identify the Coke logo.) Coke is far from the only junk food that packs on pounds, but in its classic form, a can of zingy Coca-Cola can create real havoc in our bodies, captured most memorably by the internet meme "What Happens in One Hour after Drinking a Can of Coke!" (shown in figure 0.4).[21]

Coke was not only the *market* leader, it was the *science* leader as well, the prime mover in the global effort to create a body of knowledge that

WHAT HAPPENS IN ONE HOUR AFTER DRINKING A CAN OF COKE!

① THE FIRST 10 MINUTES

Ten teaspoons of sugar hit your system. (100% of your recommended daily intake) You don't immediately throw up from the overwhelming sweetness because phosphoric acid cuts the flavor allowing you to keep it down.

② 20 MINUTES

Your blood sugar spikes causing an insulin burst. Your liver responds to this by turning all that extra sugar into fat. (there's plenty of that at this moment)

③ 40 MINUTES

Caffeine absorption is complete. Your pupils dilate, your blood pressure rises, and as a response your liver dumps more sugar into your bloodstream. The adenosine receptors in your brain are now blocked preventing drowsiness.

④ 45 MINUTES

Your body ups its dopamine production stimulating the pleasure centers of your brain. By the way, this is physically the same way heroin works.

⑤ 60 MINUTES

The phosphoric acid binds calcium, magnesium and zinc in your lower intestine providing a further boost in metabolism. This is compounded by high doses of sugar and artificial sweeteners increasing the urinary excretion of calcium.

⑥ 60 MINUTES

The caffeine's diuretic properties come into play. (It makes you have to pee.) It is now assured you'll have to evacuate the bonded calcium, magnesium and zinc that was headed to your bones, as well as sodium, electrolytes and water.

⑦ 60 MINUTES

As the rave inside of you dies down you'll have a sugar crash. You may become irritable and/or sluggish. You've also now literally pissed away all the water that was in the Coke. But not before infusing it with valuable nutrients your body could have used for things like having the ability to hydrate your system or building strong bones and teeth.

FIGURE 0.4. "What Happens in One Hour after Drinking a Can of Coke!" Created by Gary L. Sohl based on an infographic prepared by Niraj Naik from research by the health writer Wade Meredith.

takes the blame off soda for the rise of the obesity epidemic. Most of us knew nothing about this until 2015. On August 9, 2015, the *New York Times* published a front-page exposé revealing how Coca-Cola was secretly paying leading exercise scientists to establish a Global Energy Balance Network (GEBN) to promote the notion that what matters most for obesity is physical activity, not dietary restraint—a claim few experts accept.[22] What was disturbing was not just the brazenness of the company's attempt to buy science favorable to its commercial interests, but also the scale of the investment (some $20 million) and the enormous secrecy surrounding the operation. Facing widespread criticism from experts and the public alike, Coke moved quickly to contain the damage, dissolving the GEBN, retiring its chief science and health officer, and abandoning its support for exercise science. That exposé in the *Times* brought Coca-Cola, one of Americas' most admired companies, to its knees and spelled the death knell of its project to sway obesity science. Of that two-decades-long hidden history of making and spreading science, only the very end—the scrapping of the network and Coke's very public efforts to restore its reputation—has come to light. As a result, the full story of this covert science urging exercise to counter obesity—who made it and how, what it was, how it lasted so long, what difference it made—remains unknown. Indeed, the science does not even have a name.

While the history of Big Tobacco's and Big Pharma's decades of scientific deception is now part of our cultural common sense, recent work in public health discloses that the production of distorted science is exceptionally widespread. In his eye-opening 2020 book *The Triumph of Doubt: Dark Money and the Science of Deception*, David Michaels draws on more than seven years as head of the US Occupational Safety and Health Administration, and decades of research in epidemiology and environmental health, to document a corporate strategy of manufacturing doubt used by virtually *every industry* that produces products that endanger the environment or human health. While the strategy began with the corporate-driven scientific defense of tobacco in the 1950s, Michaels shows how, since the 1990s, the production of industry-friendly science has grown into an institutionalized sector of the US knowledge economy. In this uncharted arena, a small number of very large "product-defense firms" (the industry's own term) employing scientists, engineers, lawyers, and public relations experts have been creating and advancing "bad science" (Michaels's term) to defend a host of disreputable products from scientific and legal claims that they are harmful. The products protected run the gamut from diesel engines to opioids to silica dust in construction worksites, many familiar from negative press attention. Michaels's primary interest in *Triumph* lies with these highly visible, multi-industry profit-making firms, like the Silicon Valley–based Exponent or the Boston-headquartered Gradient, that openly create science for sale. Yet he briefly mentions another, less visible class of corporate-affiliated scientific organizations: the nonprofits that serve the interests of single industries. In Michaels's account, these nonprofits are minor producers of "junk science," mere "front groups" that, like the large firms, are in the business of sowing doubt about injurious products, and thus unworthy of analytic attention.[23]

Following Michaels, we can call this kind of corporate science sponsored by companies to protect their harmful goods *product-defense science*. Historians of science have written magisterial histories of the sciences created to defend a handful of notorious industries; the tobacco, drug, and energy industries are prime examples.[24] Yet as Michaels's work suggests, there are many other deliberately distorted sciences begging for attention. As noted above, today the chronic diseases of modern life—heart disease, cancers, respiratory disease, diabetes—are the major killers in the United States and globally. Obesity is a risk factor for all these, as well as a chronic disease in itself. Despite their potentially large impact on the rise of these diseases, we know almost nothing about the sciences created to protect the food industry. One kind of product-defense science—what I call soda-defense science, or more simply, soda science—lay at the heart of Coke's project to clandestinely shape the science of obesity in the United States and around the world. That science, and the organizations and people who made it, are the subject of this book.

Taking the Science in Soda Science Seriously

Since the mid-2010s, a small but growing number of critics within the public health field have been working to expose food-industry efforts to distort science for corporate benefit. The giant in this emerging field of research is the well-known nutritionist and longtime food-industry critic Marion Nestle. Her *Soda Politics: Taking On Big Soda (and Winning)* (2015) sketches the outlines of an "astonishingly comprehensive" campaign on the part of the soda industry to maintain sales of sugary soda in face of the obesity threat.[25] Nestle's *Unsavory Truth: How Food Companies Skew the Science of What We Eat* (2018) tackles the influence of the industry on nutrition professionals and associations, unraveling the often-hidden conflicts of interest involved. In another hard-hitting book, a decade earlier, in *Appetite for Profit: How the Food Industry Undermines Our Health and How to Fight Back* (2006), public health lawyer Michele Simon exposed the "shameless publicity stunts" used by the food industry to obscure its massive public relations campaign to protect corporate images and bottom lines.[26] As their subtitles suggest, these books are driven most fundamentally by political agendas. By demonstrating the food industry's culpability in biasing the science and obfuscating the truth, they aim to hold it to account and get the public to join the fight. Written in a populist style, Simon's text serves as a kind of consumer's guide to spotting and challenging corporate malfeasance in the food industry. Since around 2018, there has also emerged a small but growing journal literature by critical public health scholars, based largely outside the United States. Less overtly political (or perhaps political in a more academic way), the authors have mined company emails and internal documents acquired through public-records requests and, in a few cases conducted key informant interviews, to illuminate the motives and strategies of key food-industry actors involved in manipulating the science.[27]

Through their efforts to expose wrongdoing, these writings have contributed to a shared public narrative about the food industry and its project to bend the science of obesity to industry ends. We can call this the "PR (public relations) disguised as science" story, and it is told using a standard set of terms. Not every commentator uses all these terms, but they crop up frequently enough to provide a useful point of departure for us. In this story, the Big Food companies and their industry front groups (e.g., product-defense firms, single-industry nonprofits, and academic-industry partnerships like the GEBN) have been "co-opting" academic researchers to become "mercenary scientists" who create "junk science" (or PR-as-science) that "manufactures doubt" about mainstream accounts of the product's harmfulness, in order to protect profits and escape government regulation. The ethics are clear-cut: these actions are self-evidently wrong and deserve condemnation.

These are not just academic terms. In an era of widespread skepticism about expertise and the ascendancy of the internet as a source of information, this language has spread widely among publics around the world. A November 2023 Google search for "front group" produced 8.7 billion hits, and "co-optation of scientists" yielded 1.6 billion, while "[to] manufacture doubt" produced 228 million. What these numbers tell us is that this black-and-white, PR-as-science story, and the language in which it is told, have been reproduced again and again, becoming the dominant narrative through which many of us think and talk about the corporate distortion of science. Yet these terms are problematic.

The problems start with the outright rejection of the science. The public health critics deplore the science created for corporate ends. They've dismissed it as "corporate PR," "mercenary science," "bought and paid for conclusions," "marketing in the name of science," "so-called science," and "junk science," all signaling that the knowledge is unworthy of the name "science" and undeserving of further attention. In these accounts, the interest in the science is largely confined to discovering the myriad ways it has been twisted to achieve industry ends. Figuring out how mainstream science has been perverted is an important task, and partly as a result of this work, we now have incontrovertible evidence that the obesity science funded by Coke and created by its scientific nonprofit was biased in industry's favor. Yet we can take the story further.

Historical and broadly sociological understandings of science suggest what we might gain by taking the claims of the Coke scientists—that they *are* doing science (of some sort)—seriously. Why should we do that? For one thing, soda science is recognizably science. Moreover it represents an unusual but only too common kind of science about which very little is known. Acknowledging the "scienceness" of this industry-funded knowledge is also useful because it allows us to see the advantages that the label "science" bestows on the experts who make and promote it. Even on the political front, we cannot hold the food companies accountable for distorting the science of chronic disease unless we know exactly what they've been doing in the name of "science." In short, the scienceness of soda science is the key to its (mostly hidden) power in the world.

In this book I will argue that corporate science *is distorted* and at the same time *is (a kind of) science*. Seeing it this way opens up a host of new questions about corporate science in general and soda science in particular. Instead of asking simply how the knowledge was twisted to serve industry ends, we can ask more open-ended questions about the who, how, when, why, and so-what of soda science. Approaching the science in this way will open up a secret world of transnational science making, science spreading, and effects that no one (beyond those involved) knew existed. And by uncovering hidden pathways of power and influence, it will show that soda science's effects are especially harmful in good part because they are invisible.

A second problem with existing writing on corporate interference with nutrition science is its US focus. (Since 2020, Chile, Colombia, and other Latin American countries have attracted growing attention.) That makes sense, given that many of the key companies and their main scientific nonprofit are headquartered in the United States. But the companies funding soda science—Coca-Cola, PepsiCo, Kraft, and dozens of other peer firms—included some of the world's largest food companies, with global brands and markets that needed protecting from the critics in public health. If the companies operated within global spheres, so too did the scientists, who belonged to transnational networks of public health researchers and institutions working to find answers to the rapidly worsening crisis of obesity.

The food giants and their scientific agents were thinking globally from the beginning; to understand them and their project of science slanting, we need to do the same.

In this book we will see that, from conception to execution, soda science was a fully global enterprise designed to protect soda by making industry-friendly science about obesity in the United States, spreading it to large, rapidly growing markets, especially in the Global South, and translating it into industry-favorable policies to protect those markets from government regulation. Taking a global approach to soda science allows us to see both the full scope of the soda-science project and its most insidious effects.

The third problem with the standard story lies in the political nature of its key terms. These terms—*industry front group*, *co-optation of academic scientists*, *mercenary science*, and so forth—serve as powerful battle cries in the larger war being waged over the excessive power and greed of huge corporations in our society. What ties them together is their negative character: they are all derogatory labels for individuals and institutions doing things many of us find troubling. This approach expresses our collective discomfort and dismay at the deliberate distortion of modern science, a crowning achievement of modern society. These pejorative terms, though, stop our thinking at a fairly superficial level: we label something bad and put it out of mind. If our aim is not just to condemn, but *to understand* the how's, why's, and so-what's of corporate science, we need a very different vocabulary. In this book I draw on a new set of concepts and methods, ones that draw readers much deeper into the hidden recesses of corporate science to see who inhabits that world and what drives them. The result is a story about soda science that is both bigger (more global) and smaller (more human) than the story that now dominates public understanding.

Drawing on the methods and analytic perspectives of two allied fields—social studies of science and the anthropology of global health—I take that science created for industry ends as my object of investigation. I show how—by taking the science seriously, studying the scientists and their organizations in action, seeing them as parts of a global project, and asking big, open-ended questions of the actors themselves—we can tell a bigger story about the corporate science of obesity that has been making us as a society and people elsewhere on the planet sick. In pursuing this agenda, I will be building on the important work of the public health scholars who, as insiders in this world, are superb guides to the science and the culture of the field. I also rely on the writings of science and health journalists who have been closely following this story, submitting open-records requests, and documenting critical developments as they unfold. My findings generally support their conclusions but add to them in surprising new ways. I draw most heavily, however, on my own on-the-ground, documentary, and online research to tell a human story of soda science, and so it is to that I now turn.

Unearthing the Secrets of Soda Science

In 2013, two years before Coke's ill-fated foray into science making became public, I began pursuing the hunch that the giant food companies that are skewing our diets toward salt, sugar, and fat might also be working behind the scenes to sway the scientific and governmental responses to the obesity epidemic. Although our story in this book will start in Atlanta, Georgia, and Washington, DC, my research was launched in China's capital city, Beijing. Why China? In the late twentieth century, China was the single most important target market for the food and soda industries. Mao was long dead, and the country was on an economic roll developing a market economy and looking to the West, especially the United States, for ways to create the good life. If Coke was anywhere selling sugary soda and quietly pushing the science of obesity in its direction, I figured, it would be here in Beijing. Also, China was my area of expertise. After two decades investigating the science behind the one-child policy, I had an intimate understanding of how science and politics/policy were knotted together at the apex of the country's political system.

My plan was to look for corporate influence on the science and policy of obesity in China, and then trace those influences back to their roots in the United States, where many of the biggest food companies are based. At that time, some longtime industry critics no doubt suspected Coke was trying to manipulate science to its advantage, but no one had any idea of the phenomenal scale of the effort. To discover how corporate actors might be clandestinely tampering with the science, I combined the fine-grained ethnographic methods of anthropology (seeking the actor's or participant's point of view, and placing that in wider political-economic and cultural context) with the painstaking "deconstructionist" methods of science studies (taking apart the facts, arguments, and so forth, to see how they were created), innovating additional methods where needed.

Because corporate science turned out to be exceptionally complex and the soda industry and its agents had sophisticated strategies for keeping their secrets hidden, it would take almost ten years, four involving sustained data collection, to get to the bottom of things. By that time, I had accumulated a very substantial body of empirical material on the corporate science and policy of obesity, and the organizations, individuals, and political dynamics involved in making them in the United States and China over twenty to twenty-five years. Using that information, I then created a set of data archives and histories that made the information more readily available for teasing out larger patterns and crafting stories. The list below provides a capsule account of the methods used, the materials gathered, and the archives and histories I assembled prior to writing this book. This broad overview can serve as a guide to the kinds and depth of empirical support that lies behind my claims, some of which are rather strong.

THE UNITED STATES

1. Coca-Cola, history of obesity work: constructed from books on the firm, company annual reports, Securities and Exchange Commission (SEC) reports, sustainability reports, articles on the website Coca-Cola Journey, and other sources
2. Email archive: emails between Coca-Cola executives and cooperating academic and government scientists (2009–15)
3. ILSI history: history of ILSI-Global, ILSI Center for Health Promotion, and ILSI North America, created from materials on the ILSI website, annual reports of ILSI-Global and branches, and conference programs and papers
4. US Internal Revenue Service (IRS) tax forms: 990 forms for all ILSI entities in the United States (1997–2020)

BOTH THE UNITED STATES AND CHINA

5. Interview file: numbered transcripts from interviews I conducted with twenty-five Chinese (2013) and twelve internationally prominent obesity specialists in the United States and Europe (2015–16); includes interview with the executive director of ILSI-Global
6. Scientist profiles: professional biographies of the main soda scientists, American and Chinese, constructed from their curricula vitae and other online sources
7. Article archive: set of relevant scientific articles on obesity and its management

CHINA

8. Newsletter file: news items on obesity from ILSI-China's semiannual newsletters (1990s–2015), used to construct history of ILSI-China and its work on obesity

9. Policy archive: collection of Chinese policy documents on obesity and chronic disease, articles on Ministry of Health website, photos of Chinese public health campaigns, and so forth (late 1980s to 2020)
10. Fieldnotes: Observations from ethnographic research carried out at conferences and in restaurants, drugstores, bookstores, and other sites in Beijing (2013)

This neat, orderly list implies a straightforward process of data collection, from step 1 to step 10. The research itself was anything but linear. I began with clear questions and a detailed plan, but as I learned more about how things were done on the ground, new questions kept arising, and the project kept expanding in scope. The work took me from China to the United States, and then to Switzerland, England, and Scotland. Serendipity played a big role. It was my good fortune, for example, to meet the key figures in the GEBN project at an obesity conference in Beijing in late 2013, two years before their names were linked to the science-for-sale scandal that enveloped Coke. Though I had no inkling of what they were planning behind the scenes, I was astonished to hear one of them tell the audience of more than two hundred top Chinese obesity specialists that the real public health problem was not obesity but physical inactivity, and to hear all three claim that exercise was what mattered for obesity, a finding, they added, firmly rooted in the science of energy balance. My fieldnotes record my reaction: "OMG, ** Coke is funding research that says obesity is not the issue **." I did not know it at the time, but what I had encountered was a full-blown, Coke-funded product-defense science of obesity. The labyrinthine process by which I uncovered many of the secrets of soda science is a story in itself, and I tell it in the second appendix. In the chapters that follow, where the findings are especially striking or unexpected, I include little methodological notes in the text to fill readers in on the data or methods that produced the discovery.

Understanding Product-Defense Science

To make sense of soda science, I turn to the ideas of science and technology studies (or simply "science studies"). Science studies holds that science does not simply reflect the natural world. Rather, science is social, a product of the specific individuals, communities, and wider contexts in which it is made, and it comes to bear the marks of the circumstances in which it is produced. Science is also active; the actors are constantly working to create and advance scientific knowledge.[28] A central question in the field is how scientific knowledge is actively constructed by participants who attempt to form stable knowledge structures and networks. That is our interest here as well: how specific facts and analytic frameworks on obesity were created.

This project on soda science was inspired by science studies work on commercialized science funded by the pharmaceutical industry. The drug industry plays a huge role in contemporary biomedicine (that is, scientific or mainstream medicine) because of the centrality of drugs in treatments for modern diseases. Work by Jeremy A. Greene, Adriana Petryna, Sergio Sismondo, and Joseph Dumit, among others, has shown that Big Pharma has come to control or shape virtually every phase of the scientific process surrounding chronic disease, from the clinical trials and other research to the analysis, writing, publication, and dissemination of the results.[29] This extraordinary control of the science has given the industry a critical role not just in marketing its drug treatments, but also in defining the diseases that are treated. Further maximizing sales and profits, the industry has also been spreading new notions of the self as inherently ill and in chronic need of those drugs. Indeed, Sismondo contends that this commercialized science that is being molded by Big Pharma represents a new model of science that is distinct from academic science. He calls it simply corporate science.

This project also drew inspiration from anthropological work on global health. As giant corporations have become the predominant force shaping the health of people around the world, anthropologists, ever attuned to questions of power, have been exploring the influence of corporate power in the field of global health. Seeking to capture these ongoing shifts, Vincanne Adams has characterized global health as "a platform that binds the public and private sector interests in novel form, inheriting the common sense of neoliberalism and increasingly mobilizing the private sector and its market-based, profit-driven solutions for all health problems."[30] Anthropologists have also urged a rethinking of the notion of "health." If Susan L. Erikson is right, then today "global health" is no longer simply about eliminating disease but "has become a thing en route to, amid, and alongside making money."[31] Thinking about health promotion as fundamentally about the pursuit of money, anthropologists have been asking open-ended questions to learn what cluster of actors, institutions, ethics, and so forth are coming together to wrest money from "health," how these vary from society to society, and with what effects. This is the approach I take here.

If the pharmaceutical industry has had a pervasive impact on the medical management of the chronic diseases of modern life, what about Big Food? Very few researchers working within science studies (or anthropological) perspectives have tackled this question. Nutrition science presents special challenges for science studies work on corporate science. The science is unusually complex; in addition, the influence of food companies on health is much more difficult to measure than the impact of drug companies.[32] In this book, I build on core concepts in science studies to trace the practices by which Big Food and its agents created and advanced a science of obesity and translated it into policies on chronic disease over twenty years (1995–2015), with a look ahead to 2020. Reflecting the anthropological concern with the *effects* of corporate power and corporate science, I also track the impact the food industry has had on our ideas, practices, and public policies regarding health. Pharma products may be widely used to treat our ills, but food products are consumed every day. Moreover, agents of Big Food have been creating and spreading corporate science for decades. We will see that the industry's imprint is much deeper than anyone has suspected.

CORPORATE SCIENCE: KEY TERMS

If our current terms are too limited, what terms should we use? Let us begin at the beginning: what is corporate science? By *corporate science* I mean simply science that is funded by industry and created to serve industry ends. It often looks like academic science but differs in that its primary aim is not to increase the stock of knowledge but to advance the marketing goals of industry. Following Michaels, *product-defense science* is a distinctive kind of corporate science produced with corporate support for the express aim of defending harmful products of all kinds.[33] Soda-defense science is a product-defense science created to defend sugary soda from the cascading threats to corporate profits and reputations brought about by the public health critiques of soda. The *soda* in soda-defense science refers primarily to carbonated soft drinks, but it also stands for other sugary drinks and, further afield, junk foods that harm health, all of which gain some protection from the science created to defend sugary drinks. Coca-Cola was the dominant corporate force in this science enterprise, but other companies making highly processed food worked with it and benefited from the science.

Despite the casual use of epithets like "junk science" by some observers, soda science can by no means be considered junk science. A prime example of junk science is the set of pattern-matching techniques that are widely used in forensics but have been shown by subsequent DNA analysis to be deeply flawed and the basis of countless wrongful convictions. These problematic matching techniques include analyses of bite marks, shoe prints, tire tracks, and handwriting. As M. Chris Fabricant shows in *Junk Science and the American Criminal Justice System* (2022), these techniques have no empirical or scientific grounding and—despite the hype surrounding them, on TV shows like *Forensic Files* and in real courtroom cases—amount to mere expert opinion based on training and experience. Soda science is nothing like this; soda science is a real science that has been manipulated in certain ways.

Like all sciences, soda science is a body of knowledge that is humanly created and shaped by its institutional and wider political and cultural contexts. We will ask not whether the science developed for the soda industry is good or bad (the core concern of the public health critics). Instead we'll ask how it is made (by whom, with what practices, in what historical circumstances), how it differs from mainstream science, what effects it has on public health programs and policies, and how it is ethically rationalized. (Because it was heavily shaped by corporate interests, by definition the science was "bad" in certain ways, some of which I point out as we move along. Its "badness" is not my primary concern, however.)

Fact Making: Scientific Elements of Interest

In telling the story of soda science, I take up a classic topic in science studies: how scientific facts are made, made stable or solid, and produce effects. A common way to study science is to examine a finished product (an article, a report). If instead we study *science in action*—as a messy process of actors making facts, crafting arguments, advancing claims, and so forth over a period of time—we get a much livelier and more robust picture of who is making it, how, and in response to what institutional and political-economic conditions. By following soda science over two decades, we are able to trace it over its full lifespan, from its birth through many years of development to its sudden demise in the wake of the *New York Times* exposé.

To see the imprint of industry on the science, it's not necessary to delve into the technical details (data sets, equations, computations, and so forth). Instead, we can focus on a cluster of more general elements of the knowledge. Soda science produced one weighty fact: physical activity is the primary answer to obesity.[34] That fact was buttressed by a larger "energy balance" analytic framework, the concepts making up the framework, supporting arguments, and rhetorical (or persuasional) devices designed to convince audiences of the soundness of the claim. Facts, frameworks, concepts, arguments, and rhetorical patterns: these are the main elements of the science I will be tracking in the pages that follow. When I talk about "the science," I will be referring to these five elements.

Phases of Science and Policy Making

In the pages that follow I trace the process by which scientists associated with Coke first created and then advanced these elements of soda science.

Since soda science, like all public health sciences, was an *applied science* meant to inform concrete programs and policies, I follow the process outward to include the *translation* of the science into practical health interventions and policies.[35] We thus track the corporate science of obesity over three phases: science making, science promoting (or spreading), and science translating. Although these overlap in practice, it is useful to analyze them as separate phases, as each one involved somewhat different actors, dynamics, and effects on the overall project of spreading soda science around the world.

Much of the writing on soda science describes the work it does as "sowing doubt" on mainstream scientific views of obesity. Soda science did indeed spread doubt, but its goal was much more ambitious than that. The overarching goal of Coke and other corporate sponsors of soda science was to create an alternative scientific story about obesity, one that downplayed the role of sugary soda, and then to advance that science, spread it to major markets, and get it embedded in local policies meant to control chronic disease. Since the global dissemination and policy translation of soda science was an active aspiration of the soda industry and its scientific agents, we can talk about an intentional *project* to make soda science the dominant approach to understanding and managing obesity around the world, or at least in the parts of the world that formed important markets for sugary soda. Indeed, we can talk about a global vision, an *imagined empire of soda science* endorsed and built into public policy by countries around the world.[36] Recognizing the global ambitions of the soda-science project allows us to see international effects of the science that we would not otherwise have noticed.

Science-Making Organizations: A Dedicated Product-Defense Nonprofit

To understand the making and effects of corporate science, we need to move beyond the one-dimensional notion of a front group to specify the kinds of organizations that are making the science, how they work, and what work they do. Corporations provide vital funding for corporate science, but the successful creation of such a science lies in the hands of a set of poorly understood, dark-money organizations whose mission is to create and circulate corporate science while rendering the corporate in the science invisible. Unlike the huge multi-industry product-defense firms mentioned by Michaels, in the making of soda science, the food industry would rely heavily on a small, shadowy, secretive, and yet sophisticated *single-industry nonprofit*, the International Life Sciences Institute (ILSI). Because its overriding objective was to create science to protect the interests of the food industry, we can call it a *product-defense nonprofit*. In the absence of research on the organization and inner workings of ILSI or any other product-defense nonprofit, I draw on my own research to identify six key features of the organization's structure and operations. Ranging from organizational dimensions to personnel practices to rules on ethics and participation, these six attributes will help us understand what made ILSI such an effective scientific agent for the soda industry in both the United States and China, where ILSI had a branch.

ILSI was the most important, but it was far from the only organization that made soda science. The academic scientists working with ILSI were highly entrepreneurial and, in the fifteen years of soda science's existence, created a variety of ad hoc academic-industry partnerships to move the project forward in different ways. In tracking how these were built, how they worked, and what work they did, we will uncover an entire sociology of recruitment, cultivation, and retention of academic scientists that provides vital keys to the long life and worrying effects of soda science.

A GLOBAL PROJECT; SCIENCE IN DEMOCRATIC AND AUTHORITARIAN CONTEXTS

The story told here unfolds primarily in two parts of the world: the United States, where soda science was created and initially promoted, and China, a major destination for its spread, endorsement, and translation into official policy. China under market reform (1978–present) was a top destination for the soda industry and its scientific agents. With its enormous population hungry for the products of modern life and its government preaching the marketization of everything, China offered an irresistible destination for giant soda companies seeking to market both soda *and* soda science. Coca-Cola had long had eyes on the China market and was rewarded for its patience by being tapped to be the very first foreign company allowed to do business in China after it opened to the global economy. Less well publicized is that the very same year, Coke's scientific nonprofit (ILSI) also established a foothold in the country. China was also a pet country of ILSI's president—who was concurrently a Coke executive—and he cultivated it with exceptional care. As a result of his efforts, and the way ILSI was inserted into China's political system, the global project of soda science may have been more successful in China than anywhere else in the world.

Science is shaped by its wider context, and the vast political and cultural differences between the United States and China—one, an open democratic society in which science is largely free from direct state control, the other, an authoritarian society in which science is subordinate to the party-state and subject to its political hierarchies—offer a truly rare opportunity to observe the same science as it is made and circulated in radically different political contexts. Soda science was also molded by the very different cultures of the two nations. By following soda science as it traveled from the United States to China, we will be able to glimpse two very different ways of making science, doing ethics, and turning science into practical programs and policies. Following the science across national borders also underscores how the success of soda science depends vitally on the larger context in which it is introduced. In both societies, soda science developed rapidly, but its secrets were exposed, and it was felled, in only one of them, the United States. In the other, China, soda science, with all its industry-friendly biases, continues to shape official thinking and public policy on chronic disease to this day. At least in this case, one of world's leading authoritarian systems was much friendlier to corporate science than the world's leading democratic one.

CORRUPTION: DOING ETHICS, HIDING SECRETS

As sciences deliberately distorted to serve industry interests, soda-defense and other product-defense sciences are fundamentally tainted, or contaminated. The science studies scholar Sergio Sismondo calls this *epistemic corruption* (corruption of the knowledge, or simply "corruption"). Corruption exists, he writes, "when a knowledge system importantly loses integrity, ceasing to provide the kinds of trusted knowledge expected of it. . . . Often [it] occurs because the system has been coopted for interests at odds with some of the central goals thought to be behind it."[37]

If corruption is the central feature of product-defense science, ethics—the moral reflection on and management of that corruption—is a central practice of those making and using corporate science. For the analyst of a case of corrupted science, the issue of ethics arises on two levels: the analyst's ethical stance toward the project as a whole, and the ethics of the actors involved in creating the science. My position on the ethics of the soda-science project is clear: it is wrong and needs to be exposed and those responsible held to account. The ethics of the actors is a different question. Rather than prejudging the scientists who created soda science, I adopt an agnostic stance and seek to understand their ethical decisions and actions in the context in which they were made. In this book, I develop the notion of (epistemic) corruption further by proposing a set of knowledge practices by which the scientists who make product-defense science try to manage the ethics of corporate science to their advantage.

For corporate science to be credible and durable, the corruption at its core must be concealed. And indeed, the secretive nature of corporate science making has been abundantly documented. In the product-defense sciences, Michaels writes, "dark money rules. . . . Secrets abound."[38] In his study of corporate science in the pharmaceutical industry, Sismondo describes a "shadowy knowledge economy," a "ghost-managed medical science" in which industry presence is a ghostly apparition, everywhere busy in the background shaping the science and its consumption and yet nowhere clearly seen.[39] I discovered the ubiquity of such covert domains and furtive actions early in my research, as those involved in making obesity science in China skirted questions about funding, left gaps in the conversation, and gave answers that did not quite compute. As I peeled back layers of obfuscation surrounding ILSI's operations, I came to realize that it was devoting as much effort to protecting its soda-defense agenda from detection as to pursuing the agenda itself.

Clearly, the appearance of sound ethics was vital to the believability and durability of the science and to the reputation of its makers as upstanding scientists. But instead of labeling the scientists who made soda science *unethical*, in the pages that follow I ask what they were doing to make their (corrupt) work *appear ethical*.[40] I discovered that, in both the United States and China, achieving neutrality for their (nonneutral) science of obesity was one of the central challenges facing those engaged in making soda science. It involved two sets of practices. The first was *hiding secrets*. As the research progressed, I came to understand that the making and burying of secrets—especially regarding corporate funding—was a fundamental and structured part of making, spreading, and translating corporate science. The second practice was what we might call *doing ethics*: demonstrating scientific integrity to put one's conscience to rest, while not in fact resolving the ethical contradictions. The twin practices of ethics (performing integrity) and secrecy (camouflaging what could not be demonstrated to be ethical) became a central focus of my research, and they will be a big part of the story told below.

What Lies Ahead

Hoping to reach general readers as well as professional colleagues, in this book I keep academic discussion to a minimum, highlighting instead the human story of soda science. That is a story of how—once the soda industry came under attack for the obesity epidemic—the giant soda companies mobilized a sprawling, largely invisible apparatus of scientific nonprofits, academic scientists, and public-private partnerships to create a corporate science of obesity whose primary aim was to defend industry interests, not combat obesity, and how they then succeeded in getting it endorsed and translated into official policy in one of the world's largest countries. I argue that soda science was a real if unconventional kind of science that deserves attention because it had distorting—and largely invisible—effects on popular ideas about health as well as on the health policies (and no doubt also the health) of the United States and other nations targeted by the soda industry. Beyond the science's problematic claims—exercise is the most important answer to the obesity epidemic, soda taxes are unworthy of consideration—it had a number of unusual characteristics that make little sense until one recognizes that this is no ordinary science, but a product-defense science corrupted to the core. In telling a comprehensive story of the life, death, and afterlife of soda science, my larger aims are both analytical or conceptual and political: first, to contribute to the broader understanding of corporate science, its nature, dynamics, and significance; and second, to hold industry and its agents to account.

The book has two parts, each with four chapters. Part 1 traces the birth and advance of soda science in the United States in the wake of the mid-1990s emergence of the obesity epidemic. We follow the elaboration of the science step by step over some fifteen to twenty years, as different parties—the food industry's scientific nonprofit ILSI, ad hoc academic-industry partnerships, and the soda industry itself, especially the industry giant Coca-Cola—succeeded in quietly building a credible science of obesity that put industry interests first. Suddenly, in late 2015, just as the science was poised to go global in a big way, its corporate funding came to light, and soda science as an active project abruptly ended. A brief coda to part 1 documents the sudden death of this project and traces the aftershocks for Coke, ILSI, and the soda scientists. In this coda we learn why, despite its untimely demise, the soda-science project was still a big win for Coca-Cola.

Part 2 charts the spread of soda science to China in the years 1999–2015, uncovering how Coca-Cola, working through ILSI and its American advisers, took advantage of the peculiarities of China's political system and its admiration for Western science to get soda science accepted as authoritative. Here we follow China's top obesity scientist-officials, as they laid the groundwork, got soda science endorsed as the best science for China, and built it into China's policies on diet-related chronic disease. We see too how they reworked the discourse to make their industry-friendly science of obesity appear ethical in Chinese cultural terms. In the coda to this part, we discover that, despite its demise in the United States, soda science with its industry-friendly provisions has lived on in China. It has continued to form the basis for official policy on chronic disease, including Healthy China 2030, at least through 2020, with worrying consequences for the health of China's people.

At the center of all these stories is a remarkable cast of well-educated, highly motivated, and enterprising characters—industry executives, academic-turned-corporate scientists and, in China, state scientist-officials—all deeply committed to making the dream of soda science come true. Four Americans and two Chinese remained at the center of the action for some fifteen years, allowing me to tell much of the story through their words, writings, and actions. The interpretations, of course, are my own.

In the conclusion, I highlight the crucial role of soda science in the formation of America's step-counting, weight-obsessed fitness culture. While soda science is no longer an active project, Big Food is now aggressively defending its ultraprocessed foods and manipulating science and policy in markets around the world. The conclusion sketches out these latest developments to make clear that the work of exposing the industry's tactics and holding it to account has just begun.

Finally, the conclusion returns us to the central political question surrounding corporate science, that of accountability. As an anthropologist, I take it as part of my craft and my ethics to be responsible for what I write. In this book I have strived for accountability by basing every part of my story and every argument on empirical evidence and laying out my methods and data to show how I came to these conclusions. In presenting extensive evidence that the food industry and affiliated scientists distorted the science of obesity and transported it to other countries, my intention is to hold these parties to account. Have they taken responsibility for the science they created? We're able to answer that question because I published some initial results in medical journals between 2019 and 2021. In the conclusion we will see how Coke and ILSI responded to the publication of research documenting their complicity in the soda-science project.

A first appendix lays out the set of concepts I developed to study soda science. I hope they might be useful to readers interested in investigating other product-defense sciences. A second, methodological, appendix tells the story of the research behind the book. It includes brief profiles of the scientists I talked to as well as an assessment of the limitations of the research.

My citation practices are unusual and important to understand at the outset. In order to reduce the number of endnote markings (note numbers) in the text, I have used numbered endnotes only for essential information (such as where the material discussed can be found, usually via short citations keying to entries in the bibliography, where full information will be found) or clarification. In addition to the numbered endnotes, I've created

three sets of citations that are not marked in the text (thus "unmarked citations") but appear in the notes section of the book just after the numbered notes for the relevant chapter. These cite material from three types
of sources: my interviews, my fieldnotes, and emails located in the Food
Industry Documents collection maintained by the University of California
San Francisco Library (UCSFL). For each of these three kinds of citations,
I've created a list arranged by order of location in the text (section heading
and the paragraph number within the section). For example, when I use a
quotation from an interview, in the text I make clear that that material is
from an interview. Readers who wish to know the source will need to note
the heading and the paragraph number in the section where the quote appears, go to the notes for that chapter, find the relevant section heading and
paragraph number, and the source will be listed there. For material taken
from interviews, I provide a reference to the numbered transcript of my
Interview File (IF), as well as the date and location of the interview. For material taken from my fieldnotes, I give a brief summary of the circumstances
where the note was taken. For those three types of material, there will be
no marking in the text indicating that the source can be found in the notes
at the end of the book.

∴

Part One

MAKING SODA SCIENCE
IN THE UNITED STATES

∵

CHAPTER ONE

ILSI and the Birth of Soda Science

The first decade of this century saw the emergence of a new body of knowledge on obesity whose main purpose was not to improve understanding, but to protect the soda industry from threats posed by the worsening epidemic of excess weight and the growing calls to address it by taxing soda. The new body of knowledge would protect industry interests in two ways: by focusing Americans' attention on exercise, not dietary restriction, especially of soda and junk food, as the primary solution; and by working quietly to keep soda taxes off the government's policy agenda. Despite its significance, the story of this unorthodox science remains largely untold.

Following the mundane activities of science making over time brings to light a fascinating and consequential science project with distinctive characteristics that spring from that raison d'être of industry defense. This first part of the book unearths the story of the birth, development, peaking, and then sudden collapse of that science after fifteen years. The chapters move roughly historically, each featuring the activities of one or two academic-industry partnerships or companies. We follow the "soda scientists" in these organizations as they assemble the core elements of the new science (chapter 1), translate it into programs for the American public (chapter 2), weaponize the science to fight a historic battle between rival sciences (chapter 3), and develop it into a global force (chapter 4), before witnessing its rapid demise following the discovery that this supposedly neutral science was funded and heavily shaped by industry (coda 1).

This first chapter draws us into that hidden world of soda science in its earliest days when those charged with mobilizing science to protect industry were just coming to grips with the alarming statistics on obesity and figuring out what needed to be done. While other parties would soon join the effort, the first responder was a publicity-shy scientific nonprofit that had been established by the food industry years earlier precisely to solve such problems. In this chapter we meet the key actors associated with that nonprofit and follow them as they created the foundations of a new industry-friendly science of obesity: a fact and a framework. While charting the early years of soda science, we begin teasing out its distinctive features and unraveling the secrets to its success.

A New Threat on the Horizon: Industry Mobilizes for Self-Defense

In July 1994, the *Journal of the American Medical Association* (*JAMA*) published a newsworthy study. Analyzing a series of top-quality surveys, the authors reported "dramatic increases" in the prevalence of overweight (here, including obesity) among American adults.[1] During the most recent period, 1988–91, fully 33.4 percent of adults were overweight, and that number had grown 8 percent in the decade-plus since the period 1976–80. The authors were astonished by their results.[2] Most specialists had believed that the rate of overweight in the United States was relatively stable, hovering around 25 percent of the population. The spiking numbers "started freaking a lot of [them] out."[3]

Leading obesity experts quickly weighed in. F. Xavier Pi-Sunyer of Columbia University called the prevalence number "stunningly high" and its continued increase "alarming," portending worsening disease and death rates.[4] Yale's Kelly Brownell published an op-ed in the *New York Times* provocatively titled "Get Slim with Higher Taxes." Just as the government taxes cigarettes and alcohol and limits their advertising to children, he urged, it should tax and restrict ads for the unhealthy foods that are fattening American kids: soft drinks, candy, fast food, and sugar-coated cereals.[5] The obesity crisis had arrived, and the food industry was being assailed. Big Food would need to be defended. And so the science organization tasked with protecting it began to lay the foundation of a scientific response. In this section we see how it preemptively solved the credibility problem that faced all science made for industry by recruiting a leading academic expert to serve as key adviser.

AN INDUSTRY-FUNDED NONPROFIT ORGANIZES TO DEFEND THE SODA INDUSTRY

In 1995, the International Life Sciences Institute (ILSI), a Washington, DC–based scientific organization concerned with issues of nutritional health and food safety, established a new program to study the causes of and solutions to the newly identified problem of rapidly rising obesity rates. ILSI was driven by more than curiosity. Behind the innocuous sounding name was an industry-funded and industry-governed global scientific non-profit created and led since its 1978 founding by a senior vice president of the Coca-Cola Company, Alex Malaspina (b. 1931, PhD 1955, nutrition science).[6] Established to serve the collective needs of the food industry, this shadowy organization was funded by hundreds of companies in the food (including beverage), agricultural, pharmaceutical, chemical, biotech, and supporting industries. Its members included a veritable who's who of the global leaders in the ultraprocessed-food-and-beverage industry, from Coca-Cola, PepsiCo, and Dr Pepper Snapple to Heinz and Hershey, Kellogg and Kraft, Mars and McDonald's, Nestlé and Proctor and Gamble, all with big stakes in the obesity issue. To serve its corporate members, ILSI had evolved a complex and supple structure, with headquarters (which I'll refer to as "ILSI-Global") and a handful of entities (such as the ILSI Research Foundation) located in Washington, and over a dozen branches in major countries and world regions where ILSI member companies operated. The North America branch, which would play a central role in addressing the obesity crisis, was also based in Washington.

Though its website tells a different story, my in-depth investigations point to the conclusion that ILSI's primary mission was to create science to meet the needs of the food industry. ILSI's core agenda, carefully camouflaged by self-protective rhetoric, was to sponsor corporate science, that is, science funded by and created for the benefit of industry. Its portfolio included a variety of product-defense sciences aimed at defending harm-producing industries from threats like those unsettling the food industry in the mid-1990s. Indeed, ILSI was founded in the late 1970s for the express purpose of defending the safety of caffeine, an addictive drug that is a crucial ingredient in carbonated colas.[7]

Most of us think of science as a producer of objective, disinterested facts. Those who observe science at close range have a different view of things. Science studies scholars contend that science is in fact deeply *interested*, reflecting the interests (preferences, needs, and so on) of its makers and of the institutions and contexts in which it is made. This is so for all science, both bad and good. The interestedness of product-defense science, then, is not special. What distinguishes it from academic science is the kind and intensity of the interests that shape it. The interests that shape product-defense science are especially narrow, focusing on corporate profits and reputations. They are also unusually strong, backed by potentially enormous corporate resources. As a result, product-defense sciences are almost certain to be *corrupt*. By that I mean they are systematically biased in favor of facts, frameworks, concepts, and so forth that promote solid profits and appealing reputations for companies manufacturing injurious goods. For the science to be credible and durable, capable of warding off critique, it must be, or be *perceived to be*, neutral and disinterested. One of ILSI's core tasks was to frame its work in those terms. To this end, ILSI had developed a sophisticated set of organizational structures and everyday practices designed specifically to bury its corporate secrets and publicly present its science as ethical and uncorrupted by industry interests. These tactics would be used to great effect in its project on obesity.

An essential first step was to recruit as advisers prominent academic scientists whose expertise, reputation, and institutional independence from industry would lend both authority and credibility to an ILSI-sponsored science of obesity. As supposedly disinterested scientists free from pressures to produce certain kinds of science for their employers, university-based experts could perform critical tasks of making and promoting science that industry scientists (below called "corporate scientists") could not effectively carry out. Strategically selected and incentivized with exciting professional opportunities, these putatively neutral academic advisers could turn into loyal supporters of the corporate-science project, *quasi-corporate scientists* who could be relied on to create science that meets corporate needs and to defend the industry whenever required.

AN INDUSTRY-FRIENDLY ADVISER

To lead the scientific effort on obesity, in 1995 ILSI tapped James O. Hill of the University of Colorado School of Medicine, naming him chairman of the scientific advisory board of the ILSI Research Foundation's program in childhood obesity and, a year later, of a similar program in adult obesity. One of the nation's leading authorities on obesity, Hill was an inspired choice.[8] Trained in physiological psychology (PhD 1981), Hill had an impressive academic record. By the mid-1990s, he had published nearly one hundred scientific papers and been elected president of the Obesity Society (service 1996–97), a professional association of experts. He was highly sought after in Washington, having served on advisory panels for several federal government agencies charged with developing the nation's response to the emerging public health crisis. Happily for ILSI, Hill had an entrepreneurial flair. Just a year earlier, he had cofounded the National Weight Control Registry to track the strategies of individuals achieving long-term success in losing weight and keeping it off.

Hill was friendly and outgoing, with a genuine interest in people. When I first met him in Beijing in 2013, he was eager to strike up a conversation, expressing great interest in knowing more about what I, an anthropologist, had written on his subject. Linguistically gifted, Hill had a knack for translating complex scientific ideas into easy-to-understand language. As I tracked down and read more and more of his work during my research, I came to understand that he was also a brilliant rhetorician who had rare powers of persuasion. He greatly impressed Greg Critser, journalist and author of *Fat Land: How Americans Became the Fattest People in the World*, who introduced him this way: "a vigorous, intellectually engaged fellow with an agile debating style and a wide-ranging presence in the field. . . . Hill is the dean of obesity studies."[9]

The Colorado-based scientist had one more attribute that ILSI's leaders would have known about: a demonstrated record of working with industry (which was not uncommon then—or now).[10] According to his curriculum vitae (CV), by 1995, when ILSI brought him on board, Hill had received not only eight grants from the US National Institutes of Health (NIH), but also two grants from pharmaceutical firms developing anti-obesity drugs, and six more from food companies, including Proctor and Gamble, Kraft, and Kellogg, all longtime corporate members of ILSI. Far from being an academic naïf co-opted by industry, a notion central to a popular public narrative about corporate science, he would be a savvy and congenial participant. Through sheer force of personality and entrepreneurial genius, Hill, the university-based outsider, would remain at the center of this sprawling corporate-science enterprise until the very end.

In later years, countless studies would show that industry-funded research is much more likely to produce industry-friendly results than research funded by nonconflicted researchers. This is known as the *funding effect*. For example, a 2018 systematic analysis of 133 studies of health risks associated with consuming sugary beverages published between 2001 and 2013 (the years of greatest interest to us) reported that industry-related papers were overwhelmingly more likely to conclude that the drinks had weak or no detrimental health effects than did independent studies.[11] (Sugar-sweetened beverages include regular soda, fruit drinks, sports drinks, and other beverages with added sugar.) There are many such studies, and their findings are strong and consistent. Although industry-funded research does not always produce industry-friendly results, it very often does, and it certainly gives the appearance of conflicts of interest.

Academic scientists who serve as advisers to industry or industry-funded scientific nonprofits are likely to face charges from colleagues of ethical impropriety and industry influence on their conclusions. Yet most researchers who accept corporate funding, while often willing to disclose their funding sources, steadfastly maintain that any conflicts of interest are negligible or easily managed. They claim too that their scientific findings are not affected by those conflicting interests or by industry funding.[12] Hill was no exception. Over the years he worked with ILSI, Hill would continue to insist that industry funding did not influence either the conduct of his research or his conclusions. Again and again, he would espouse the many benefits of academic-industry partnerships, while saying little about potential interest conflicts.

How can we understand these responses to the ethical questions swirling around industry funding of science? Is the science that Hill would create working with ILSI and industry simply *mercenary science*—science for sale—as some writers have suggested, in which a scientist with a complex relationship to the truth "sells out" and simply says whatever is required to protect his reputation? A more generous and, I think, more plausible interpretation emerges when we see corporate science as a real science. In this view, comments such as Hill's reflect how he *does ethics*, how he strategically protects his reputation as a good scientist with close industry ties. In this account, this ethics stance is the result not of industry co-optation but of what some public health scholars call motivated reasoning.[13] I use the more precise term *incentivized reasoning*. The two methods of recruitment differ in subtle but important ways. According to any number of dictionaries, co-optation means including someone in something, often when they don't want to be included. Incentivization refers to the provision of a positive motivational influence. The person may be willing or unwilling, but the incentive stimulates greater interest and investment in the new group. The language of co-optation tends to shut down discussion, while the language of incentivization opens up new kinds of questions about what incentives were offered, what bargain was struck, and so on.

Hill was not co-opted or pressured into working with ILSI; as we will see, he was extremely enthusiastic about the collaboration. Offered substantial research funding and professional advantages—hobnobbing with corporate executives, global travel, opportunities to realize his entrepreneurial ambitions, and much more—he was well incentivized to join and maintain the partnership. In incentivized reasoning, funding relationships and the professional advantages flowing from them often subtly affect the scientific process—how questions are asked or results interpreted, for example (how well the research is conducted appears less affected)—but in ways that tend to remain imperceptible to the researcher. Starting in the mid-1990s, when Hill was recruited to advise ILSI, he would enjoy countless benefits that would be the envy of many academics. This steady stream of professional perks, as well as spending significant time in the world of corporate science, where industry support was valorized and normalized, could easily have colored his views, leading him to miss the subtle shaping of his thinking and making him see the benefits but not the risks of collaboration. In any case, Hill would serve ILSI as a loyal quasi-corporate academic scientist for decades, advancing industry-friendly scientific positions, all the while performing a self-protective ethics. That performance involved three routine practices, declaring his science bias-free, extolling the benefits of partnerships, and defending industry interests when the occasion demanded. Surrounded by an aura of neutrality lent by his academic base, he would be vital to ILSI's success.

The CEOs Speak, and a Soda-Defense Science of Obesity Is Born

In the late 1990s, the mounting public alarm about what the US Centers for Disease Control and Prevention (CDC) were now calling an "obesity epidemic" was deeply concerning to ILSI member companies in the sugary-drink and highly processed food industries.[14] CDC mapmakers understood well that colorful maps, as well as line drawings, bar graphs, and other visual displays of scientific data, possess a remarkable ocular power that communicates scientific messages more emphatically than mere words or numbers. In everyday language, we say: "a picture is worth a thousand words." To convey the seriousness of the obesity threat, CDC cartographers began creating sets of maps of the United States showing the rapid increase over time in the number of states with high obesity rates (10–14, 15–19, and over 20 percent of adults; the highest rate kept rising over time). The maps for 1990, 1995, and 2000 are reproduced in figure 1.1.[15] Widely circulated by the media for years, such maps made an unforgettable impression. From then on, obesity in America would be an "epidemic" (technically, the rapid spread of a disease to large numbers of people within a population), an object of national dread and fear.

At the same time, soda was being fingered as a major culprit, bringing the threat to corporate profits close to home for the industry. In his hard-hitting 1998 report *Liquid Candy: How Soft Drinks Are Harming Americans*, the pioneering anti-junk-food scientist-crusader Michael F. Jacobson of the Center for Science in the Public Interest (CSPI) called attention to the rise in consumption of sugar-laden, calorie-rich "soda pop" in parallel to rising rates of obesity, especially among teens.[16] Soda was nothing but liquid candy, Jacobson declared, and posed major threats to the health of growing bodies. The public health community was also concerned that calories from soft drinks were replacing calories from more nutritious foods and drinks in the American diet. In *Food Politics: How the Food Industry Influences Nutrition and Health* (2003), Marion Nestle decried how, by the late 1990s,

Obesity Trends* Among U.S. Adults
BRFSS, 1990
(*BMI ≥30, or ~ 30 lbs. overweight for 5' 4" person)

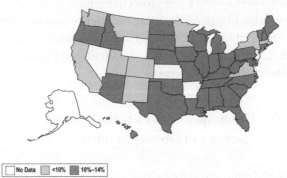

No Data <10% 10%–14%

BRFSS, 1995

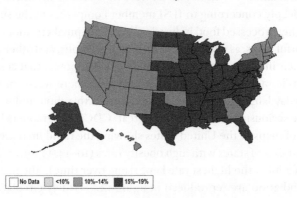

No Data <10% 10%–14% 15%–19%

BRFSS, 2000

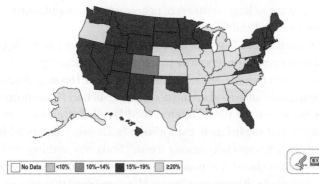

No Data <10% 10%–14% 15%–19% ≥20%

FIGURE 1.1. The spread of the obesity epidemic in the United States, 1990–2000.
Source: Behavioral Risk Factor Surveillance System, CDC.

soft drinks were no longer occasional treats but were being produced and consumed in vast quantities. Between 1970 and 1997, the annual production of sugary sodas almost doubled, while the production of milk declined by almost a quarter.[17] As the major producer of sugary soft drinks in the United States, accounting for 44 percent of the market, Coca-Cola must surely have felt the heat.[18]

Around this time, growing numbers of specialists, echoing Brownell at Yale, were calling for taxes and advertising restrictions on soda and other junk foods.[19] The public was getting the message. After climbing rapidly for decades, in the late 1990s soda sales in the United States began to flatten and fall, as health-conscious consumers cut back. Meantime, the CDC was organizing a national effort to address the emerging crisis.[20] Word was out that the surgeon general was preparing a major report on the epidemic of excess weight looming on the horizon. Fueled by memories of the US surgeon general's landmark 1964 report on smoking and health, which sparked a wave of antitobacco legislation and led millions of Americans to quit smoking, there was a palpable fear in corporate quarters that soda would become the next tobacco. In this section, we see how ILSI member companies moved to protect their prized products by setting in motion a new soda-defense science that looked like ordinary science, but differed from it in crucial ways.

A SECRET CEO DINNER AND A NEW SCIENTIFIC FACT

In 1999, influential members of ILSI's board of trustees, acutely aware of these threats, took action. In the early months of the year, members of the board sat down with experts on nutrition and obesity to brainstorm how the food industry should respond. Hill, a member of the ILSI (and ILSI North America) boards since at least 1997, was most certainly involved in these discussions, the details of which have never been made public. Then in April 1999, a prominent member of the ILSI and ILSI North America boards, a top Pillsbury Company executive, invited about a dozen food-industry chief executives to Minneapolis for a secret "ILSI CEO Dinner" (his label for the event) to consider a collective ILSI response.[21] What they decided would determine the shape that a product-defense science of obesity would take for years to come.

In the main presentation at the meeting, Michael Mudd, senior scientist and vice president at Kraft, and Hill, his scientific adviser, took turns in laying out the sobering news about the menacing environment and the urgency of taking action.[22] Mudd began: "The food industry is being portrayed as a major cause of an epidemic of obesity and all its disease-related effects. The proposed remedies are troubling—taxes to control consumption, and regulations to restrict marketing and advertising, especially to kids. . . . There are no easy answers. . . . But this much *is* clear[:] . . . the one thing we *shouldn't* do is nothing. Stonewalling the issue can destroy credibility, intensify hostility, and ultimately may produce more damaging consequences."

Mobilizing a rhetoric of warfare—industry versus its critics—Hill described the formidable array of opponents lined up against the food giants: "The forces aligning on the issue include public health and obesity experts, activist groups, professional associations, politicians, and the media. . . . While virtually all the experts blame changes in both diet and exercise for the obesity problem, the more hostile critics, lifting tactics from the tobacco war, paint a far darker picture. They've developed sensationalized charges of a 'toxic food environment,' an emphasis on diet rather [than] exercise . . . where big corporations can be targeted . . . [and] a focus on kids . . . where emotions are highest."

What could be done? Deeming an industry response imperative, and knowing the CEOs would reject a proposal involving major modifications in their food business, Mudd and Hill offered a solution that, while not ignoring food, promised significant attention to another facet of the problem, physical activity. While acknowledging that the industry bears responsibility—"no credible expert will attribute the rise in obesity solely to decreased physical activity"—they sought to reassure the CEOs that, since activity "is a big part of the issue," the debate would never focus *solely* on food. Under their plan for a national program to promote healthy lifestyles—involving the dissemination of the most promising diet-and-exercise interventions, research using the best science, and an aggressive public education campaign—the activity side of the equation would receive sustained attention. All of that could be had for $15 million over the first three years, with ILSI managing the funds and leading the effort on the scientific front. Wrapping up, Mudd sought to coax his peers into action while filling them with fear of the dangers of collective inertia: "By being part of the solution, we hope to deflate the growing criticism of the industry and avoid the drastic measures our critics are proposing. . . . By speaking out we can make sure the debate focuses, as it should, on physical activity as well as food. . . . By acting now, we may *head off* any stampede to demonize the food industry before it gathers broad-based momentum."

The empirical support for an emphasis on exercise (or physical activity; I use these terms interchangeably)[23] was weak, however. No data demonstrated that exercise worked to prevent or reduce obesity (or even that a decline in activity was a major cause of rising obesity levels). In an article in *Pediatrics* the year before, Hill had emphasized the scarcity of studies of childhood obesity.[24] Precious little was known about how it develops. Even such basic questions as how it should be defined and measured had not been settled. Hill needed to give the exercise recommendation scientific authority, and he did so in Minneapolis by drawing on his own credibility as a university expert: "I personally believe that the decline in physical activity has been a key factor in the increase in obesity and I also believe that this provides a good target for intervention." The CEOs may have accepted Hill's reassurances (or, more likely, they cared little about the lack of data), for they dismissed the broad diet-and-exercise program and endorsed only the meagerest of efforts: the promotion of physical activity among children, a plan that kept their business models intact.[25]

This plan—encouraging exercise as the most promising solution to obesity—was a novel and even radical idea. It would be a turnaround for Hill, whose writings as ILSI's adviser on obesity during the late 1990s, as well as findings from the National Weight Control Registry, had underscored the importance of both diet and exercise, *with neither given priority*.[26] Nonetheless, the CEOs, the ultimate powerholders in ILSI, had spoken, and their word evidently was final, for Malaspina quickly adopted the exercise solution and made it the cornerstone of ILSI's new project on obesity science. On that single scientific "fact"—that physical activity is the key to the obesity question—an entire edifice of corporate science would be built.

SCIENCE MADE BACKWARD AND OTHER FEATURES OF SODA-DEFENSE SCIENCE

With food executives establishing the foundational element of the science— the framing of the obesity problem and solution as one of activity—and scientific basics like the need for data to support scientific claims brushed aside, from the beginning it was clear this would be no ordinary science. Instead, it would be a product-defense science of obesity, whose raison d'être was to protect the profits and other interests of corporate actors from the threat that public health critiques posed to their financial well-being and public image. Because sugary soda was singled out for its outsized contribution to the obesity crisis—it was targeted far more than, say, breakfast cereal or candy, which were also harmful to children's health—and because the Coca-Cola Company would be the main corporate force behind the creation of a science to defend soda from the critics in what was already being depicted as a "war on industry," I call the science we will track in this book soda-defense science.

In this inaugural episode, which gave birth to the new science, we can see three of the features that marked it as soda-defense science. First, as a science funded by industry and created to defend harmful products from critique, all the elements of the knowledge served industry ends in some way. In most cases, product-defense science protects companies making harmful products by creating doubts about the claims of mainstream science that those products pose risks to human health. The central fact associated with this science (activity comes first) protected the soda industry not only by casting doubt on mainstream (diet-restricting) solutions but also, and even more so, by diverting attention from diet to an entirely different solution: physical activity.

Second and more curiously, the science would be made in reverse order. Though there is no standard order of scientific tasks, we can recognize a logical sequence to the activities that yield a scientific fact, and the makers of soda science inverted the sequence. In that logical order of scientific activities, ideally, scientists start with a hunch or hypothesis, gather data, test the hypotheses, and gradually build to a conclusion (a fact, a narrative), which may then be translated into concrete policies or interventions. The makers of soda science, however, *started with the conclusion* (exercise is the answer). They would go on to sketch a very simple conceptual framework, then translate a still minimally developed science into real-world interventions, and later backfill with methods, data, more concepts, and a strengthened analytic framework to support the prescribed conclusion. In following this order of procedures, they would invert the logical order of science making in two important ways: they would put the conclusion first, and they would develop practical interventions or programs before solidifying a rationale for them.[27] That *inverted sequencing of activities* will become increasingly visible as we observe the science making over longer periods of time.

A third distinguishing feature of this new science stemmed from the corruption of its knowledge base. To maintain the appearance of credibility, the influence of corporate actors on the science would be wrapped in layers of secrecy and silence. The CEO meeting in Minneapolis was kept a tightly held secret. According to the journalist Michael Moss, who learned about the meeting while reporting for his book *Salt, Sugar, Fat: How the Food Giants Hooked Us* and who spoke to a few in attendance, there were no minutes taken, no recordings made, and of course no reporters allowed.[28] The meeting came to light—and to my attention—only with the publication of Moss's book in 2013, which made sense of a clue to such an event I had encountered on the ILSI website.[29] I located a copy of the Mudd and Hill talk in the Truth Tobacco Industry Documents Archive, a resource built by the University of California San Francisco Library to house the internal documents of the tobacco industry that were produced through litigation. Why was a food-industry document in a tobacco archive? Because Kraft, Mudd's employer, was part of Philip Morris. The making—and burying—of corporate secrets would be a fundamental part of every stage of the construction of soda-defense science.

That same year, 1999, ILSI-Global took the big step of asking all its international branches to put obesity on their agendas. Over the next decade and a half, three different US-based ILSI entities would engage in a major effort to make, disseminate, and translate into concrete interventions a science of obesity whose main mission was to protect the soda industry from the threats laid out so clearly at the CEO meeting. What then was ILSI?

ILSI: A Product-Defense Nonprofit for the Food Industry

When I visited on a bright September morning in 2016, I found ILSI's Washington headquarters on the second floor of an office building near Thomas Circle. Its executive director, Suzanne Harris, had kindly agreed to talk to me. Located in the heart of DC, ILSI is a few blocks north of the White House and southeast of the trendy Dupont Circle, and a short walk from its sister organization, the IFIC (International Food Information Council). Responsible for creating and globally promoting science beneficial to industry, ILSI is a sprawling organization that touches many lives and appears to have inordinate, largely hidden, and uncharted influence around the world. A few numbers will show what I mean. In 2015, ILSI had fifteen branches (fourteen regional branches as well as the Health and Environmental Sciences Institute) with a total of (precisely) five hundred member companies among them.[30] The branches were remarkably active. For example, in that same year, the North America branch (ILSI-NA for short) and its sixteen scientific, project, and other committees sponsored thirty-four projects, including workshops, conferences, sessions, and posters, and published twenty-three scientific articles and other works in mainstream scientific journals. The North America branch may be one of the more active ones, but these figures offer clues to the likely reach and influence of the organization.

Like the corporations that form its membership, ILSI does not welcome probing inquiries into its operations. Responding to a spate of published critiques, in recent years it has revised its bylaws, revamped its website again and again, and undertaken significant reorganization. My account here is based on in-depth research into ILSI's work on the obesity issue between 1995 and 2015, primarily in the United States and China. Because of that focus, I have used the bylaws, code of ethics, and branch structure that were in effect at the end of 2015.[31] The organization may work somewhat differently in other topical areas or branch sites. Certainly, some things have changed since 2015; whether fundamentally or superficially is impossible to say. In this book, when I write "ILSI," I will be referring to ILSI as it functioned in these specific domains.

The term *front group* is commonly applied to all organizations that create science for industry benefit (other than formal product-defense firms). ILSI is often labeled an industry front group, and indeed it does serve as a "front" or cover for industry by doing industry's shady work. While the term serves as a useful cudgel to denounce organizations covertly working for industry, it's a blanket term that cannot capture the differences among the various organizations creating corporate science. In this book we will encounter a remarkable array of configurations—combinations of industry and academic actors working in partnership—each of which made a distinctive contribution to the larger soda-science project. To see these differences, we need to use a more expansive vocabulary and pay attention to the inner workings of these organizations, something neglected to date. A more useful label for ILSI, *product-defense nonprofit*, conveys one of its primary objectives as well as its organizational form, the nonprofit. I discussed the product-defense goal in earlier sections. I turn here to its nonprofit status and to the many financial and reputational benefits that status conferred.

My research reveals ILSI to be a complex, sophisticated, and highly secretive entity brilliantly designed to accomplish its product-defense goals. Its effectiveness is rooted in six features of its structure and operation. These key attributes of the organization will crop up again and again as our story unfolds:

1. A dual structure and mode of operation (visible/invisible, formal/informal)
2. A welter of hidden pathways that enabled industry to quietly influence the science and ILSI leaders to discreetly influence public policy
3. A nominal rule purporting to ensure scientific integrity (the *ethics rule*)
4. A recruitment and incentivization system by which key personnel (academic advisers, branch heads) were turned into reliable allies
5. Exclusionary rules of participation that created a quasi-closed scientific world in which only or mostly ILSI-approved voices could be heard, and
6. A hierarchical (center-branch) global structure that concentrated power in DC and a set of mechanisms that facilitated the top-

down flow of ideas and practices to ILSI branch countries around the
world

Of these features, only the ethics rule and global structure (3 and 6)
are mentioned on ILSI's website, and there the description is only partial.
The rest I teased out by following the organization in action. I begin, in
this section, by mapping out ILSI's basic structure and mode of operation.
Then, stripping away the legalese, I uncover its dual character and the hidden workings of its ethics rule, showing how together they solved the organization's most fundamental problem: benefiting private industry while
appearing to benefit the public. I introduced ILSI's recruitment and incentivization practices earlier, when introducing its main obesity adviser, and
will describe the other attributes as the story moves along.

CORPORATE MEMBERSHIP AND FUNDING

According to its mission statement, ILSI brings together scientists from
three sectors—academia, government, and industry—to collaborate in
"providing science that improves human health and well-being and safe-
guards the environment."[32] As noted above, ILSI's headquarters and key
affiliated entities (such as the Research Foundation) are in Washington, DC.
From its establishment, however, the organization's outlook has been international, with attention focused on the roughly fifteen geographic branches
located in key markets around the world. With this branch structure, ILSI
science, which is created largely by US-based researchers, can be readily
disseminated around the world. This center-branch structure turned out to
work wonders for the organization.

Corporations play a key role in ILSI's funding, governing, and science-
making activities. ILSI is funded by several hundred member corporations
of all the ILSI branches. (Other ILSI entities, such as the Research Foundation, are not membership based.) Representatives of the member companies form an "assembly of members," which elects the board of trustees.
That board, which is composed of both public (academic or governmental)
and private (industry) trustees, is responsible for supervising, controlling,
and otherwise managing the affairs of the organization.[33]

The great bulk of ILSI's revenue comes from these member companies.
In ILSI North America, 91 percent of revenue came from member dues and
committee assessments (45 and 46 percent, respectively) in 2015.[34] In the
organization as a whole (ILSI corporate, Research Foundation, and the
branches), 66 percent of revenue came from members, while another 24.4

percent came from grants and contributions (including government grants; again in 2015). Not surprisingly, ILSI rejects the view that industry funding biases science. Its position, posted on its website, is that the quality of science should be judged on its technical merits:

Q: Is industry-funded science reliable?
A: ILSI believes good science can be funded by a wide variety of organizations, in both the public and private sectors. ILSI maintains all science should be judged on the merits of study design, methodology, and validity of the conclusions regardless of funding source.[35]

While this stance may sound sensible, it fails to consider the findings that research funded by industry is more likely than research funded by other entities to produce industry-friendly results. And the gap is hardly trivial. To my knowledge, the organization has never fully addressed the problem those findings of bias pose. Instead, it has restated this position on its website and created a set of conflict-of-interest guidelines for food companies sponsoring scientific research. (We delve into these later.)

For many years, the Coca-Cola Company was a predominant force in the organization, if not the foremost. Coca-Cola was one of six founding companies. (The others were General Foods, Heinz, Kraft, Pepsi-Cola, and Proctor and Gamble.)[36] ILSI's founding president, the MIT-trained food technology specialist Alex Malaspina (whom we met earlier), was concurrently a vice president of the Coca-Cola Company (1969–ca. 1998). Malaspina remained at the helm for nearly twenty-five years (1978–2001), and much evidence suggests he looked after the company's interests. Malaspina formally retired as ILSI-Global president in 2001, but he remained actively involved in ILSI affairs for at least fifteen more years, reportedly phoning the executive director several times a week to keep a tab on things. That astonishing nugget of information—which suggests that even when Coke was not in charge, it continued to guide, direct, or just meddle in the activities of this supposedly neutral scientific organization—was shared with me by the executive director herself. After his retirement, other Coke executives took his place at the helm. From 2009 to 2011, Michael E. Knowles, Coca-Cola's vice president of Global Scientific and Regulatory Affairs (based in Brussels, Belgium, at the time), served as ILSI president. In 2015 Rhona Applebaum, vice president and chief science and health officer (based in Atlanta) served one year (before resigning in the wake of the scandal over Coke's support for the Global Energy Balance Network). The Atlanta-based company thus called the shots at ILSI for twenty-eight of its first thirty-eight years. Beyond its prominence in leadership positions, Coke was a major ILSI funder. In 2015, the company (or a subsidiary) was a fee-paying member of every one of ILSI's fourteen international branches, giving it a presence throughout the entire network.

BUSINESS LEAGUE OR CHARITABLE NONPROFIT? "DOING ETHICS" AND "SPURNING ADVOCACY"

Presented on its website and in its core documents, ILSI's official aim is to provide science for the public good: "All ILSI scientific activities have a primary public purpose and benefit."[37] A clue to a rather different purpose can be found in the historical record. A letter to ILSI from the US Internal Revenue Service (IRS) dated March 28, 1985, indicates that, at the time of its founding, in 1978, ILSI was incorporated under the IRS tax law (that is, the Internal Revenue Code), as a "business league" (also known as a chamber of commerce or board of trade) (501[c][6]).[38] A business league is a tax-exempt organization that promotes common business interests and may engage in lobbying but is not organized for profit. ILSI leaders at the time affirmed that agenda, stating that it was established "to unite the food industry" and "to permit companies to pool resources to support research programs of common interest."[39]

By the early 1980s, the organization had found a more advantageous legal category that would allow it to cloak more distasteful aspects of its work—potential conflicts of interest, mercenary agendas, aspirations to influence policy—in the language of "science for public benefit." Since that time, all US-based ILSI entities have been incorporated as nonprofit, tax-exempt "charitable, scientific, and educational organizations," commonly known as "charities" or "nonprofits" (501[c][3]). This desirable designation not only surrounds ILSI with a halo of public service; it has financial benefits, permitting member companies to deduct contributions to ILSI and exempting ILSI from paying federal income taxes on funds it receives. To qualify for this favorable status, an organization must meet two major criteria: its activities must be primarily for public benefit, and it may not engage in political or legislative activities (such as lobbying, the attempt to influence legislation) as a substantial part of its work.[40]

Though ILSI would quietly violate both rules in the service of creating obesity science, it maintained a public narrative in which it met both requirements and had procedures available to ensure they were followed. Here we see that dual character of the organization (public narrative / private reality, the first attribute in the list above) in operation. Set out in ILSI, "Bylaws, Conformed as of January 17, 2015," and ILSI, "Code of Ethics and Organizational Standards of Conduct," this narrative was utterly crucial to the credibility of ILSI's science. According to that narrative, the organization's main strategy for achieving neutrality (or, more accurately, minimizing corporate bias of its science) was to maintain a public-private "balance of perspectives" by specifying that at least 50 percent of all ILSI trustees and members of scientific committees be "public-sector representatives" (university or government based), with the rest "private-sector representatives" (those from industry, both determinations based on primary place of employment). This 50 percent rule was ILSI's nominal *ethics rule* designed to demonstrate its neutrality, what it called "primary public benefit." While the strategy may have appeared solid on paper, this unusual approach to demonstrating neutrality was problematic. Here is where science studies allows us to see things in a new and compelling light.

The public/private distinction at the heart of ILSI's good-ethics strategy was based on the questionable assumption that academic scientists (for readability, let's just call them professors here) are all largely free of conflicts of interest and immune to industry influence. To be sure, professors' compensation from their employers is not based on their success in making science serve their employing organization, as is the case for corporate scientists. Professors are generally free to pursue topics of their own choosing. Yet working at a university hardly guarantees that one creates science primarily for the public good. Far from it! What matters is whether the scientist is offered and accepts incentives to create industry-friendly science. Academic scientists offered significant professional and other benefits for producing industry-friendly science can be expected to act like corporate scientists in creating science aligned with corporate interests and not recognizing the conflict of interests involved. In practice, many professors are likely to produce a mixture of public-benefit (mostly government- and foundation-funded) and private-benefit (industry-funded) science. As long as they are working extensively with and for industry, though, the label "public scientist" would hardly seem appropriate.

Because of his university position, from the beginning Hill was labeled a public-sector scientist, even though industry had funded a substantial amount of his research before he joined ILSI. Hill would become a long-time ILSI insider, serving as a designated "public representative" on the board of ILSI from 1999 to 2002, and on the board of ILSI's North America branch from 1997 to 2015 and beyond. This is despite the fact that for all those years, and especially after the 1999 CEO meeting fixed the scientific agenda, he would spend a substantial amount of time working with ILSI and in other academic-industry partnerships to create company-friendly science. How can we make sense of that?

Surely, we can assume, ILSI was fully aware of these logical inconsistencies and of the practical difficulties of fitting the actual careers of ILSI's scientific advisers into these two tidy boxes, "public" and "private." In building their careers, many health scientists are opportunistic, combining university and industry work and funding, sometimes moving back and forth between sectors, in the belief that industry funding has no influence on their work. How then could ILSI expect more strictly "either/or" careers among its advisers? To answer that, we need to think about rules like the 50 percent private-sector criterion in less conventional ways. Science studies suggests that formal rules like this one are useful to organizations because they appear to provide hard-and-fast solutions to problems, but on closer inspection turn out to be no more than loose guidelines because they are open to flexible interpretation. Rules are better understood not as prescriptions for behavior but as rhetorical resources actors use to achieve various ends.[41]

The science studies scholar Sergio Sismondo follows this line of reasoning in his work on corporate science in the pharmaceutical industry. In that field, he suggests, formal ethics rules are just that: a formality, designed not to produce ethical behavior, but to insulate the organization from charges of *un*ethical behavior.[42] ILSI's formal ethics rule appears to have served the same objective. In this interpretation, the point of the public/private distinction and the 50 percent rule was not to correctly sort and label ILSI scientists, but to have a formal ethics rule in place that staff could point to as showing they did ethics. I call this doing (or even performing) ethics because it involved an active and ongoing process of claiming ethical purity (virtue, uprightness) by citing the rule and flexibly interpreting it to fit changing circumstances. In this view, the rule's principal purpose was to enable plausible deniability ("we've got it under control"), allowing ILSI's work to proceed. ILSI associates could deny wrongdoing and proceed to create industry-friendly science with their consciences clear. The public/private scientist distinction proved highly useful, providing ILSI legal cover while allowing it to engage the services of academic (nominally "public") scientists who were comfortable working closely with industry. By giving industry-friendly academic scientists such as Hill important seats at the table (of governing boards, scientific committees, and a CEO meeting, among others), ILSI member companies could ensure that industry's interests would be taken into account when decisions on the science were being made.

On the second IRS requirement for nonprofits—no lobbying or policy advocacy—ILSI's code of ethics clearly forbad attempts to influence policy and legislation by "strictly limiting [advocacy] to promotion of the use of evidence-based science as an aid in decision-making. ILSI does not conduct lobbying activities."[43] Although I was unable to discover concrete enforcement mechanisms for this rule in ILSI's primary documents, in the mid-2010s ILSI took the extraordinary step of temporarily suspending the Mexican branch for engaging in lobbying-style activities. This suggests that the *threat* of suspension was a de facto means of enforcement for major violations (or just violations that had become publicly known).[44]

Behind this public face of the legitimate nonprofit pursuing the public good, however, in-depth research reveals the existence of a host of carefully hidden informal channels by which member companies could shape the science ILSI made and ILSI executives could influence policy in branch countries. One such channel was the secret CEO meeting, whose company-chosen solution to the obesity problem would become the foundation of a new ILSI-fostered science. More such channels will come to light as we follow ILSI and its academic advisers in action.

ILSI Center for Health Promotion: Science Made in Reverse

ILSI was a supple organization that could spin off new divisions when pressing problems arose. In April 1999, the CEOs of nearly a dozen of ILSI's most powerful member companies had declared the promotion of exercise the only solution to the obesity crisis they would accept. ILSI's task now was to create the science to support their solution. Finding (industry-friendly) solutions to the childhood obesity epidemic was one of Malaspina's obsessions. He would throw his energies into it, persuading ILSI to create a new entity to address the problem just as he was about to retire. In early 1999 ILSI established the Center for Health Promotion (ILSI-CHP), an offshoot of the Research Foundation. Located in Atlanta and headed by Malaspina, whose home was in the city, the CHP was formed to jump-start the formation of a corporate science of obesity.[45]

In creating science for industry defense, the CHP would reverse the standard sequence of scientific activities in the two ways outlined earlier. First, it would start with the conclusion. Malaspina was not coy about this upside-down, last-first agenda. The CHP's goal, he declared, was "combating childhood obesity by promoting physical exercise."[46] The charge to the new organization was not to conduct research to discover the most effective solution to the epidemic; it was to defend the food industry by encouraging exercise rather than dietary restraint as the key approach. Reversing the standard order of science making in a second way, the CHP would move ahead with creating and disseminating a model childhood exercise intervention before the scientific justification for it was firmed up. This agenda was clear in the CHP's tax filings, which describe its mission as translating science into practical tools. In the staid language of form 990, it was "[to] function most actively in the area of technology transfer, moving science from the laboratory to practical application in communities to benefit the public." [47] This statement of CHP's mission presumed that the science had already been established, when clearly it had not.

While others within the CHP created the exercise program, Hill, who continued to advise ILSI on obesity, began developing two components of the exercise-first science: the analytic framework and an assortment of arguments rationalizing exercise as the approach of choice. Acting already as a quasi-corporate academic scientist, Hill's contributions would defend industry interests in both subtle and not-so-subtle ways.

THE ENERGY BALANCE FRAMEWORK:
GOOD FOR SUGAR, GOOD FOR SODA

In May 1999, just a month after the history-making meeting of corporate heads, ILSI-CHP cosponsored an international conference in Atlanta to establish a research and programmatic agenda on childhood obesity. The slim, paper-bound conference volume was ILSI's earliest major scientific statement on obesity that I have been able to locate. Malaspina was proud of the publication, describing it as "an authoritative monograph on the causes and consequences of obesity in children."[48] The bright-blue volume was edited by ILSI staff and published by the organization's in-house press.

In the book's introduction, Hill began elaborating an underlying conceptual framework to support the preferred solution. To understand the factors responsible for rising obesity, he drew on an analytic tool that not only was conventional wisdom in the nutrition field, but had served the sugar industry remarkably well since the 1920s: the energy balance framework. According to this simple scheme, obesity is a disorder of energy balance that occurs when energy consumed through eating exceeds energy burned by moving (both measured in calories), resulting in positive energy balance and thus weight gain. With all calories assumed to be the same, regardless of their macronutrient (carbohydrate, fat, protein) content and degree of processing, obesity can be managed by simply adjusting the number of calories consumed or the number expended to achieve numerical balance. For this introductory chapter in the blue book, this basic formulation of the model was sufficient; Hill and his associates would develop it more fully as time went by. Before moving on, though, we should take a little detour into that history of Big Sugar's scientific subterfuge.

During the early to mid-twentieth century, the sugar industry spent heavily to sponsor sugar-friendly science to help protect its market.[49] Scientists working with the industry mobilized the energy balance concept to argue that sugar had a place in a healthy diet and to defend sugar from claims that it was a major cause of the chronic diseases of modern life. The concept served the sugar industry in at least two important ways. First, it depicted obesity and diabetes as problems simply of caloric imbalance, and especially excess intake through overeating. The emphasis on the number of calories consumed allowed them to set aside a rival explanation for diet-related chronic diseases—high-carbohydrate and especially high-sugar diets, leading to hormonal dysregulation—and to ignore the question of the biological mechanisms that promote weight gain. Such mechanisms were irrelevant to the simple calories-in-calories-out model. Second, the concept treated all calories as the same—"a calorie is a calorie," or so the mantra went—allowing industry-affiliated scientists to argue that since sugar was no worse than any other food, obesity should be addressed by reducing calories from *all* foods, not just from sugar. Both arguments got sugar off the hook as a cause of obesity.

In the early twenty-first century, a different generation of scientists working with and for the soda (and junk-food) industries would put the energy balance concept to new uses. Despite surface similarities, soda-defense science was a new and relatively distinct science. Though both soda- and sugar-defense science built on the energy balance framework, they developed it in different ways, leaving few substantive overlaps between the two bodies of knowledge. Moreover, there is no evidence of continuity between sugar- and soda-defense science, no record that I found of the soda scientists reaching back and borrowing ideas or citing work from earlier in the twentieth century to give their work credibility or heft. Although it was a small step from the basics of energy balance to the new emphasis on activity, it was a novel and creative move that should be recognized as such.

ENERGY BALANCE, EXERCISE-FIRST, AND OTHER ELEMENTS OF SODA SCIENCE

In the late 1990s and early 2000s, the energy balance framework would serve the interests of the soda and food industry in a host of ways that will occupy us throughout this book. In his chapter in the blue book, Hill took advantage of several features of the framework to press industry-friendly ideas. One was the model's inability to specify which side of the equation was more to blame for the recent rise in obesity (eating or physical activity). That indeterminacy provided an opening for Hill and other industry-affiliated scientists to cast doubt on the argument of mainstream science and industry's critics that high-fat, high-calorie, high-sugar diets were contributing disproportionately to the obesity crisis. Without naming Brownell or the others calling for junk-food taxes, Hill tackled their critique of the ultraprocessed-food industry with a series of logical assertions pointing to the conclusion that the sheer complexity and unmeasurability of the factors behind the energy imbalance that was causing obesity to rise made it impossible to say which factors are most responsible:

> This complexity . . . makes it difficult—in fact, inaccurate—to identify a single cause that can be remedied to fix the problem. . . . One target that has been blamed . . . is the food industry. It is cited for marketing high-calorie, energy-dense foods in large portion sizes and for focusing advertising efforts toward children. . . . These are certainly important issues . . . but food intake is only part of the equation that produces obesity. Physical inactivity is certainly also a contributor. . . . Is it not logical [then] to single out the computer industry, [or] the manufacturers of video games . . . as contributors to inactivity and sedentary lifestyles? Automobile manufactures might also be blamed for discouraging physical activity. . . . Other observers may blame parents . . . [but] this is unlikely to be a useful process. . . . If "blame" is to be placed . . . it rests broadly across many segments of our society.[50]

Hill's simple, straightforward prose, which carried readers along from one point to the next, drew their attention away from a clever rhetorical strategy he was using to protect the food industry. By stringing all these industries together in one long list (food, computer, video games, automobile), he made them seem *equally culpable* for rising obesity rates. By refraining from ranking them by degree of likely responsibility, he was able to suggest that since *all parties* are to blame, *no party*—especially not the food industry—bears special blame. And if the food industry is not the main cause, the passage quietly implied, taxing its products should not be a priority solution. In this short passage, penned in the early stages of the creation of soda-defense science, we can see the rhetorical talents that made Hill such an asset to ILSI and the food industry.

Hill not only protected the food industry from being singled out for blame; he declared industry an essential partner in solving the obesity problem, arguing that the complexity of the issues signaled the need for cross-sectoral partnerships that included business leaders in the food and some other industries. With this, he moved the food companies belonging to ILSI from one side of the ledger (part of the problem) to the other (part of the solution), using purportedly scientific logic to secure their new roles as defenders of the public's health. Of course, as the producers of much of the food Americans consume, the food industry did have a critical role to play in improving food quality or shrinking portion size, for example. In his writing, however, Hill stressed the *benefits* of industry involvement while remaining silent about the *costs*. Here again we see the quasi-corporate scientist openly defending the industry that welcomed him into its fold.

In his writing, Hill protected the food industry not only by casting doubt on mainstream, diet-centric science, but also by energetically advocating exercise as the priority solution. Here again his rhetorical skills would serve him and the food industry well. To support the new emphasis on activity, Hill offered a few commonsense arguments. A successful strategy requires a variety of approaches, he suggested; "promoting physical activity is likely to be a key component."[51] In contrast to his enthusiasm for activity, he found dietary strategies were merely "potential interventions that need to be further explored." What we might call *diet tokenism*, the most minimal possible recognition of diet's role, allowed him to claim his views were balanced, when in truth they were not. Another tack was to stress the urgency of doing something. "Although more knowledge . . . [is] critical, we also need to move forward with the knowledge we have." Hill acknowledged the data problem but quickly finessed it, relying on his status as a well-known academic expert to whisk away any doubts.

These simple arguments—exercise should be part of any strategy, some solution is urgently needed—would be the first of many that Hill and his associates would offer for why exercise should be the approach of choice. Because it prioritized physical activity, I call this centerpiece of the emerging soda-defense science the *exercise-first solution*. It would not ignore diet; indeed, diet was a necessary component since moving had to be in balance with eating, and no one could reasonably argue that diet had *no* role in obesity reduction.

This question of diet would be especially troublesome for the soda scientists. For while they needed to offer some dietary recommendations, they had to protect the products of the sugary-drink and junk-food industry that was paying the bills, a difficult assignment given junk food's undeniable contribution to the obesity epidemic. This *diet problem* was the fourth distinctive feature of soda-defense science. Diet tokenism would be the first of many solutions.

It bears noting that in his chapter summarizing the opportunities available for addressing the obesity crisis, Hill failed to mention soda taxes or regulations on corporate marketing of junk food to children, strategies that some in public health deemed highly promising. Although the Kraft executive who spoke in private at the CEO dinner had highlighted the fearsome possibility of taxes and marketing limits, in this public venue Hill maintained a stark silence about these subjects. This would become the soda scientists' go-to approach over the next fifteen years. To my knowledge, they did not openly argue that soda and other junk foods should not be taxed or regulated; instead, they worked to keep the subject off the policy agenda by saying nothing about it, evidently treating it as a taboo topic. That silence did important work, though, producing the second of the two policy prescriptions that would be associated with soda science: taxation and regulation of the junk-food sector are not necessary. We can call this pair of policy ideas *exercise-first (soda-tax-never)*, placing parentheses around the second part to indicate that it remained largely unspoken.

EXERCISE-FIRST VERSUS EXERCISE AS ADJUNCT, PER THE SURGEON GENERAL

The exercise-first solution—which would be ILSI's (and Coke's) preferred approach for the next fifteen years—was not supported by the weight of scientific opinion. At the time, the role of physical activity in weight management remained poorly understood. (Even today the issue remains controversial.) Since the 1960s it was widely understood by experts and the public alike that regular exercise was good for health generally and an important component of a healthy lifestyle. Yet there was little agreement on the possible value of exercise for weight loss.[52]

In the late 1980s and the 1990s, new scientific evidence began to link regular physical activity to a wide array of physical and mental health benefits. Concerned that the vast majority of Americans were sedentary, in the mid-1990s the CDC and the American College of Sports Medicine (ACSM) convened an expert panel to review evidence on the health benefits of regular, moderate-intensity physical activity and prepare recommendations for promoting fitness. The article, published in *JAMA* in 1995, made no mention of any benefits for obesity reduction.[53]

At the time of this initial ILSI conference on obesity, basic ideas about energy balance and weight management were widely accepted. The 1995 Dietary Guidelines for Americans of the US Departments of Agriculture and Health and Human Services, for example, informed citizens that: "In order to stay at the same body weight, people must balance the amount of calories in the foods and drinks they consume with the amount of calories the body uses."[54] A year later the Office of the Surgeon General issued its first-ever report on physical activity and health.[55] This historic document clarified the relative role of diet and exercise, concluding that physical activity is a useful supplement to methods of weight loss that should be primarily dietary. "Physical activity is important for weight control. By using energy and maintaining muscle mass, physical activity is *a useful and effective adjunct to dietary management* for avoiding weight gain or losing weight" (emphasis added).[56] Elaborating, the report continued: "Weight loss resulting from increased physical activity [alone] is relatively slow, [but] the combination of increased physical activity and dieting appears to be more effective for long-term weight regulation than is dieting alone."[57]

In 1998, Hill himself published two articles in prominent professional journals asserting that both approaches were important. While stressing how little was known about the causes of and solutions to the obesity epidemic, an article in *Science* with John C. Peters framed rising obesity levels as a product of broad shifts in the food and exercise environments that were producing unhealthy lifestyles.[58] Citing the spread of high-fat, energy-dense foods, the supersizing of food and beverage portions, cutbacks in school physical education programs, and technological changes fostering sedentary lifestyles, the authors called for broad measures to restrict caloric intake and encourage activity. These were fairly mainstream views on the obesity question. Data from the National Weight Control Registry Hill had cofounded had shown that the secret to the success of long-term weight losers was to pursue both strategies, *with neither given priority*: consume a low-fat, low-energy diet and maintain a high level of regular physical activity.[59] Among the successful long-term weight losers, 89 percent used both diet and exercise, while only 1 percent relied on exercise only.[60] Hill's own data showed the limits of a strategy prioritizing exercise.

Now that the CEOs had spoken, however, the scientific messaging had to be adjusted. Although Hill and his associates would routinely state that diet was an important part of their program, following the new mandate, from then on virtually all the scientific effort would be devoted to supporting and elaborating the activity-first solution. Rhetorical tactics like dietary tokenism would allow them to claim interest in both sides of the equation.

TAKE 10! PROGRAM FIRST, EVIDENCE LATER

ILSI-CHP also reversed the ideal order of science making by translating the science into a practical tool before solid evidence was in hand demonstrating that the tool would work. In doing so, the CHP seems to have pushed the core scientific tasks of data gathering, hypothesis testing, and so forth into the position of afterthought, mere justification for conclusions already reached. One of the first steps Malaspina took was to hire an exercise specialist to create a model exercise program that could be used to spread exercise-first ideas far and wide. Developed in late 1999 and officially launched in 2001, Take 10! was a classroom-based physical activity program for elementary school students that built short activity breaks into the academic curriculum.[61] Although Take 10! was not designed as a weight-loss intervention, according to its creator it was considered one important part of a "whole cafeteria of programs" that together would add up to a solution to the childhood obesity crisis.[62] Clearly, ILSI wanted Take 10! to be seen as a worthy contribution to solving the childhood obesity epidemic.

On its website, ILSI boasted that the organization had begun addressing the obesity issue before the wider community took it up. The implication was that ILSI was so sincere about tackling this problem that in its ardor it had jumped ahead of the federal government and everyone else. ILSI did indeed anticipate the problem early, but there is another interpretation that merits consideration. In this view, ILSI wanted to get its exercise-first approach widely accepted *before* the mainstream science on obesity, which would certainly focus heavily on dietary approaches, was stabilized. In other words, it wanted to win the first battle of the sciences that Hill had referred to in Minneapolis by getting its conclusions published first. But in the rush to get them out quickly, Take 10! was introduced with virtually no data on its effectiveness. "We don't necessarily have any data collected on it," the specialist who developed it is quoted as saying in an enthusiastic article on the program, "but teacher reports indicate that the students love to get out of their chairs."[63] Measures of effectiveness would come a decade later and be equivocal.[64] Meantime, in the next few years Take 10! would be adopted in over a dozen US states and would become a model for physical activity programs in countries throughout the ILSI network, including China.

⁘

Its main tasks accomplished, at the end of 2004 the CHP board voted to dissolve the organization. The next year the Take 10! program, along with the CHP's financial assets, were folded into the DC-based Research Foundation. As the center and its energetic head took their leave, the larger project of promoting soda science would get a boost from two bold new partnerships that took the mantra of exercise-first directly to the American people.

CHAPTER TWO

Ad Hoc Partnerships

Taking Soda Science to the American People

In the chaotic climate of the early 2000s, when the science of obesity was still in its infancy, real-world solutions remained elusive, and ILSI's CHP was shutting down, entrepreneurial scientists saw an opportunity to do critical work in public health—with industry support. The Colorado team of James Hill and John Peters, who had served as key advisers to the CHP, were at the forefront of this movement in health entrepreneurship, and they will remain front and center here.

This chapter tells the story of two ad hoc partnerships, one an academic collaboration, the other an academic-industry partnership, and a pair of brilliant projects they launched in the early years of the new century. In these projects—a count-the-steps diet book and a walk-for-health nationwide campaign—ILSI's advisers-turned-entrepreneurs created a novel body of translational science promoting exercise-first and made it the basis of concrete weight-loss programs for use by individuals and communities nationwide. Translational projects such as these are rarely thought of as part of the science. But in a public health science whose ultimate aim is practical—slowing the rise of obesity—they form the essential third phase of the making-spreading-translating science dynamic.

In this chapter we follow the development of these two ventures to see the strategic ways in which Hill and Peters solved the problems associated with soda science—concealing industry involvement, finessing the diet question—and the larger effects of those solutions. While both projects worked to keep soda science alive, making it available to untold millions of Americans at a time when it had no institutionalized sponsor, they had other, mostly invisible effects as well. We will see how the new translational science placed psychology at the center of soda science, introducing new tenets to guide weight management: weight control must be fun, and cravings for junk food must be indulged. In promoting these ideas, the two linked projects fostered the creation of a new society of pedometer-wearing, weight-obsessed citizen-consumers. These new quantified selves were to count their steps, translate all their foods and activities into step equivalents, and then calculate their energy in and out to achieve energy balance. The approach was indeed creative, but it was also consequential, reinforcing the notion that individuals are responsible for their weight and health. The food industry was off the hook.

The Step Diet Book: Count Steps, Not Calories, and Eat 75 Percent of Your Doughnut

In the early 2000s, as the Center for Health Promotion was winding down, Hill was busy on the side working on ways to get the exercise-for-obesity message beyond the professional audiences reached by ILSI and into the hands of the ultimate target audience: the American consumer. The short-term partnerships he created—an academic collaboration and a university-industry partnership—would put him at the helm and provide an opportunity to realize his vision of making energy balance ideas the centerpiece of a large-scale social movement to stop the obesity epidemic. His chief collaborator in these ventures was John C. Peters (PhD 1982, biochemistry), director of the Nutrition Science Institute of the Proctor and Gamble Company in Cincinnati, Ohio. Hill and Peters went way back. They had worked together on numerous research grants and articles since the mid-1980s, and Peters had served as president of ILSI-CHP in its final years (2002–4). Launched in those early years of national panic about the obesity epidemic, these complementary soda-science activities that Hill and his associates developed would act as force multipliers for ILSI's efforts, intensifying the effects of its work.

Hill and Peters hit on a potentially lucrative idea: to repackage the basics of energy balance science as a fun, easy-to-manage diet, and market it to a public being inundated with urgent warnings about the ever-worsening obesity crisis. The two scientists teamed up with a nutritionist to offer a bold, new, "scientifically based, people-tested, easy-to-follow program that will take you step by step toward successful weight management."[1] This ambitious program was laid out in *The Step Diet Book: Count Steps, Not Calories, to Lose Weight and Keep It Off Forever* (figure 2.1). We are not accustomed to reading our diet books as scientific texts, but researcher-physicians working on nutrition and chronic disease often use diet books to introduce new weight-loss ideas. By far the best known of these is *Dr. Atkins' Diet Revolution: The High Calorie Way to Stay Thin Forever*, which fifty years ago introduced a potentially limitless readership to a model of

❝This is THE diet for the new millennium. It's a whole new lifestyle that is perfect for our hectic lives.**❞**

—MIRIAM E. NELSON, PH.D., author of *Strong Women Stay Young* and Associate Professor, Nutrition, Tufts University

the STEP

BOOK DIET

JAMES O. HILL, PH.D.,

JOHN C. PETERS, PH.D.,

with BONNIE T. JORTBERG, M.S., R.D.

Foreword by

PAMELA M. PEEKE, M.D.,

author of *Fight Fat After Forty*

DB23320537

COUNT STEPS, NOT CALORIES, TO LOSE WEIGHT *and* KEEP IT OFF FOREVER.

FIGURE 2.1. *The Step Diet Book* (cover)

weight loss very different from that promoted by Hill and Peters. Instead of blaming excess weight on energy imbalance and telling readers to count calories or steps (energy in and out) to lose weight, Atkins wrote that fatness was a metabolic disorder that could be managed by lowering the carbohydrate content of our food. Hill and Peters certainly anticipated making money from their step diet book, but, like Atkins, they had scientific objectives as well. Their primary goal was translational: turning their emerging but still unproven ideas into practical weight-loss strategies. In this they continued the pattern of science making in reverse we saw with Take 10!: application first, solid science later. But they also had to solve two problems always associated with soda science. They had to finesse the tricky question of diet (that is, junk food), and they had to find a way to reassure readers that their ideas were credible and scientifically robust, and would work in practice. Reading the diet book as a vehicle for scientific communication and persuasion allows us to see the clever ways the authors addressed these issues while turning the diet-book genre into a powerful vehicle for industry defense.

The diet book and the entrepreneurial ventures described in the next section return us to the larger question of how the story of industry-funded obesity science is told. As an alternative to the well-known account of that science as merely corporate PR, I've been spelling out a story that treats it as real—and really significant. Which better fits the facts? In *Appetite for Profit*, Michele Simon heaps scorn on the food companies for co-opting scientists into their nefarious projects. In her hands, scientists come across as naïve dupes of industry who sign on to projects that are virtually always initiated by industry. "Industry wolves in scientists' clothing," she calls them sarcastically.[2] Complaining about "how low . . . health . . . professionals are prepared to stoop in exchange for a hefty cash payout from Big Food," she writes as though their sole motive was money.[3]

While Simon's critique of industry tampering with science is well placed and much needed, my investigations of the ties between key soda scientists like Hill and Peters and the food giants show that, although the scientists' thinking was indeed shaped by corporate interests, the rest of this picture oversimplifies a much more complex situation. In getting university scientists to do their shadowy work, the companies relied not on co-optation but on incentivization, the provision of enticing benefits of all sorts. The evidence I uncovered also shows that the party taking the lead was just as often the scientists as the companies. Hill for example was a scientific entrepreneur par excellence. At many points in the twenty-year history of making a science of obesity that supports industry, it was Hill who, perhaps confident in the knowledge that corporate executives had his back, seized the initiative and took the story of soda science in different directions. The extracurricular projects explored in this chapter illustrate those talents and the kinds of contributions such short-term partnerships made to the larger project of defending soda with science.

THE NEW SCIENCE OF ENERGY GAPS
AND SMALL CHANGES

The Step Diet Book incorporated the insights of a new body of scientific work on energy balance. In a major article published in *Science* in 2003, Hill, Peters, and two colleagues argued that the key to arresting the rise in obesity is to offer a concrete and feasible goal for behavioral change at the individual level.[4] In that article, the authors introduced the notion of an *energy gap* of one hundred excess kilocalories per day. That small gap, which explained why Americans gained an average of 1.8 to 2.0 pounds a year, could be closed in most people by increasing lifestyle (that is, everyday) physical activity, by reducing portion size, or by doing both, without chang- ing the *type* of food consumed. By simply walking fifteen more minutes a day (roughly two thousand steps or about a mile) or eating a few bites less at each meal, the authors argued, one could close the gap, restore energy balance, and stop weight gain. And in that project, impressive results can come from making *small changes*, a term that would become one of Hill's signature slogans. To the authors, the small-change model had many things going for it. For most people, the entry barrier was low (who can't walk a little more and eat a little less?). The daily goals were easily achievable and could add up to substantial benefits over time, a claim that, in the absence of concrete evidence, remained conjectural. Cinching their argument for small changes was the glaring fact that the big-change models tried so far had produced little success. While a few critics dismissed the approach as grossly inadequate to the challenge of containing a rapidly rising obesity epidemic, in the persuasive language of the *Science* article there was much to gain and little to lose.[5]

What made such a program thinkable was the development of the modern pedometer. Human societies have long counted steps to measure distance, but it was only in the mid-1990s that health researchers became widely interested in using steps per day to measure physical activity and study its connections to health. Around 1995 this research got a big boost from the introduction of fairly accurate spring-levered pedometers with digital displays.[6] Building on this new line of research, in the mid-2000s Hill and Peters made step counting the centerpiece of their new diet, including a pedometer with each book purchased.

The new focus here on individuals calculating their personal gaps and fine-tuning their daily routines was a clever and easy-to-swallow proposal for solving the nation's obesity problem. It had a larger effect that should be noted, though: it quietly reinforced the notion that obesity was a problem for us as individuals to solve. Indeed, as though purposely protecting industry, the authors explicitly set aside the idea of trying to get larger forces in the environment to encourage healthy lifestyles. On the one hand, they wrote with passion that "we must mount a social-change campaign that will, over time, provide the necessary political will and social and economic incentives to build an environment more supportive of healthy life-style choices." Yet at the same time, they actively discouraged that project by lamenting that such campaigns take a long time to work, have not succeeded so far, and may never work anyway given innate biological drives for eating and inactivity. The focus now must be *"to help individuals manage better within the current environment"* (emphasis added).[7] That would be true of virtually all soda-science projects. Obesity is *our* problem. And we had the experts we needed to guide us to the answers. In writing the diet book and leading the nationwide movement campaign, Hill and Peters would become two of the century's most prominent gurus spreading the gospel of exercise as the number-one solution to excess pounds.

THE SELF-CULTIVATING, HEALTHIEST CITIZEN IN A WORLD DOMINATED BY BIG BUSINESS

This focus on the individual as the agent of weight management and health care more generally reflected wider trends in American culture since mid-century. One was a growing concern about new, so-called lifestyle diseases. In the 1950s, a cardiac crisis in which millions of middle-class men were suddenly dying of heart attacks brought home the importance of a class of diseases caused not by pathogens, but by the unhealthy eating and (in)activity habits of a society enjoying the fruits of postwar prosperity, including the single-family suburban home, the family car, and new processed convenience foods such as General Mills' Cocoa Puffs, Kraft's Cheez Whiz, and Swanson's TV dinners, all invented in the 1950s. With poor lifestyle habits deemed risk factors for the chronic, degenerative diseases of modern life, modifying diets and taking up new fitness routines were seen as forms of preventive medicine for which individuals themselves needed to take responsibility.[8]

Second was the growing power of big business in an era of financial crisis. In the 1970s both the country and the medical field faced serious financial crises whose ideal solution, according to fiscally conservative politicians and thought leaders, was to displace some of the costs of health care onto individuals by promoting self-help and self-health. Anyone who lived through the 1970s will remember the steady stream of get-healthy crazes—cycling, jogging, dieting, fitness, natural foods, vegetarianism, the self-help movement, the women's health movement, and many more—that washed over America, or at least middle-class America, in that decade. The problem of health, once a collective matter, came to be relocated to the sphere of individual lifestyle, and the individual was responsible both for detecting threats to their health and for addressing them. That decade saw the birth of a new morality, called *healthism* by the political theorist Robert Craw-ford, which is still with us today.[9] In healthism personal health became the key site for defining well-being and demonstrating one's good citizenship. Health was attained primarily through lifestyle modification and, because the environment was full of health hazards, could be achieved only by mak-ing health a daily preoccupation.

The emergence of the self-cultivating healthiest citizen in the 1970s went hand in hand with the rise of the corporate-dominated (or neoliberal) po-litical and economic order outlined in the book's introduction. In the late 1970s and 1980s, individual preoccupation with health both contributed to and, Crawford argues, provided political cover for corporate and political strategies aimed at deregulation, freeing big companies from the burden of proconsumer regulations. With individuals responsible for themselves, government health and safety regulations were no longer needed, or so the free-market ideology went.[10] This is not the place for a lengthy discussion of the larger political economy, but it's important to remember that this is the larger context in which all this became thinkable. With huge corporations dominating the national and global economies, product-defense sciences could proliferate, restrictions on unhealthy products could be quickly dis-missed, and individuals could be tasked with taking care of their health on their own.

By the early 2000s, when Hill and Peters were coaching Americans to take small steps to prevent weight gain, obesity was being depicted as an alarming, ever-worsening threat to the nation. In the society-wide war on fat that had sprung up, ordinary Americans, blamed and shamed for "not taking care of their weight" (the flip side of responsibility for fixing is blame for breaking), were desperate for solutions that promised to give them the ideal body of the good thin-and-fit American.[11] In advancing soda science as the answer to the obesity crisis, Hill and Peters were tapping into these deep currents in American culture and politics, dragging the tenets of big-business-ascendant neoliberalism into the twenty-first century. When Hill and Peters described the goal of soda science as "helping individuals manage better within the current environment," what they meant was they would leave the soda industry alone—no soda tax needed—while asking you and me to become entrepreneurs of the self who devote countless hours a week to keeping track of steps, miles, bites, portions, and other measures of lifestyle and consumer behavior in an arduous effort to keep our bodies from packing on pounds.

A NEW RECIPE FOR WEIGHT MANAGEMENT

To the student of product-defense science, *The Step Diet Book* makes for absorbing reading. The central problem the authors address will be familiar to anyone who has tried to lose weight. Most people can easily shed pounds by limiting how much they eat. But most quickly gain them back because their metabolism drops, but they haven't compensated by adjusting their energy intake or expenditure. One underlying difficulty, the authors explain, is that few people can keep accurate track of the calories they consume. So they provide an alternative way to compute progress: by translating everything into steps. Dieters can largely ignore calories and just count steps, using the pedometer included with the book and the tables provided to translate calories eaten and activities undertaken into step equivalents. Once they've mastered basic energy balance principles, dieters begin making small changes (eating less, walking more) to stop weight gain, then lose a sustainable number of pounds, and finally adopt a new lifestyle to keep the weight off permanently. For authors seeking to upgrade the role of exercise in the management of obesity, this simple move—which reframed the entire weight-loss project in the discourse of steps—was a stroke of genius, for it got readers to see steps as the basic unit of weight-related behavior and activity as an essential key to achieving a healthy weight. While promoting their exercise-first diet, the authors were also coaxing readers to become weight-obsessed healthist citizens, citizens who saw weight as a key to being a good American and were willing to follow the multistage process of lifestyle modification to achieve the good body.

The diet question was always a vexed one for the creators of energy balance science—some dietary solution was needed, but it could not hurt the (junk) food industry—and they developed different positions on diet at different times. In the new diet book, Hill and Peters offered a novel approach rooted in human psychology. Food restriction, they contended, is simply too difficult for people. So too is asking them to give up their favorite sugary and salty foods. Adding exercise to their daily lives is much more fun, especially with the pedometers provided with their book. Arguing in the name of consumer psychology, they stressed the importance of limiting changes to those that would fit relatively easily into people's lifestyles. The formula they hit on was to ask people to make small adjustments on both sides of the equation, while promoting the whole package as easy and enjoyable. Instead of worrying about good and bad foods, all readers needed to think about was balancing daily energy consumed and burned. (Whether calculating energy in and out is less trouble seems debatable.)

Following the logic of the energy balance framework, which sees all calories as the same, they told readers: "It doesn't matter if you eat all carbs or all fats or all sugars or all Twinkies (though we wouldn't recommend this!). . . . If you are taking in the same amount of energy as you burn each day, you will not gain weight."[12] So, the authors coached, instead of one doughnut at 210 calories, you eat three-fourths of a doughnut and save 53 calories. The same three-fourths principle applied to other foods and beverages, including soft drinks. Reducing your portion size by 25 percent works, the authors assured their readers, because it's a specific target, not just a vague suggestion to "eat less." By following the energy-in-balance rule, readers could also "'buy' an occasional fun food."[13] If you're craving a Baskin-Robbins chocolate-fudge ice cream, for example, you can either subtract calories from other parts of your diet or add calories to your program of activity, measuring both in step equivalents. In this new solution to the diet problem—which can only be called *dietary coddling*—Americans were told it was fine to consume junk food and sugary soda—in moderation. While readers today, attuned to the dangers of too much dietary sugar, might find these recommendations outlandish, they followed directly from the all-calories-are-equal energy balance framework. By claiming "on scientific grounds" that junk food is OK, exercise is the key, and obesity is simple and fun to fix, the step diet acted both to protect the ultraprocessed-food industry from critics and to extend the reach of the exercise-first message to a very broad public.

A Rhetoric of Science: Making Soda Science Credible

In the previous chapter I identified four features of product-defense science that distinguished it from academic science. (Product-defense knowledge is industry friendly, corrupt, and made backward, and it faces challenges concerning diet.) A close reading of the diet book brings out a fifth notable attribute, one I began noticing only after I started thinking of the book as a text in soda science. Precisely because product-defense science is skewed toward industry interests, its makers often indulge in an unusually strong *rhetoric of science* to present it as solid, precise, rational, and authoritative. The diet book is full of science talk. In this section we see how the authors mobilized the discourse of science for two ends: to highlight the trustworthiness of their work and to promote their figure of the step-counting citizen-consumer as the answer to the practical problems of weight loss.

A NEW DIET ROOTED IN SCIENTIFIC TRUTH

Rhetoric of this sort appears on the very first page of *The Step Diet Book*. In their acknowledgments, the authors present the new diet as the product of scientific truth: "This book is our attempt to communicate the truths we have learned about weight management over the past 25 years."[14] And those truths, they write, came from two sources. The first was the National Weight Control Registry, whose participants, because they "have learned how to take control of their own energy balance and their own weight . . . should be great inspiration for everyone who reads this book." The second source of truth was the authors' federal-government-supported research: "We owe a great debt to the National Institutes of Health for funding 25 years of our weight management research." No mention was made of the other grants Hill had received from food and drug companies (some nineteen by 2004, the year the book was published, compared to thirteen from the NIH), some as co–principal investigator with Peters. Nor did the authors mention the years Hill and, to a lesser extent, Peters had spent helping ILSI come up with industry-protective scientific ideas on obesity.

At that time, corporate sponsorship apparently was commonplace in nutrition research, and disclosure of potential conflicts of interest was not yet a professional norm. (Nutrition has lagged far behind medicine in its concern about the distorting effect of industry funding.)[15] The role of corporate support in Hill's and Peters's careers was not completely out of the norm. Yet they snipped industry out of the story of the career histories that had led to the new book. To be sure, they thanked Proctor and Gamble for graciously allowing Peters to work on the project. But that raises other questions. Skeptical readers might wonder why a company would release a senior scientist to work on a time-consuming project such as this unless it hoped to benefit in some way. The authors' decision to acknowledge only one type of funding (the more prestigious NIH grants) and research experience (the more academic projects) can be read as a form of rhetoric intended to establish their scientific authority by conveying that their science was unassailable and that they were free and clear of conflicts of interest. Looked at differently, mentioning their longtime connections to Coca-Cola and other food giants would certainly have raised eyebrows and led to concern about possible industry influence on their thinking. Telling a partial story about their careers served to distance them from the companies they had worked with. For readers who had not studied the authors' CVs—virtually everyone, since they were not available online at the time—the book's acknowledgments assured them their new diet guides were eminent scientists at the top of their game. The message was reinforced in a foreword by Pamela M. Peeke, MD, author of *Fight Fat after Forty*, who touted the authors as "two of the most powerful voices in obesity research in the country . . . the real scientists who have done the hottest, most current research."[16] To reinforce the point, the authors peppered their text with reminders of their grasp of the best science—"we bring you the latest scientific knowledge," "studies from researchers at Harvard," "scientific research . . . has shown over and over"—without citing the actual sources so curious readers could look them up. No reader could miss the importance the authors attached to the scientific grounding of their work.

THE SCIENTIFICALLY MINDED CITIZEN-CONSUMER

The rhetoric of science was quite versatile, and Hill and Peters found another productive use for it in their book. In promoting their new diet, the authors created the figure of the scientifically minded *citizen-consumer* to show us, the readers, how by consuming their diet advice we could remake ourselves to "lose weight and keep it off forever," as their title promised. Vignettes of everyday Americans like Brittany, Jill, and Paul, who had remade themselves and succeeded in losing weight, told us we could do it too. The key was to become a self-cultivating healthist citizen who thinks and acts scientifically. As citizens learning for the first time a "logical" approach to our weight, we needed to create a long-term plan, break it down into six stages, and follow the rules for each stage. Before proceeding, we had to translate our lives into the language of science: numbers, charts, graphs, and the like, colorful examples of which filled the pages of the book. One of our first concrete tasks was to learn energy balance concepts. In order to apply them to our lives, we then needed to become self-surveilling, self-disciplining consumers who maintain detailed records of the foods we eat and activities we engage in. Using the conversion tables in the book, we then translate everything into step equivalents so we can calculate calories in and out. The goal was to become a weight-preoccupied quantified self whose numbers on both sides of the equation balanced.[17] When in doubt, the authors instructed, we should follow the "masters."

The "masters" were the more than three thousand people who had joined the National Weight Control Registry (NWCR), lost an average of sixty-seven pounds, and kept them off for six years. "Pretty impressive!"[18] Yes, indeed! In the absence of large-scale studies demonstrating that the energy balance approach worked, the authors mobilized the experiences of the registry participants to serve as proof that it did. "Much of what we know about successful long-term weight management," the authors write confidently, comes from these "'masters' of weight management," whose strategies for losing weight and keeping it off they had analyzed in a series of scientific papers.

In truth—though readers would not know this unless they read the scientific literature—the NWCR was a wobbly source of evidence. The information on weight-loss strategies and success was self-reported and based on recall; unfortunately, most of us are not very good at remembering such things, especially if we had cheated on our diets. And rather than a random sample of Americans, the participants were self-selected volunteers who were overwhelmingly white, college-educated, middle-class women.[19] The authors nonetheless held up the participants as models, telling readers that if the registry participants could lose weight, "so can you." *So can you*: These words of encouragement clearly overstate the generality of the experiences of the mostly white, well-educated women in the registry. Abundant research on weight and weight-loss success by income level, education, race/ethnicity, gender, and other factors shows how those who occupy privileged statuses along these lines are more likely to succeed in losing weight than others. For readers who do occupy such statuses and who try and fail at the step diet program, such upbeat words may sound like false promises—as indeed they were. The authors addressed some of the concerns in the scientific literature, but the book stood.[20] There the tactic of subtle exaggeration, of stretching the meaning of science to include what they wanted it to include, served them well. The rhetoric of the scientifically managed citizen-consumer guided by prominent, government-funded scientists created a larger story of how science had led the reader to energy balance concepts, which would lead them to weight-loss success.

America on the Move: Teaching Energy Balance to the Masses

The step diet was but one of the more visible products of a massive social movement that Hill and his associates, especially Peters, set in motion in those initial years of alarm about the exploding levels of obesity in the country. This larger movement, brilliantly named America on the Move (AOM), was a national initiative to slow the rise in obesity by getting Americans to make exercise more central to their efforts to lose weight and keep it off. Another translational project, AOM sought to "inspire Americans to choose healthy lifestyles" by galvanizing change at the community level.

America on the Move was the product of years of planning and discussion. In 1998 Hill and his collaborators formed a new nonprofit, the Partnership to Promote Healthy Eating and Active Living, to tackle the growing epidemic of obesity.[21] Unlike ILSI, where the role of industry was deliberately obscured to veil the interested character of ILSI's science, here we find a full-fledged academic-industry partnership in which industry's participation—in funding, deciding on, and implementing the solutions—would be described quite openly as essential and helpful. Industry's relatively visible role in the new partnership allows us to glimpse, in a way not possible with ILSI, the kinds of influence the food giants wielded in making and spreading soda science. In this section we take an up-close look at the partnership and the AOM it gave rise to, to see how industry gained its influence and what difference it made to the scientific messages AOM would take to Americans around the country.

AN ACADEMIC-INDUSTRY PARTNERSHIP:
HOW INDUSTRY GOT ITS WAY

Industry partnerships for health are inherently contradictory projects. While including industry may enable a large-scale initiative to promote much-needed shifts in lifestyle, participants are often constrained to make compromises to accommodate industry interests. This partnership would be no exception. Describing obesity and related diseases as products of an "epidemic of poor [individual] lifestyle," in 2000 it brought together nearly 190 experts from all concerned sectors—academia, government, industry, advocacy groups, health professions—in a high-profile industry-funded summit meeting.[22] In starting this new endeavor, Hill and Peters would draw heavily on their ILSI ties. Half the partnership's board of trustees were drawn from ILSI-CHP, while the funding for the conference was provided in good part by food companies (and a few drug companies) associated with ILSI North America.[23] In return, the partnership and the campaign it produced would take ILSI's scientific agenda and Take 10! program to ordinary Americans. In this way, the various organizations involved in making and spreading soda science would become more deeply entangled as time went by. These ties had invisible effects. For example, over time groups pursuing the soda-science project in different ways would come to form a self-referential circle in which the parties promoted each other's work as some of the best available, spreading the word to unknowing consumers.[24] After the summit ended, ILSI's journal *Nutrition Reviews* published a veritable treasure trove of materials on the summit: the major papers, a summary, a transcript of (at least some of) the discussions, and a list of everyone who attended along with their affiliations.[25]

The main goal of the summit was to reach consensus among all the parties on which lifestyle behaviors should be prioritized for future research and inclusion in action programs. While many different voices were present at the summit, the way the event was organized may have given one voice—industry's—an outsized influence on the outcome. Before the meeting, participants were divided into three working groups, each charged with creating one part of a multidimensional conceptual framework aimed at identifying and sorting out factors on every level (from individual to community to societal) affecting individual activity and eating behavior. The organizers' hope was that, through wide-ranging discussion at the meeting, participants would agree on a shared framework, then take it home and use it to guide research and action programs fostering healthy lifestyles. Industry representatives, roughly one-fifth of the attendees, would of course be full participants in the discussions.

The way the summit was structured may have constrained the discussion by giving companies subtle veto power over which issues went into the framework. With each of the working groups including both public- and private-sector members, one can surmise that solutions that were unacceptable to the food industry would not be recommended, or perhaps even discussed, by the group. Think of it this way: If a representative of Dole Food, PepsiCo, or the Sugar Association (all of whom were present) is in your group, would you suggest taxing sugar or soda to reduce obesity? At a time when the junk-food industry was under attack for contributing to the obesity epidemic, it took little effort to grasp that if you wanted to work with the industry you could not promote a measure that would directly threaten its profits. Here again we see how, in discussions of soda science, the subject of soda taxes would operate as a (mostly) silent taboo.[26]

Published transcripts of the summit discussions suggest that corporate sponsorship, as well as the industry-friendly views of the summit planning group, especially its chair, James Hill, may also have limited the lifestyle options considered. In a wider political environment in which growing numbers of public health researchers were calling the food industry to task for the obesity epidemic and urging governmental action to curb soda sales and marketing, Hill, by now an adroit quasi-corporate scientist, seems to have taken pains to protect his funders. Remember that, in his work with ILSI creating a proindustry science of obesity, Hill did ethics by praising academic-industry partnerships and reassuring doubters that industry had no effect on his thinking. He would follow those same reputation-protecting practices in his work with his partnership. In their background paper for the summit, Hill and the others on the planning committee offered his standard defense of the food industry, asserting that "there is no clear single villain; everyone owns a piece of the responsibility."[27] Those comments by the organizers, as well, perhaps, as a concern about biting the hand that fed them, seem to have deterred participants from criticizing the industry. In transcripts of summit discussions, they tiptoed around the question of what to do about the ultraprocessed-food industry. Some offered arguments that mimicked those of the organizing committee. Others put a more positive spin on things, suggesting that, rather than pointing fingers, "we need to build on the positive."[28] Still others reported rather vapidly that "what [our working group] learned was not to blame, but to try to understand the context in which all of these different groups [in American society] are operating." In the materials I reviewed, no one is on record suggesting that the food industry take responsibility for its part in creating the epidemic or recommending options unfavorable to industry interests. This is quite remarkable given the wide range of people involved, some very prominent figures in the field with strong commitments to producing high-quality, neutral science. It's possible, of course, that some did speak out, but their views were not carried in *Nutrition Reviews*. If the censoring was in fact imposed by *Nutrition Reviews*, the ILSI journal, that would still be interesting, because the upshot would be the same: industry remained free of public critique.

AMERICA ON THE MOVE: "A FUN, EASY WAY TO STOP WEIGHT GAIN"

The consensus reached at the meeting became the basis for the next, even grander scheme. In 2003, the Partnership to Promote Healthy Eating and Active Living created the America on the Move (AOM) Foundation and now, leaving academic discussion behind, launched a national movement, led by Hill and Peters, to galvanize lifestyle change at the community level. AOM was the lucky beneficiary of one very generous partner: PepsiCo, a longtime member of ILSI.[29] Like its rival Coca-Cola, PepsiCo was keen to be seen as working tirelessly to address the obesity problem. This highly visible nationwide program provided an attractive opportunity to send that message, and the company jumped on it, reportedly volunteering its support.[30] The company's sponsorship totaled nearly $6.3 million, paid over the five years from 2004 to 2008. That bounteous support allowed AOM to keep its programs running for many years and to extend its reach across the country. In the late 2000s and early 2010s, PepsiCo funding dropped off precipitously, declining to virtually nothing in 2012, and the organization folded. As with all product-defense projects, when corporate funding ends, the project expires.

America on the Move took the ideas presented in *The Step Diet Book* and built them into messages and programs aimed at sparking lifestyle change and weight loss nationwide. Under the mantra "small changes = BIG results," AOM was a "fun, easy approach to stop yearly weight gain that easily integrates into our busy lives."[31] Schools would teach energy balance concepts, teachers would add ILSI's Take 10! program to their curricula, and Girl Scouts would earn "healthy lifestyle badges." Families and neighbors would form walking clubs, while pizza chains would give away free pedometers with the purchase of healthy toppings (not described). AOM named September the "month of action for a healthy you," calling it STEPtember to remind people what they should do all month (see figure 2.2). By going to the AOM website and pledging to make two small changes each day—take two thousand more steps (about one mile) and eat one hundred fewer calories (about a tablespoon of butter), the promotional materials explained—they were eligible for a chance to win a prize, from an iPod to a gift certificate for Dick's Sporting Goods. The solutions were as imaginative as they were appealing. But the larger product-defense effects of this effort should not be missed. By claiming "on scientific grounds" that junk food is OK, movement is the key, and obesity is simple and fun to fix, AOM worked both to protect the ultraprocessed-food industry and to extend the reach of

FIGURE 2.2. "STEPtember 2010: A Month of Action for a Healthy You"

the energy balance message to the broad public. Indeed, these simple ideas and appealing messages ("it's easy," "walk away the pounds," "let's do it together"), along with energy gap numbers that could easily be recalculated to fit different contexts, would travel remarkably well, enabling this science of obesity to be translated into concrete programs around the United States and the world.

Partnerships like AOM not only served as force multipliers for Hill's and ILSI's science; they also worked as scale multipliers, vastly expanding the numbers and kinds of people who heard the OK-to-junk-food, exercisefor-obesity message. AOM had an elaborate structure designed to deliver the message to as many Americans through as many channels as possible. According to a 2004 informational presentation, AOM engaged individuals through its website and 800 number, and through partnerships with an extraordinary array of grassroots affiliates, nonprofit national delivery partners, and corporate sponsors. These sets of partners were responsible for creating AOM-related programs and events, bringing them to local communities, and communicating AOM messages to consumers all over the country. These complex arrangements are summarized below.[32]

AOM PROGRAMS

A. Worksites/organizations: PepsiCo, Mayo, Guidant, Cargill, and others
B. Health professionals: AOM kit designed to reach patients through health-care professionals
C. Schools: Lesson plans teaching children energy balance concepts
D. Walking groups: Family and neighborhood walking clubs

VEHICLES BY WHICH AOM ENGAGES AMERICANS

1. AOM website and 1-800 number
2. AOM affiliates
   Located in twenty counties across the United States (2004)
   Spotlight: Tennessee on the Move (state version of AOM), engaged in legislative work, media events, education, and programs at worksites, as well as programs for children, including development of a Girl Scout badge promoting healthy living
3. AOM national delivery partners
   Academy of Family Physicians
   American Diabetes Association
   American College of Sports Medicine
   National Coalition for Promoting Physical Activity

Two more in development (as of 2004)
4. Corporate sponsors
    PepsiCo: AOM integrated into PepsiCo's 2005 launch of SmartSpot
       products
    Quaker Oats: AOM registration included in Quaker Oats Quakes
       walking promotions (pedometer giveaways, walking prizes)
    Nick N Willy's Pizza Chain: Colorado on the Move (state version of
       AOM), promoted through pizza chain that offers step counter
       with purchase of healthy toppings and other incentives

At least on paper, the effort seems to have been remarkably comprehensive, and the impact may have been as well. Hill states on his CV that AOM reached over three million Americans between 1999 and 2004. While there is no way to know how many lives the movement touched over its roughly ten-year life, the reach of just one campaign by a single delivery partner—a 2007 YMCA campaign reportedly involving over fifteen hundred YMCAs countrywide—suggests the number reached may indeed have stretched well into the millions.[33] Yet while the millions of Americans reached by AOM programs were told that industry was an AOM partner, they were left in the dark on how its involvement silently affected the weight-loss advice they received. Once again, the makers of corporate science had largely invisibilized the corporate in the science.

⁙

The step diet and AOM were important movements to spread soda science, but they were small potatoes compared to what was to come. With its products under attack by obesity experts in public health and its profits at serious risk, in the mid-2000s the Coca-Cola Company, the biggest seller of carbonated soft drinks in the world, launched a major campaign for self-defense. Its weapon of choice was science, more specifically, soda-defense science, and over the next decade it would undertake a series of initiatives bolder by far than anything the leaders of ILSI or these partnerships had imagined.

CHAPTER THREE

Coca-Cola

Fighting Science with Science

From the late 1990s, when the epidemic nature of obesity could no longer be ignored, the world's largest soda company had a target on its back. The threat to profits was only too real. With public health critics leading the charge and urging the institution of soda taxes, the battle to protect corporate profits would need to take place on the battleground of science. Through the efforts of Coke vice president (VP) Malaspina and his academic advisers at ILSI, by the mid-2000s the basics of a science to protect sugary soda had been worked out and were being spread through projects such as *The Step Diet Book* and AOM. To save the soda industry and its corporate giants, though, an effort on an entirely different scale would be required. Coca-Cola, the industry leader, would head it up.

In this chapter, we follow Coke as it tackled this unprecedented threat to its profits and reputation. Springing into action in early 2004, its first step was to hire an entrepreneurial vice president and chief scientific officer to take charge. Drawing on a treasure trove of emails exchanged between Coca-Cola's top science officer and her partners in academia, we see how she framed the project of corporate self-defense as one of open warfare. Weaponizing obesity science, she poured large sums into a series of extramural projects aimed at advancing the science of exercise while undermining the science of governmental regulation.

But Coke's science officer had more than industry-friendly science in mind. In her campaign to defend the company, she turned its grants program into a springboard for a secret project of cultivating a handful of major grantees and other top scientists to join Team Coke, an inner circle of advisers loyal to their benefactor. Through a silent logic of exchange—my corporate favors for your service to the company—she transformed them into virtual corporate scientists, willing to do the company's bidding, whether that meant providing scientific advice, giving proindustry talks around the world, or endorsing its paper ethics. This broader campaign for scientific defense, never before brought to light, is important because it had hidden effects that matter—on the scientists, on Coca-Cola, and on the wider field of obesity science.

Delving into these emails also gives us a chance to get to know some of the scientists on Team Coke. By reading their messages, we discover what exercised them, what amused them, and, most generally, what they found email worthy. Reading about the fun they had planning trips to conferences in far-off places, say, or scheming to silence their adversaries in public health, we can easily understand what led them to stay in the game despite the risks to their reputations. These more personal communications topple the narrative that reduces motivation in corporate science to simple material greed.

Profits at Risk, Industry Turns to Science

In the early 2000s, the environment facing the soda industry was becoming outright hostile. In 2001, the US surgeon general's Call to Action urged all sectors of society to join the fight against obesity and called for providing low-fat, low-calorie, low-sugar foods and drinks in schools and other sites where people congregate.[1] The "pouring rights" that since the mid-1990s had given the soda giants exclusive rights to advertise and sell their drinks on school campuses were being challenged as never before. In 2002, the Los Angeles Unified School District, the second largest in the nation, voted to ban soft-drink sales during school hours, inspiring similar moves elsewhere.[2] Nor did the bad news end there. That same year David S. Ludwig and his colleagues at Harvard University published the first scientific evidence tying consumption of sugar-sweetened drinks to childhood obesity: with each additional soda consumed a day, they found, a child's risk of obesity increases 1.6 times.[3] Why would sugar-sweetened drinks contribute to weight gain and obesity more than other categories of high-energy food, such as solid foods that are high in sugar and fat? For one thing, they are loaded with added sugar. (A twelve-ounce can of Coke, for example, contains thirty-nine grams of sugar, all added through processing.[4] Added sugars have none of the healthful qualities of naturally occurring sugar in fruit and milk, among other foods.) And drinking calories also fails to elicit the same fullness signals, or satiety, in our bodies as eating them. Rather than compensating by eating less later on, when we drink our sugar we can end up feeling hungrier and taking in more calories.[5] Sugary soda, it turns out, is a particularly insidious beverage.

In 2003, international banking firms issued a pointed warning to the food-and-beverage industry that the worsening epidemic of obesity posed a major risk to corporate profits.[6] In *Appetite for Profit*, Michele Simon, who was doing research for her book in 2003–4, describes companies tripping over each other trying to spin themselves as part of the solution to the obesity problem.[7] Many of the nation's largest food companies—including PepsiCo, Kraft Foods, General Mills, and McDonald's—launched programs in what Simon dubs "exercise philanthropy" to associate their brands with "active healthy living," while deflecting attention away from the role their food played in the obesity crisis.[8] Virtually all those companies would join ILSI in promoting "energy balance," "exercise," and "active healthy lifestyles" as the answers. Simon derides these efforts as mere public relations stunts. They were indeed PR pushes—Simon is absolutely right—but there was more to it than that. In order for the exercise-for-obesity project to be persuasive and effective, it had to be more substantial than PR; it had to be—or appear to be—grounded in *science*. Here is where ILSI and one of its most powerful member companies, Coca-Cola, working in tandem, performed outsized services for the industry. In this section we see how Coke turned to science to protect profits, and how its dynamic new science officer turned science into a weapon of corporate warfare against the critics in public health.

FUNDING ACTIVE HEALTHY LIFESTYLE PROGRAMS AROUND THE WORLD

Coca-Cola took the bankers' warnings seriously. In December 2003, outgoing chairman and CEO Douglas Daft told colleagues that obesity represented the biggest challenge the industry had faced in half a century. Yet the solution some were proposing—governmental legislation against soft drinks—was "absurd and outrageous." Instead, he insisted, the soda companies should take things into their own hands, fighting obesity by offering more beverage choices and promoting healthy, active lifestyles.[9] From at least 2003, Coke's annual reports to the Securities and Exchange Commission (SEC) listed challenges associated with "obesity and inactive lifestyles" as the first and most serious risk to profitability. The report for the fiscal year ending December 31, 2003, reflects the heightened concern and the company's plan of attack:

> Increasing consumer and regulatory awareness of the health problems arising from obesity and inactive lifestyles represents a serious risk [to our company]. We recognize that obesity is a complex and serious public health problem. Our commitment to consumers . . . includes adhering to the right policies in schools and in the marketplace [and] *supporting pro-*

grams to encourage physical activity and to promote nutrition education. . . .
We are committed to playing an appropriate role in helping to address
this issue in cooperation with governments, educators and consumers
through science-based solutions and programs. (emphasis added)[10]

Under new CEO Neville Isdell, in 2004 Coca-Cola announced it would
join the battle against obesity by sponsoring "active healthy lifestyle programs" as part of the company's widely publicized corporate social responsibility agenda. "Active lifestyles" was an especially appealing catchphrase
for Coke, which had been marketing its signature beverage as the drink of
Olympic athletes since 1928, when it sponsored its first Olympic Games in
Amsterdam (see figure 3.1). Beyond the Olympic Games, Coke had been
sponsoring sporting events—especially for global youth, its main target
market—for decades.[11] Visible on billboards around the world, the ubiquitous images of strong, healthy young athletes smiling broadly while enjoying Coca-Cola sent the unmistakable message that the drink made people
healthy. Certainly, it did not make them fat or metabolically diseased!

FIGURE 3.1. Coca-Cola ad for the 1928 Olympic Games in Amsterdam

Now committed to being part of the solution (as it defined it), in 2004 Coke began energetically investing in the programs it had announced in the SEC report. That year it funded a school program for American kids called Step with It!, distributing Coca-Cola-red pedometers (it's an actual color) and encouraging youngsters to take at least ten thousand steps a day.[12] The next year, it began supporting active healthy living programs in every country where it operated. Under the new banner, the company funded a huge range of activities for youth, including Copa Coca-Cola in Mexico (soccer), the Coca-Cola School Cup in Kazakhstan (football), Live It! in North America (fitness), and many more.[13] By 2014 Coke was funding over 330 such programs, reaching teens the world over with its message of good health and fun with Coca-Cola.

A SCIENCE STRATEGY TO FIGHT ENEMIES IN PUBLIC HEALTH

In order for the corporate message on active lifestyles for obesity to be persuasive, it had to be supported by solid science produced by credible scientists. Given the known bias of company scientists, Coke would have to recruit external scientists from academia or government or both to create the science for it. In 2004, it hired Rhona S. Applebaum (PhD 1981, food safety and microbiology), a longtime scientific affairs officer in the food industry, to serve as vice president and chief scientific and regulatory officer (from 2013, vice president and chief science and health officer). Although I met her only once and then very briefly, she struck me as an articulate, energetic, and effective professional. As head of the company's Global Scientific and Regulatory Affairs, Food Safety, Health and Nutrition division, with a staff of forty and an annual budget of $21 million, Applebaum was responsible for developing and executing Coke's global health and well-being strategy on issues of food safety, diet, and health.[14]

In her new post, Applebaum took charge of a new project on obesity, one that would serve as the scientific companion to the active healthy lifestyle programs. More covert than the lifestyle programs—this was nowhere to be found in the annual sustainability reports—the obesity project was a massive, multifaceted initiative to legitimize the activity solution scientifically, pushing forward the work Coke vice president (and ILSI president) Malaspina had begun in the mid-1990s when he recruited Hill to join the ILSI team. Because this work involved a deliberate, sustained effort over many years to promote "facts and evidence"—Applebaum's term for science—I call it a *science strategy*, or even *campaign*, because at times it took the form of a crusade.[15]

Aiming to counter criticism of sugary drinks, Applebaum weaponized science, framing her efforts as part of a larger battle to defend her company against the attacks of the public health critics. In an early 2010s talk at the International Sweetener Symposium, she declared that, in the obesity discussions, "misinformation is rampant; too often science and facts are MIA," likening absent facts to soldiers missing in action. In that same talk, she called for the industry to "fight back" by funding "defensive and offensive science and research" that "balances the debate" by countering the "agenda-driven science" of the industry's critics.[16] In her battle plan, the "evidence-based scientists" working with industry would demolish the "agenda-driven scientists" in public health. The term *soda-defense science* thus had a very literal meaning for Coke. Over the next decade, the language of warfare, far from "mere rhetoric," would be critically important to Coke and affiliated scientists. That language would create a rationale for their actions, a discourse of "extremist them" versus "rational us," and the moral foundation for a collective fight against their foes. And the company's chief science officer, who would do more than anyone else to secure the vital funding for the project and mobilize a team of scientists to join in, would be the *corporate scientist in chief.*

Coke's science strategy to neutralize the obesity threat unfolded as a series of opportunistic investments in people and programs where Applebaum saw opportunity to advance the exercise science of obesity. In rationalizing its scientific work on obesity, the company adopted the industry's stock-in-trade framework, energy balance. As a sustainability report explained: "There is widespread consensus that weight gain is primarily the result of energy imbalance. . . . No single food or beverage alone is responsible for people being overweight or obese. But all calories count. . . . We believe all foods and beverages can have a place in a sensible, balanced diet combined with regular physical activity."[17] Boiled down to essentials, the message was: eat and drink all you want, then exercise it off. Given this statement of the problem of healthy weight, it is not surprising that most of the company's investments in science went into the activity side of the equation, that is, finding ways to encourage consumers to exercise off what they ate and drank. Only one went into the dietary side of the equation, and that was undertaken before Applebaum joined the firm. Let's take a look.

In early 2004, the company established an online platform, the Beverage Institute for Health and Wellness, to support nutrition research and education on beverages. Hosting web-based information, webinars, conferences, and other activities, it provided "evidence-based science" to "educate" health professionals and consumers about the benefits of active healthy lifestyles to health and well-being, the role of beverages in delivering hydration, and other industry-friendly points.[18] To ensure its science reached important global audiences, it created country-specific websites for major markets from Brazil and Mexico to India, Indonesia, China, and Russia. Applebaum served as executive director of the institute, but Maxime Buyckx (MD 1986), Coca-Cola's director of health and wellness programs (1995–2015), often served as its public face.[19]

As an early effort to mobilize science for corporate ends using in-house means, the Beverage Institute approached its tasks in a rather crude and clumsy way. By foregrounding scientific issues that were blatantly self-serving (the importance of soda in providing hydration, for example), the Beverage Institute quickly invited the scorn of critics like Simon, who called its label "education" "Coke-speak for shameless marketing."[20] Indeed it was, but instead of simply dismissing the institute as a failed initiative, as many do, we should register its significance as part of a longer history of corporate obesity science and seek out its uncharted effects. The creation of the Beverage Institute just months after Daft's warning about the historically unprecedented threat on the horizon marked a turning point when Coke itself seems to have embraced (some notion of) science as its fundamental answer to its obesity problem.[21] The Beverage Institute had long-term, invisible effects as well. For the next eleven years, its staff, especially Buyckx, a corporate scientist par excellence, would be actively involved in spreading the new science (including the notion of hydration) far and wide, sometimes in cooperation with ILSI. We will meet him again in Washington, DC, and Beijing.

FUNDING EXERCISE MEDICINE AND SCIENCE

Though Coca-Cola could not create a credible science of obesity in-house, it had ample financial resources it could tap into to persuade external researchers to assemble a science for it. As head of scientific affairs, Applebaum invested heavily in two sprawling science projects that advanced the science of exercise for obesity in powerful ways. In the first, in 2007 the company became the first founding corporate partner of Exercise Is Medicine (EIM), a now massive program to encourage health-care providers to prescribe exercise as medical treatment on a routine basis. This remarkable new program promoting exercise in the clinic was the creation of the American College of Sports Medicine (ACSM) with the American Medical Association. The idea originated with Robert E. Sallis (MD 1987), a California-based family and sports physician, who was inspired by a lecture given by Steven N. Blair, a passionate advocate of exercise for health, based since 2006 at the University of South Carolina. As Sallis explained: "Regular physical activity has been shown to be essential in achieving and maintaining weight control.... [Steven] Blair ... has dedicated his amazing career to proving scientifically that exercise is medicine and is responsible for much of the evidence base behind it."[22] Coke's enthusiasm for EIM is not hard to understand. Let's listen to Sallis again: "The key is to shift some of the public health focus off obesity and onto physical activity. People need permission to be fat and still be healthy. The way to do it is by getting them more active."[23] One critic described the hidden agenda this way: "EIM aims to turn public health's focus away from obesity entirely, and towards physical activity."[24] That, the evidence suggests, was precisely Coke's agenda: let's play down obesity and highlight exercise.

For Coke, partnering with EIM offered an opportunity to extend the exercise-first message beyond the sphere of public health into the huge field of clinical medicine. The company presented the sponsorship, several million dollars in the first year, as a key part of its contribution to solving the global obesity problem. Coke continued to support the program after that initial donation. Between 2010 and 2015 it invested some $785,000 in the EIM program at the ACSM alone (that is, not counting investments in foreign programs), an average of more than $130,000 annually.[25] Every year, the company's annual sustainability report celebrated the partnership, charting the growing number of countries with active EIM programs contributing to the battle against obesity (there were forty-three by 2014–15). For Coke, EIM was a near-perfect solution to the obesity problem: a highly visible program that said "we are part of the solution" while redefining the central problem of public health as inactivity.

In the second investment in Coke-friendly science, the company introduced a grants program to support scientists whose research was friendly to corporate interests. Two of the projects were awarded multimillion-dollar grants. The first was the ISCOLE study (International Study of Childhood Obesity, Lifestyle and the Environment) managed by the Pennington Biomedical Research Center at Louisiana State University, Baton Rouge (circa 2010–15, $6.4 million). The second included the Energy Balance and Energy Flux Studies conducted at the Arnold School of Public Health at the University of South Carolina (2010–15, about $4.7 million).[26]

By any measure the grants program was a smashing success for the company. Critical public health scholars who have studied the output found that research funded by the company dealt largely with Coke-preferred topics and produced Coke-friendly results. By one conservative count, the program resulted in some 389 publications between 2008 and 2016, written by 907 researchers and published in 169 different journals.[27] Over half the studies examined focused on topics in energy balance, physical activity, diabetes, and obesity, all potentially helpful to Coke's project of defending soda.

Most of the articles concluded that physical activity is more effective than diet in weight control and that soft drinks are largely harmless. Many were published in influential journals, giving the work both visibility and credibility within the scientific community. This remarkable output—the sheer number of articles with industry-friendly conclusions—reveals the power of deep corporate pockets to shape the scientific truth.

While it's impossible to measure its impact precisely, the grants program was clearly instrumental in building a body of published research supportive of industry-friendly solutions to the obesity problem. Coke-funded research would lay the groundwork for two ambitious projects in exercise science described in the next chapter. But the grants program would have another benefit as well. In Applebaum's hands, the program would serve as a springboard for a secret project of cultivating scientists, and especially the recipients of large research grants, to become special friends of Coke. We turn to that in a moment.

Paper Ethics, Performing Integrity

The grants program brought many benefits, but it also carried extraordinary risks. In offering direct company-to-investigator research funding, without the kinds of institutional layers that separated funding companies from co-operating academic scientists in an organization like ILSI, Coke risked the public perception that it was trying to buy favorable findings. Of course it *was* trying to do just that, but that had to be hidden, or the company could face serious reputational damage. In this section we see how Coca-Cola, working with peer firms, sought to solve the ethics problem by creating a much-touted set of guidelines that amounted to paper ethics: principles that looked good primarily on paper. Turning to ethics in action, we then take up one concrete case—the development of a research protocol—to show how easy it was for Coke to quietly ignore those guidelines and leave its mark on the research it funds.

SCIENCE ETHICS BY AND FOR INDUSTRY: THE ILSI PRINCIPLES

In the late 2000s, as wider debates over conflicts of interest in industry-funded research grew more heated, ILSI's North America branch set out to address the issue on an industry-wide basis. ILSI-NA leaders were deeply concerned about the emerging literature on conflicts of interest in industry-funded work. In that literature, "conflicts are typically treated as disqualifying factors in scientific papers. . . . Even with complete and open disclosure," scientists who accept industry funding "are regarded, at least to some extent, as of suspect scientific credibility."[28] If industry-funded science and scientists are deemed suspect, the entire project of corporate support of research that ILSI embodied would be at risk. Once again, industry had to be defended. This time the defense encompassed its collaborations with academics.

In 2007, the organization convened the broadly multisectoral Working Group on Guiding Principles. Its mission was to find solutions to potential problems of scientific integrity and conflicts of interest in industry-supported research in nutrition and food science.[29] Applebaum, a trustee of the branch, joined the effort on Coke's behalf. The working group, which included distinguished academic and industry researchers, was supported by educational grants from ten of the biggest food companies, suggesting how far the concern about bias in nutritional science had spread throughout the industry. Those firms were Cadbury Adams USA, Coca-Cola, ConAgra Foods, General Mills, Kraft Foods, Mars Snackfoods US, PepsiCo, Proctor and Gamble, Sara Lee, and Tate and Lyle. The discussions culminated in the 2009 publication of a multiauthored article outlining eight principles for minimizing the potential for bias in research funded by food companies. These principles, listed below, were offered as a first set of ground rules.

In the conduct of public/private research relations, all relevant parties shall:

1. Conduct or sponsor research that is factual, transparent, and designed objectively, and, according to accepted principles of scientific inquiry, the research design will generate an appropriately phrased hypothesis and the research will answer the appropriate questions, rather than favor a particular outcome;
2. Require control of both study design and research itself to remain with scientific investigators;
3. Not offer or accept remuneration geared to the outcome of a research project;
4. Ensure, before the commencement of studies, that there is a written agreement that the investigative team has the freedom and obligation to attempt to publish the findings within some specified time frame;
5. Require, in publications and conference presentations, full signed disclosure of all financial interests;
6. Not participate in undisclosed paid authorship arrangements in industry-sponsored publications or presentations;

7. Guarantee accessibility to all data and control of statistical analysis by investigators and appropriate auditors/reviewers;
8. Require that academic researchers, when they work in contract research organizations (CRO) or act as contract researchers, make clear statements of their affiliation; and require that such researchers publish only under the auspices of the CRO.[30]

Reflecting the group's ambitions—and the seriousness of the threat—the article was published simultaneously in the top six peer-reviewed food science and nutrition journals, an unprecedented move. Of the sixteen authors listed, five were corporate scientists, seven were academic researchers, two were ILSI-NA staff, and two were advisers based in a corporate strategy firm. In other words, just under half the members were academics, some of whom had close connections to ILSI. The two advisers, who were listed first, were affiliated with the International Food Information Council (IFIC), an industry-funded sister organization of ILSI whose mission was to disseminate (often industry-friendly) nutrition information. Applebaum took part in the deliberations and was listed as an author. The document she helped prepare would form the centerpiece of her company's ethics strategy going forward.

Despite the seriousness of the effort, given the strong industry involvement in the development of the principles, we should not be surprised to find they were kind to industry. On the face of it, they appeared tough, listing eight practices that were not to be allowed. These eight practices were designed to form a solid boundary between industry and science that would block efforts by company staff to intervene in sponsored research. Yet the supposedly impenetrable boundary line was pierced by tiny gaps. Closer inspection reveals that the rules restrict practices that are likely to be rare while omitting from consideration those likely to be common. The guidelines are concerned primarily with blunt forms of influence, such as the payment of money in exchange for industry-friendly findings (principle 3) or the structuring of research to yield a particular outcome (principle 1). These science-for-sale sorts of ethical violations are easily detected and for that reason likely to be rare. At the same time, more common forms of corporate influence on sponsored science are not covered. Like most (if not all) sets of guidelines, the ILSI principles have no provision for the more subtle and pervasive forms of influence that operate through the bestowing of largely nonmonetary incentives. Professional and personal benefits (opportunities to spread one's ideas globally, trips to exotic destination) can have as much persuasive power as money itself, quietly leading scientists to ask questions, choose interpretations, or reach conclusions that are consistent with the preferences of their sponsors.

The principle that all financial interests or paid authorships must be disclosed in talks and publications (principles 5 and 6) is also a feeble demand. Disclosure is a necessary but far from sufficient measure to safeguard research from undue influence exerted by funding organizations. As scholar-critics like Marion Nestle have argued, disclosure alone is not enough, since simply naming the funders reveals nothing about how they might have shaped the questions asked or interpreted the findings.[31] The principle of disclosure is also a flimsy safeguard because it allows for plausible deniability. By disclosing their funders, researchers can declare they've "done ethics" without having to address any of the tough questions about unconscious bias on their part. On top of that, the ILSI principles skirt the question of enforcement, leaving it to the parties involved to create mechanisms to ensure compliance. Without an enforcement mechanism, the principles are just that: abstract norms with no bite.

Taken together, these features of the guidelines suggest that their main purpose was to *perform* integrity, to *demonstrate* a sincere effort to handle conflicts of interests and minimize industry bias. Just as ILSI's 50 percent public-private rule on participation was designed not so much to produce ethical behavior as to insulate the organization from allegations of *un*ethical behavior, the ILSI principles on research funding seem devised to demonstrate compliance with standard norms on science ethics, while leaving gaps through which industry influence can slip. The guidelines allowed Coke and its academic grantees to plausibly deny wrongdoing ("see," they can say, "we follow these widely agreed-on principles"), while flexibly interpreting the principles to fit their needs.

ETHICS IN ACTION: DEVELOPING A PROTOCOL

Corporations trying to shape the research they sponsor leave few paper trails, making it difficult to discover how such practices work or how common they are. Yet email exchanges between grantee Steven Blair of the University of South Carolina (USC) and Coca-Cola staff, in the months before they finalized the initial agreement for the 2010–13 Energy Balance Study, provide tantalizing hints about the ways suggestions of corporate actors can slip across the industry-science boundary to leave corporate fingerprints on the research. Principle 2 of the guidelines "require[s] control of both study design and research itself to remain with scientific investigators." While that sounds like an effective deterrent, the fuzziness around the meaning of "control" leaves room for corporate funders to tamper with the design of a proposed study. A company could make any number of suggestions on study design, while still leaving the ultimate "control" of the project design in the hands of the investigator. Such suggestions would be hard to reject by researchers hoping for generous financial support.

Such a dynamic appears to have played out in the preparation of the research protocol in mid-2010. To track that process we begin drawing on the email archive. In the spring of that year, Applebaum, Blair, and one or two technical/scientific representatives of Coke (KC and SR) met in person in Toronto to begin fleshing out what a study of energy balance and obesity might look like. (Here and below, citations to each email used can be found in a section following the notes to this chapter.) On June 1 Applebaum wrote to her partners sheepishly apologizing for nagging, but reminding them she needed a short, paragraph-length proposal and budget ASAP if the project was to start in 2010. That demand drove the planning process quickly forward. The emails reveal the development of the study design to be a collaborative effort in which, over a series of face-to-face meetings, email conversations, and teleconferences in June, July, and August, Blair sketched out a draft protocol and solicited input from Applebaum and Coke's two scientific representatives. At the initial meeting, the group had discussed a prospective observational study, but questions of sample size, the nature of the study population, and the duration of the project remained unsettled. (In a prospective study, researchers follow a group over time, looking for outcomes as they unfold.) After talking to one of the Coke experts (SR), Blair increased the proposed length of the study, creating a one-paragraph project description and asking the other expert (KC) to "please modify and improve." Adopting a deferential tone, he replied to Applebaum: "Please let me know if the current document is sufficient at present, or if you want me to make changes."

In late July, SR at Coke informed the USC team that she had added "a few comments to the draft Study Design / Methods Outline" and, in a separate message, had "a few comments on the Main Study Questions document." USC participants thanked her for "continuing to refine the documents" and "mak[ing] some additional great comments . . . in the protocol document." The project coordinator at the university input the changes and updated the documents. All parties joined a conference call on August 11 and then met in Atlanta for three days to develop the final protocol. Blair seems to have accepted the corporate input into the design of his study as just how things were done, a routine part of doing business / making science with Coke. While the details of the additional changes are not included in the emails, the emails show decisively that Coke influenced the duration of the project. It mostly likely affected the size and nature of the study population and may well have influenced other things during private conversations for which there is no record. There is no evidence of the use of external reviewers to evaluate the document.

The ILSI guidelines were attached to the formal grant agreement signed in late 2010, a semipublic demonstration that ethical issues had been taken care of. But had they? The formal guidelines left enforcement to the parties involved. Coke seems to have set the issue aside. There was no discussion in the emails that I reviewed about the propriety of Coke meddling in (or was it simply perfecting?) the science. Nor is there any evidence from the document-signing stage that anyone checked what happened (that is, who actually developed the protocol) against the guidelines, to make sure the ethical rules had been followed. (I checked the correspondence from August 20, when the group meeting ended, to November 4, when the document was signed.) Instead, Coke's technical staff and USC scientists developed the protocol together, as if the rule requiring that control of the study design must remain with the investigators had no bearing on their work.

A Soda-Defense Squad for the Industry

Coke's ethics document sought to create a strong. unbroken boundary between industry and science, the company and its grantees. As the story of the protocol suggests, however, that boundary had breaches and breaks that permitted crossing by parties on both sides. The thousands of emails exchanged between Applebaum and her research partners during the early 2010s give us an extraordinary opportunity to inspect these border crossings to see what was going on behind the scenes. Reading these communications, one quickly discovers that the relationship between grantor and grantee involved far more than a simple transfer of funds. In these conversations we can tease out a hidden project of corporate cultivation of academic scientists in which Coke's chief scientist took advantage of her grantees' reliance on her company's support to ask favors of them. By building multidimensional relationships with these scientists, Applebaum evidently sought to cultivate them to contribute to the broad campaign to save Coca-Cola from the obesity threat. The apparent goal was to create a team of *virtual corporate scientists* willing to serve Coke's needs with few questions asked.

How did Applebaum turn her grantees into virtual corporate scientists? So far I've discussed the transformation of an academic scientist like Hill into a quasi-corporate scientist in terms of incentivization. I've argued that the corporate distortion of science operates through the provision of financial and professional incentives that lead university-based grantees to subtly adjust the way they conduct research (the questions asked, data used, interpretations offered, and so on) to produce industry-friendly conclusions. The same argument applies to Coke's grantees, who were motivated perhaps first and foremost by the desire to please their funder and receive support for future projects. On both sides, the influence would have worked semi- or unconsciously, allowing both parties to deny corporate influence on the science with consciences clear. The availability of that vast email archive allows us to see how these incentives operated in everyday scientific practice. But the emails also permit us to go further and uncover additional pathways of corporate influence, and how they worked not just to produce industry-friendly scientific results but also to transform academic scientist-grantees into allies of Coke.

The Coke VP exerted influence over her grantees not merely by dangling the promise of more research money and support before their eyes (the incentivization strategy). Her communications were much more sophisticated than that. She also deployed a carrot-and-stick technique that involved making outright demands of her grantees, often in highly directive, impatient language that could not be ignored, and then later, after the demands were met, showering them with effusive praise. Applebaum also wielded influence by making small asks (and some big ones), conveying subtle or not-so-subtle expectations that the grantees should help her out with something. These requests for favors were often couched in voluntaristic language ("only do it if you want") that belied the subtly coercive logic of "our money for your help." There was also the lure of personal charm. In a style akin to cheerleader, Boy Scout troop leader, and army drill sergeant all rolled into one, the Coke VP often wrote to her "gentlemen" (as she sometimes called them gaily) in an informal, lighthearted way that introduced an element of playfulness into their interactions. Finally, the rhetoric of warfare that often animated her messages created a charged atmosphere and a reminder of what was at stake in the battle over obesity science. Instead of one, then, we can discern at least five pathways by which Coke's corporate scientist in chief sought to win over her grantees for Coke.

In the next two sections, I draw on the email exchanges to show how Applebaum used these pathways of persuasion, how they worked to create friends for Coke, and what effects those dynamics had on the scientists and the company. First, though, I want to introduce the scientists who were the object of so much corporate attention.

TEAM COKE

As she rolled out the grants program in the late 2000s and early 2010s, Applebaum began nurturing an inner circle of trusted grantees. Based on how often they communicated by email, we can identify a core team of nine researchers. Displayed in table 3.1, the group included scientists Coke had supported with

TABLE 3.1. Team Coke, 2010–15

Name	Institution(s)	Connections to Coca-Cola and ILSI, Illustrative Only	Email Exchanges with Applebaum, 2010–15
Steven N. Blair	Arnold School of Public Health, University of South Carolina	Major Coca-Cola grantee (Energy Balance and Energy Flux Studies), with Hand	2,020
Gregory Hand	Arnold School of Public Health, University of South Carolina (2010–13); School of Public Health, West Virginia University (2014–15)	Major Coca-Cola grantee (Energy Balance and Energy Flux Studies), with Blair	1,949
James O. Hill	Anschutz Health and Wellness Center, School of Medicine, University of Colorado	Long history of collaboration with ILSI and Coke, ILSI adviser since 1995, longtime member of the board of ILSI North America	1,364
John C. Peters	Nutrition Science Inst., Proctor and Gamble (2010–12); Anschutz Health and Wellness Center, School of Medicine, University of Colorado (2013–15)	History of collaboration with ILSI, president of ILSI-CHP	2,703
Peter T. Katzmarzyk	Pennington Biomedical Research Center, Louisiana State University, Baton Rouge	Major Coca-Cola grantee (ISCOLE, International Study of Childhood Obesity, Lifestyle and the Environment), with Church	1,921
Timothy Church	Pennington Biomedical Research Center, Louisiana State University, Baton Rouge	Major Coca-Cola grantee (ISCOLE, International Study of Childhood Obesity, Lifestyle and the Environment), with Katzmarzyk	2,031
Carl (Chip) Lavie	John Ochsner Heart and Vascular Inst., Ochsner Medical Center, New Orleans; also Ochsner Clinical School, University of Queensland, Brisbane, Australia	Consultant and speaker on fitness and obesity for Coca-Cola	759
David B. Allison	School of Health Professions, University of Alabama, Birmingham	Longtime member of the board of ILSI North America	569
Michael Pratt	National Center for Chronic Disease Prevention and Health Promotion, US CDC	History of research funded by Coca-Cola and backed by ILSI	309

Source: UCSFL Food Industry Documents.

Notes: The emails often had multiple recipients. The number of emails is skewed upward in 2014–15 by the extensive communications about the GEBN. Although Hill and Blair were the senior partners at their respective universities, Peters and Hand played important administrative roles in the GEBN, which directed more communications their way. The numbers include only the emails in the UCSFL Food Industry Documents archive, not all the emails exchanged between the parties. Count taken on October 20, 2023.

big multiyear grants (Steven Blair and Gregory Hand at South Carolina, Peter Katzmarzyk and Timothy Church at Pennington), two longtime advisers of ILSI at Colorado (Hill, Peters), and a handful of other old friends of Coke and ILSI (Michael Pratt at the CDC, David Allison of the University of Alabama, and Carl [Chip] Lavie at the Ochsner Medical Center in New Orleans).

During these years Applebaum and her team were in constant touch via email about matters big and small. By rough count, during the six years between 2010 and 2015, Applebaum and Blair exchanged a total of 2,020 emails, roughly one a day. In 2014–15, they emailed one another 1,605 times, more than twice a day. (These numbers include emails between the two of them and emails with multiple recipients. The numbers count not all emails, but only those collected in the archive.) The conversations with Hand were equally frequent, followed by those with Hill and Peters at Colorado, and the two researchers at Pennington. (The number of email exchanges between Applebaum and Peters is most likely skewed upward because Peters served in key administrative roles in collaborative projects being developed at the time. As we will see later, Hill worked more closely with other Coke officers, perhaps explaining his lower number of communications with Applebaum.) Because of their vital role in advancing Coke's science agenda, I call this group of nine Team Coke. There was also an outer circle of researchers with whom the Coke VP was in regular but less frequent touch.[32]

A WARRIOR FOR COCA-COLA

Of this outstanding team of exercise specialists, Steven N. Blair stood out in ways that would have endeared him to Coca-Cola. Unlike Hill, whose field was obesity research, Blair was an expert in physical education, earning degrees in that field (1939–2023; BS 1962, PED [doctor of physical education] 1968) and spending twenty-six years at the Cooper Institute (founded by "father of aerobics" Kenneth H. Cooper) in Dallas, rising to president and CEO of the institute, before joining the Arnold School of Public Health at USC in 2006.[33] By the time he arrived at the Arnold School, Blair had become one of the world's premier experts on physical fitness and activity and their benefits for human health. He was the first author of a landmark study published in *JAMA* in 1989 establishing the association between physical fitness and mortality from all causes.[34] He played a central role in preparing the two major governmental reports on the health benefits of physical activity mentioned in chapter 1, the 1995 CDC/ACSM recommendations and the 1996 surgeon general's report. In a career that combined research with advocacy, Blair was an outspoken advocate for exercise and fitness. (As a physical activity specialist, his primary interest was in advancing the cause of exercise and fitness; diet was of secondary importance.) As we have seen, he was the intellectual father of the Exercise Is Medicine program, which Coke backed with large sums. Blair not only was a prolific scientist who had received countless federal grants and honors; like his colleague at Colorado he had a history of working with industry as a grantee (Body Media, Technogym) and scientific board member (Jenny Craig).

What made Blair really stand out from the rest was his position on, and passion for, physical activity. The South Carolina scientist was known for his view that inactivity was not just important but was the major public health problem of our day—more important even than obesity. Promoting activity, he argued, was the answer not just to obesity but to chronic disease more generally. A main basis for this claim are data, first published in 1995, showing that individuals with obesity who are at least moderately fit have half the risk of dying in the next several years when compared to normal-weight individuals who are unfit. That article noted that the evidence for a role of activity in weight loss and control is inconsistent, making the topic controversial, but "an active way of life may have important health benefits, even for those who remain overweight."[35]

Within a few years that suggestion would develop into a claim that inactivity is a more important public health problem than obesity. For Greg Critser, author of *Fat Land,* Blair had become a "gladiator" in a new "crusade" to convince the American public to forget about losing weight and learn instead to be fat and fit.[36] But the data underlying his claims were problematic. Those data were gathered from Cooper Institute clients, who were overwhelmingly rich, white, and male, and thus could afford the time and money required to become fit. Just as Hill's National Weight Control Registry data were classist, relying on the reports of mostly white, well-educated, middle-class women, Critser argues that Blair's new fitness-over-fatness crusade was eminently classist, suggesting that everyone can be fat and fit, when few working Americans had the time and money for fitness. Critser points out that the project was in part a personal one for Blair. "He is, as he likes to say, 'Fat, fit, and bald—and none of those things are likely to change.' . . . With his confident, engaging manner, mile-long vita, and persuasive debate style, Blair is his own best advertisement for his fit and fat campaign."[37]

Blair's argument that fitness, not thinness, is the key to good health sparked outrage in London in 2001, with a spokesman for a leading UK-based anti-obesity organization retorting: "No amount of exhorting fat people to become fit is going to solve the obesity problem."[38] Undeterred, Blair grew bolder over time. In a typical provocation, in early 2009 he published a short article in the *British Journal of Sports Medicine* with the attention-grabbing title "Physical Inactivity: The Biggest Public Health Problem of the 21st Century." Although the article's text said only that inactivity is *one of* the most important problems and *may even be* the biggest problem, because its crucial importance was underappreciated in public health and clinical medicine, Blair appears to have decided to right a wrong by staking out a more extreme position in the title. "I have been often criticized on this point," he said in a 2019 interview, "but my response is: do some research with good measurements of PA [physical activity], obesity, and health outcomes, and show that we are wrong!"[39] In Steven Blair, Applebaum found a veritable warrior for her cause—a blunt, outspoken, risk-taking academic expert who was a fierce advocate for physical activity, and evidently had few qualms about working closely with industry. The email communications suggest that he would become her right-hand man in academia.

Quid Pro Quo: Small Favors

During the early 2010s, Applebaum stayed in close touch with the members of her team. She emailed them frequently, praising their work, passing along news items, expressing her happiness or outrage at the latest developments in the war on obesity, doling out corporate goodies or, on rare occasion, just sharing an amusing observation. Reading the long email chains, one discovers that many if not most of Applebaum's messages to her team took the form of quid pro quos, in which the Coke VP offered the grantee a benefit or perk and then subtly (or not so subtly) asked for a favor in return. Through this process of give and take, her grantees became more deeply entangled in a relationship with their grantor that pulled them further into the company's orbit. In this way, the awarding of grants became a springboard for a larger project of cultivating scientists for Coke.

Focusing on the exchanges between Applebaum and Blair during 2009–15, in this and the next section we follow several interactions in which Applebaum bestowed benefits on one of her prized grantees and then asked for favors in return. We start in this section with a relatively innocuous request: for assistance with scientific questions. In tracking their conversation over time, we are interested in how Applebaum sought to win over her grantee (that is, which pathways of influence she used), how he responded, and what effects these back-and-forths had on the grantee and the company.

FUNDS FOR RESEARCH

The tight bond between Applebaum and Blair was rooted in their shared interest in research demonstrating the value of physical activity, especially in addressing the obesity crisis. In the fall of 2010 Applebaum, Blair, and Greg Hand (his collaborator at USC) signed a fifty-three-page research agreement to guide the ambitious Energy Balance Study. (This is the project whose protocol we just considered.) The project, which would measure the contribution of energy intake and expenditure to changes in weight and fat over a full year, would run from late 2010 to the end of 2013, with a budget of $2,520,722. By moving more of the research focus to energy expenditure (that is, activity) and measuring its contribution to obesity, the project promised a big payoff for Coca-Cola.

Coke was evidently pleased with the initial results, for it responded positively to requests for more funds. Before long the USC team was on the Coca-Cola gravy train. In 2011, a new agreement was signed to undertake the energy flux study to determine whether energy balance works differently at high and low levels of energy flux. This project would last two years (2011–13), with a budget of $718,499. Meantime, the Energy Balance Team requested continuation of their project to follow their research subjects for a second and then third twelve-month period (to 2014 and then 2015) in order to obtain even more robust data. These proposals were approved, bringing the total for the Energy Balance Study to nearly $4 million.

Beyond arranging for major grants such as these, the Coke vice president seems to have taken care of her partner's funding needs in more personal ways, as suggested by this genial 2014 message to him: "Finances tight. 2015 will be very tight. Thankful . . . to have had the op and foresight to work with good folks like you and others and spread the 'gospel' on health and well-being. Not sure what lies ahead re support. Makes no sense. Anyway— pls know I'm working to squirrel you more before the end of the year. Stay tuned/fingers crossed." (Applebaum often made up abbreviations for common words like "please" and "opportunity.")

For the USC team, the benefits of the arrangements with Coke went far beyond the generous funding for research. During the years Coke sponsored their work, they were able to support a large research staff. The budget permitted them to invite leading experts to serve as advisers. Recruiting seven top experts on energy balance—four of them members of Team Coke—allowed Blair and Hand to further strengthen their ties with Coke partners. The large-scale Coke funding also made South Carolina one of the leading centers of research on energy balance and obesity worldwide, if not the leading center itself. With two major projects underway, Blair, Hand, and their colleagues could foresee a steady stream of publications for years to come. These were benefits most academic scientists could only dream of, and they offered powerful incentives to stay on good terms with the Coke VP.

FACTS ON DEMAND

There were subtle quid pro quos, however. In exchange for this largesse and the promise of more to come, Blair and his colleagues on the sponsored research team were expected to respond to requests for scientific assistance at any time. This kind of reciprocity is of course routine among scientists; what's noteworthy here is the inclusion of a powerful industry official in the network of scientific exchange. One type of request was for *facts on demand*. Applebaum would sometimes ask Blair to look into scientific questions that piqued her curiosity, usually because they suggested additional ways in which physical activity benefited human health. One day an article on heartburn came to her attention, and she saw a promising research opportunity. She sent a message to Blair with this in the subject line: "Heartburn on the rise—and scientists aren't sure why—TODAY Health." Dispensing with small talk, she dashed off this note: "Wonder what the prevalence of HtBn is in those Physically Active vs. inactive/sedentary," including a link to the article. Especially with the link enclosed, nonresponse was not a real option. Blair replied a couple hours later, cc'ing his junior colleague Xuemei Sui (known as Mei): "I have never seen any data on this. . . . Mei, do we have any data on heart burn in [the] LEAN, ACLS, or EB [studies]?" Here Blair, the full professor, directs an assistant professor in his department to find the answer for Coke, drawing her into the penumbra of Coke-serving scientists. Professor Sui, eager to please, responded to them both within a few minutes saying: "Yes, we do have frequent heartburn data" and offering her help: "I can take a quick look at the data to try to answer Rhona's question." Six days later she reported the results, which showed that active men and women were slightly *more* likely to report heartburn than inactive men and women—the wrong result. "This is probably unexpected?" she wrote and asked if it was worth pursuing the question with a different method.

While the fate of the heartburn research remains unknown, this correspondence illustrates a dynamic that was common in the exchanges. Blair was asked to draw on his professional resources (time, technical knowhow, social network) to serve as a scientific consultant to the Coke vice president. Blair was undoubtedly happy to be of use to his funder, especially because he too was a strong proponent of activity, but over time these exchanges of funding for scientific advice began to produce effects. The emails suggest that doing favors for Applebaum came to be normalized, part of Blair's routine professional life. As he responded to more and more requests for scientific assistance, the line between Blair the academic scientist and Blair the virtual corporate scientist would begin to blur. Blair's service as Coke's go-to scientist had spin-off effects as well. As he recruited colleagues

COCA-COLA › 95

to assist him, he drew them into the Coca-Cola circle, expanding the number of scientists comfortable with doing the company's work. When those colleagues were junior professors, Blair's actions modeled the professional behavior of a successful senior scientist for them. The model of success he embodied included accepting money from and working closely with the food industry. Such a dynamic would have spread a proindustry mind-set among his colleagues. Applebaum's requirements for "facts and evidence" had unseen effects that mattered.

Quid Pro Quo: Big Favors

As Applebaum's relationship with her grantees deepened, her requests became more demanding. She urged them to join her in promoting exercise science at conferences around the world. As concern about industry funding of science spread, she beseeched them to endorse Coke's position on ethics. When the calls for government regulation of the soda industry grew insufferable, she appealed to them to challenge those demands in print. These were big favors and they had big effects. In this section we drop in on two more conversations between Applebaum, Blair, and some of his colleagues, to see how Blair, showered with the kind of benefits only a company like Coke could offer, would be gradually molded into a poster boy for EIM, a propagandist for Coke's ethics, and a defender of corporate science against the "extremists" in public health.

POSTER BOY FOR EIM, PROPAGANDIST FOR COKE

From all indications, Coca-Cola loved EIM because it focused discussion of obesity on exercise, leaving food and drink out of the conversation. After investing heavily in the program's launch, Applebaum began energetically promoting it by sponsoring scientific sessions on EIM theory and practice at conferences around the world. This was a labor-intensive project. As Applebaum tried to put together sessions in which the best speakers would be available to lecture on the key topics on the right continent on the specified date, she could be highly demanding in her calls for help. She would often issue last-minute requests to her inner circle of experts. "Need your help asap," she would write, "We got the slot for the ICN [International Congress of Nutrition] Luncheon [in Bangkok in 2009]. . . . I've sketched out a brief of what is needed. We need a confirmed name(s) asap." Or she would write: "I'm thinking EIM needs to have a presence here [at the World Congress of Cardiology in Dubai in 2012]. . . . I'm thinking a booth in the exhibit area.

Views? We can try to help here." For Applebaum's grantees, these "sticks" in the carrot-and-stick ploy were hard to ignore.

As the scientist whose ideas inspired the program and a world-famous authority on activity and health, Steven Blair was the ideal champion for EIM, and he often said yes to Applebaum's appeals, willingly traveling the world as a spokesman for the program. On several occasions, Blair helped Applebaum set up the sessions. While she located promising conference opportunities, sketched out the topics to be covered, and supplied a steady stream of funding and encouragement, Blair facilitated collaborations and contacted people in his network, coaxing them to join the group in Canada or Scotland—all expenses paid and a good time guaranteed. In some cases it was a Team Coke effort, with Chip Lavie and Michael Pratt, especially, pitching in to help.

To show her gratitude, Applebaum showered the South Carolina scientist with lovely things. In addition to the all-expense-paid trips to exciting foreign destinations and honoraria for the talks—usually paid to a university research fund "so as to avoid the antis from making mischief," as Applebaum put it conspiratorially—he enjoyed special enticements. In Moscow, he and the other speakers and their spouses were treated to a specially arranged tour of the Kremlin. Beyond the material rewards, Blair's participation in the sessions earned him effusive praise (you did "a fantastic job in Moscow") and promises of more trips to come ("you two [the other was Kenneth R. Fox of the University of Bristol] are such a dynamic duo that you can both expect more trips to . . . spread the 'active, healthy living gospel'" around the world). For Blair, undoubtedly the biggest reward for his contributions to the company was the opportunity to carry the Olympic torch near Oxford as part of the 2012 London Summer Olympics. (As a sponsor of the Olympic Games, Coke had the honor of arranging a section of the torch relay.) For Blair, a runner who had devoted his life to promoting sports and fitness for health, bearing the torch must have felt like an extraordinary career achievement, and he declared himself thrilled by the invitation. Such "carrots" surely offered more powerful inducements than the sticks.

Did Blair express any concern that he might be working too closely with Applebaum and Coke, that others in the field or the media might see a conflict of interests? He may have had private concerns, but the emails I read contain no hint of such doubts or of worries that his integrity might be compromised. Quite the opposite, he seemed to revel in the praise Applebaum heaped on him. For Blair, the constant accolades seem to have acted as an enticement to keep using his talents to defend the company that treated him so well. Just as Hill was incentivized by ILSI to *not see* corporate influence on his work, Blair was cajoled, incentivized, and manipulated into *not notic-ing* any influence Coke might have had on his thinking.

At the same time that he was receiving this corporate largesse, Blair was subject to demands that, though veiled, could be quite taxing. To see these subtle forms of manipulation in action, let's consider the Q&A document on Coke's scientific ethics that Applebaum shared with Blair and other experts speaking at a big meeting on activity and public health in Sydney, Austra-lia, in 2012. Marked "Classified—Internal Use," the document defended the company's funding of science in the strongest possible terms. The compa-ny's goal was to ensure that "there is absolutely no conflict of interest and . . . the highest level of scientific integrity is maintained." These were promises any company would find hard to meet. When supporting scientific meet-ings, the document stated, Coca-Cola took a hands-off approach: "We have no involvement in deliberations about the agenda, topic areas, and speakers involved in the meeting. . . . Our role is to invite the experts and reimburse them for their travel and expenses only." The Q&A section of the document was equally emphatic, as this sample of back-and-forths suggests:

Q : So when your detractors say "industry money corrupts" are they wrong?
A : When it comes to money we provide, they are WRONG. We helped author and fully endorse the ILSI Guiding Principles.
Q : What about the experts taking part in your symposia? Isn't it a conflict of interest to pay them to speak at [this meeting]?
A : Absolutely not. . . . The information they present and the opinions they offer are their own.

The claims of noninterference notwithstanding, in email exchanges with her team about the sponsored sessions, Applebaum often offered sugges-tions that violated the spirit and sometimes also the letter of the ethics state-ment. She had strong opinions about which speakers should be invited. She would also offer her opinion on how the science should be done or pre-sented, couching her comments in self-deprecating language so as to not appear overly directive (which would be an ethical violation). On one occa-sion when asking a scientist to give a talk, she wrote: "*My POV*—it's time the 'calories out' side of the eqn was given more prominence at these nutrition/ health mtgs" (emphasis added), knowing full well that a casual comment like that coming from her could not be taken lightly. Or she would suggest research topics that "would be of interest—at least—*my humble opinion*" (again, emphasis added). On occasion she also edited PowerPoint presen-tations that would be used at a session. When Blair's work was edited, he seems to have okayed the changes with little apparent dissent or discomfort, at one point offering to adjust the content and length of a talk to fit Applebaum's needs.

In pressing her grantees to take the company's official ethics as their own, Applebaum was asking them to stretch the truth, to tell what might generously be described as an idealized story about Coke's ethical practices. In her message to Blair and other speakers, Applebaum offered the document "for your use in any way you choose," adding in an offhand, even humorous way: "we are providing *a bit of* guidance for you to use—*if you so choose* should you be asked *some wicked questions* regarding your session and why you are here. . . . Here you go. Pls share with your panelists if you so choose" (emphasis added). The casual language implied that the grantees' use of Coke's language about its ethics was fully voluntary, when in truth refusal to use the specified language might well jeopardize their relationship with the company. Whether Blair saw the "offer" of the document as an official request or took it as simply one of the give-and-takes that was part of his ongoing relationship with his sponsor the emails do not say. These sly requests matter, though, because they altered his relationship to the company in ways he may not have recognized. Essentially, Applebaum was calling on the scientists to ignore the gap between ethics as promised and ethics in practice, and to serve as propagandists for Coke's (supposedly) irreproachable ethics. In endorsing the suggested language and using it with reporters, Blair and other grantees would become accomplices in a "classified—internal access only" project of protecting Coke (and themselves, as grantees) from ethical censure. Through these kinds of interactions, the grantees were quietly maneuvered into becoming defenders of—some might say apologists for—the company's ethically dubious ways of doing things.

KELLY MUST BE CONFRONTED: TEAM COKE FIGHTS BACK

As Coke's chief science and health officer, Applebaum was constantly monitoring the science press and the scientific literature, looking for work she found distasteful ("agenda-driven"). Often mobilizing a rhetoric of warfare, she would send her thoughts, along with the links, to her team asking for suggestions about how best to fight this new battle. In this project of defending corporate science against criticisms and attacks, Steven Blair would prove invaluable. He quickly joined the cause and before long began drawing to Applebaum's attention articles that provoked his ire. Whatever his motives—staying in her good graces, keeping his funding coming, enjoying a good scientific fight—for work he deemed particularly outrageous he went further, taking it on himself to organize colleagues to collaborate in writing a commentary or even undertaking new research to combat the unfriendly work. As he energized his partners on Team Coke to counter one wrong-thinking article after another, Blair would become something of a bad-news scout and fixer for Applebaum and a vital asset to the company.

In the public health community, a top contender for enemy number one was Kelly Brownell, whose calls for soda taxes going back (at least) to the mid-1990s were a constant irritant. In 2012, around the time Mayor Bloomberg of New York City proposed a sixteen-ounce cap on soda size, Brownell and a colleague published a carefully reasoned article in the *New England Journal of Medicine* arguing that, despite the outrage directed at Bloomberg's plan, state and city governments possessed the legal authority to regulate the food environment to protect citizens' health. Though the food industry would no doubt challenge such measures in the courts, their review of possible arguments suggested such challenges were unlikely to prevail.[40] Steven Blair quickly spotted the offending article and sent it on to Applebaum and six others on Team Coke, writing in exasperation: "Kelly won't let up!" Hill and Allison chimed in, leading to a fast-paced conversation and, in no time at all, a concrete plan of action emerged to counter Prof. Brownell:

JH: [Sarcastically] Great . . . lets regulate [everything!] . . . Won't we all have fun living this way. . . . Seriously it might be fun to write a letter to *NEJM* suggesting all of this. . . .

DA: [Humorously] Could be a modern version of Jonathan Swift's "Modest Proposal."

JH: exactly !!!! should we do it?

JH: I think the 3 of us should develop a piece—either letter or editorial—along the lines of Swift's Modest Proposal—this could be fun and if NEJM doesn't publish it, someone will

DA: I'm game.

JH: I am happy to do a first draft and sent [*sic*] it to both of you

DA: Terrific. Here are a few things that might be useful to mention or draw on. [Includes ten fairly well-developed ideas.]

SB: As usual, David has many great ideas, and I am sure Jim will add to them. I will struggle to make any contributions to the outstanding work that will result from the two of you! I do have one idea for you to consider. . . .

JH: I had initially thought of this as a tongue in cheeks [*sic*] short piece saying that obesity is so complex that Kelly has not gone far enough. . . . Do we want just a tongue in cheek piece or do we want to suggest alternative approaches. I could go either way.

DA: I favor a stand-alone commentary as opposed to a reply to a specific paper or a letter to the editor. Is there an upcoming meeting where the three of us ... could sit down and brainstorm?

Team Coke was embattled, to be sure, but Applebaum's "men," as she also called them, were having fun fighting the bad guys together. Applebaum was deeply grateful for her team's help and encouraged them with unstinting praise, often packaging it in humorous language. After a favorable article appeared in a leading newspaper she sent it along to her team, writing playfully, "nice article in *USA Today* all you quotees." When the *New York Times* carried an article with the delicious title "Is Sitting a Lethal Activity?" and a Coke executive forwarded it to eighteen colleagues noting that a company grantee, Steven Blair, had been quoted in a new book on the subject, Applebaum forwarded the entire exchange to Blair, gushing: "I just have to share this internal email. I feel like a rock star—I actually know the one, the only—Steven Blair!!" Silly? For sure, but endearing at the same time.

CORPORATE DEFENSE: A BATTLE FOR HEARTS AND MINDS

This deep dive into the conversations between corporate America and some of the country's most prominent obesity experts sheds new light on the project of corporate science. Academic researchers working closely with industry face serious reputational risks. Why do they do it? The standard answer centers on financial incentives: the anticipated "hefty cash payout from Big Food." These emails tell a different story. Reading the back-and-forths, one comes away with the sense that the scientists' devotion to Coke's cause stemmed less from the material incentives than from the close personal relationship they had developed over many years with a savvy executive who combined tough demands and harsh reminders of the battle underway with endless personal charm. And Applebaum drew them into something bigger: a team of like-minded experts all fighting to win the war over obesity science. The scientists she cultivated clearly enjoyed the rush being part of Team Coke, strategizing with their teammates to spread the gospel of EIM, undermine their enemies in public health, and defend their benefactor, all the while keeping corporate secrets secret. The scientists on Team Coke were complex figures with many motives for joining in, not the least of which was their belief that their science was correct.

The email exchanges also reveal that the efforts to save Coca-Cola from the obesity threat entailed far more than sponsoring a few articles—or even a few thousand articles—with industry-friendly conclusions. Applebaum seems to have realized the challenges and mobilized her many talents in the service of a wide-ranging campaign to cultivate highly respected external experts to become virtual corporate scientists willing to deliver for Coca-Cola. Whatever her winning formula—I've laid out a few key elements of it here—her project of cultivating university scientists paid off for the company in critical ways that have not been brought to light. Working singly and jointly, the inner circle of grantee-advisers provided scientific advice and conducted scientific research on demand; promoted Coke-favored science and programs, including EIM, around the world; publicly supported the company's ethical practices; and mounted a spirited defense of Coke science. Furthermore, through their interactions with colleagues and students, they created a larger circle of researchers who came to learn that accepting industry support for research was normal and desirable, and called for doing corporate favors in return. These varied services provided by Team Coke must have helped protect Coke from criticism, perhaps delaying the time the secrets of soda science would be revealed by several years.

∵

After assembling a team of loyal academic advisers, Coke's VP and chief science officer was ready for bigger challenges. Digging deeper into corporate pockets, in the early 2010s Applebaum and her colleagues on Team Coke launched the two most ambitious soda-science projects of all time. That story comes next.

CHAPTER FOUR

Soda Science at Its Peak

After a decade of gradual development, in the early 2010s the time was ripe to take the science of energy balance to a higher level. Two new organizations shared that grand vision: a "technical committee" of ILSI's North America branch and a new ad hoc academic-industry partnership sponsored by Coca-Cola. Behind these new vehicles for the advance of soda science were the same experts and corporate executives we have been following all along: Rhona Applebaum, Steven Blair, James Hill, and even, in a bit role, Alex Malaspina.

In the first venture, the Energy Balance and Active Lifestyle Committee within ILSI North America developed a relatively sophisticated biological science of energy balance offering fresh arguments for the activity solution to obesity. In the second, an outgrowth of Team Coke, a new partnership called the Global Energy Balance Network (GEBN) created a worldwide network of experts to promote exercise-first solutions to the obesity epidemic around the world. These two projects were potentially transformative. If they succeeded—that is, if their corporate secrets remained secret—they would place corporate-friendly ideas at the center of the understanding and management of obesity on a global scale. The ILSI committee would succeed brilliantly. The GEBN would fail just as spectacularly, bringing the entire soda-science project crashing down around it.

In this chapter we follow the buildup of these two vehicles for soda science to the very peak of their development, just before the GEBN collapse of late 2015. We track Hill, Applebaum, Blair, and their colleagues on Team Coke as they went about creating the organizations and trying to solve the ethical, dietary, and other problems inherent in soda-defense science to see why one venture succeeded while the other failed. We will see that the Energy Balance and Active Lifestyle Committee benefited from the brilliant institutional design of the larger ILSI organization. Its multicompany model enabled it to create high-quality corporate science while veiling the role of the corporations involved. The GEBN was a much riskier endeavor. Instead of funding science indirectly through a nonprofit like ILSI, Coke sought to fund this project directly and largely on its own, in a political context in which the ethics of industry funding of science was becoming increasingly contentious. The organizers struggled mightily to overcome the problems. But the GEBN lacked mechanisms for concealing corruption and, in the end, was taken down by a tweet and an article in the *Times*.

ILSI North America: Energy Flux and the New Science of Energy Balance

As America on the Move was winding down and Coke was searching for new ways to invest in soda science, another member of the ILSI family was gearing up to take the science of obesity to a new level. Established in 1993 and based in Washington, DC, the North America branch had several dozen corporate members, the vast majority in the food and restaurant industries.[1] In 2005, longtime trustee James Hill was joined on the governing board by Rhona Applebaum, representing Coca-Cola. Applebaum was soon elected to the officer ranks, becoming secretary-treasurer (2006–7) and vice president (2008–9) of the board before taking over the presidency during 2010 and 2011. (She remained on the board through 2014, when she became ILSI-Global president.) The presidency of this important branch was a powerful platform from which to pursue her agenda. A decade after a loose exercise-first prescription for obesity had been translated into interventions like the Take 10! classroom activity, Applebaum and Hill mobilized the branch's committee structure to create a newly robust science of energy balance to buttress the case for physical activity. In this section, I uncover the mechanisms by which this skillfully designed product-defense nonprofit enabled a powerful new collaboration between industry and a few quasi-corporate academic scientists to take the corporate science of energy balance to a new level, while keeping the corporate contribution out of sight.

A TECHNICAL COMMITTEE ON ENERGY BALANCE AND ACTIVE LIFESTYLE (EBAL)

Under Applebaum's leadership, the North America branch would establish a new committee to give energy balance work on obesity scientific substance and weight. Although obesity rates were continuing to rise rapidly, the impetus does not seem to have come from any new threat in the corporate environment. Instead, the Coke executive's ascension to the presidency of this key branch appears to have given the two soda scientists an opportunity to use the ILSI apparatus to advance their agenda in a bold new way. In 2010, they seized the moment by forming the Technical Committee on Energy Balance and Active Lifestyle ("EBAL committee"; active 2010–15). The new initiative would combine Coke's favorite anti-obesity slogan ("active living") with Hill's ("energy balance") into a powerful force for corporate-scientific advance.

As the sites where ILSI science was made, technical committees formed the heart of the organization.[2] The new energy balance group joined about ten other issue-specific committees in the North America branch dealing with such topics such as carbohydrates, hydration, low-calorie sweeteners, food microbiology, and dietary lipids, all of obvious importance to member companies in the food industry. The name "technical committee" makes the groups sound specialized, esoteric, apolitical, and of little general interest. Behind that innocuous name, however, lay another of ILSI's secrets, this one buried in the branch bylaws: technical committees are *by design* dominated by profit-making corporations. (Key excerpts from the bylaws can be found in appendix 2.) Established to foster research, technical committees are composed of member companies whose obligations include funding the work through self-assessments, setting the membership format and program, appointing scientific advisers, and managing the dissemination of information produced.[3] The bylaws state that the scientific advisers ("scientists from companies, in conjunction with qualified non-industry scientific advisors appointed by the Technical Committee") shall guide the committee's (scientific) activities.

The responsibilities just listed offered member companies means to mold the science to corporate needs even before committee researchers began their work. For example, the members might appoint scientists known to be friendly to industry or to favor a particular model of disease causation. Both were the case with the EBAL committee. The committee's charge was to define "the state of the science" on energy balance and physical activity, its relationship to active/healthy living, including weight management, and to identify research gaps.[4] Once again, the agenda was not to find the best solution to the obesity epidemic. The agenda was to build out the activity side of the equation. Put another way, it was to amass scientific evidence demonstrating that the industry's preferred solutions were the keys to weight management.

In 2012, the EBAL committee was made up of five soda and ultraprocessed-food giants: Coca-Cola, Dr Pepper Snapple, PepsiCo, Kraft Foods, and Mars.[5] By 2015, three more had joined the committee: General Mills, Hershey, and Mondelēz International. These big and powerful food companies would pay for the work and control the information produced.

Their scientists would also be active participants in routine decisions about which projects to undertake, when, and how.[6]

To advise them—to do the concrete work of organizing sessions, conferences, and so forth—the committee's corporate members relied on a handful of well-known academic and government researchers with expertise in energy balance science. Most had worked with industry or ILSI or both over many years and knew what was expected. The committee's main advisers were James Hill (service 2010–15) and John Jakicic (service 2010–15). Jakicic (PhD 1995, exercise physiology) was a well-published exercise physiologist at the University of Pittsburgh with expertise in physical activity and weight management.[7] Two others, both core members of Team Coke, served for two or three years. David B. Allison (service 2010–11) was a statistical and study design expert at the University of Alabama, Birmingham (PhD 1990, psychology). Allison had been on the governing board of ILSI-NA since 2002 and served in leadership positions on the board during the years the EBAL committee was in operation (vice chair 2011–13, chair 2013–15).[8] Michael Pratt (service 2013–15) was an exercise scientist at the CDC with interests in physical activity and public health (MD, MPH [master of public health], dates n.a.). Pratt had a long history of conducting research funded by Coca-Cola and backed by ILSI.[9] Three others served as adviser for one year.[10] With five to eight soda and food giants securing the guidance of at least three demonstrably industry-friendly scientists (Hill, Allison, and Pratt), this "technical committee" had a clear political agenda: making science to protect the ultraprocessed-food-and-beverage industry. This close look at the mission and membership of the EBAL committee shows how the very structure of the technical committee was designed to produce science that was beneficial to industry interests. But that remained secret, hidden behind the dull name and the dry bylaws that few are likely to ferret out in search of hidden industry influence.

BACKFILLING AND STRENGTHENING THE SCIENCE

From the earliest days of soda-defense science, Hill and others recruited by ILSI had worked backward, beginning with the conclusion given them by ILSI's CEOs, and then translating the basic ideas into concrete interventions such as Take 10! for promotion to the public. Now, ten years later, Hill and a few colleagues, authorized by Applebaum as ILSI-NA president, and funded by Coke, PepsiCo, Kraft, and other wealthy food companies, would create a robust science to support the conclusion and interventions. Under the auspices of that committee, they would now backfill the science with methods and data, strengthen the analytic framework, augment the supporting arguments, and promote an enhanced science of energy balance as a new paradigm to multiple professional audiences. The EBAL committee was extraordinarily active, sponsoring fifteen events between 2011 and 2015 (set out in table 4.1). Thirteen were scientific activities aimed at advancing the science and publicizing its policy applications. The other two were webcasts designed to translate the science into action programs. Held in North America, the Caribbean, Europe, and Asia, and addressed to professionals in nutrition, exercise science, and general health, the EBAL activities spread the exercise-first message to obesity experts and policy makers far and wide.

Now, a decade after a loose energy balance framework had been rather prematurely turned into concrete interventions like Take 10! and AOM, the scientists promoting this body of work began developing more sophisticated models of the components of what they called the "energy balance system," thereby strengthening the case for activity. Arguing (once again) that because measures of key inputs were simply not accurate enough, it was impossible to say whether diet or exercise was more responsible for weight gain, EBAL researchers proceeded to develop the exercise side of the equation, leaving diet for others to study.[11] Perhaps anxious to gain ground in the battle over obesity science, the committee described its work as field redefining. One conference issued a "consensus statement" on core questions of energy balance. Another established energy balance as "a new paradigm."[12]

What made their formulation a new paradigm and constituted the major conceptual advance was to define energy balance "in terms of a biological system in which energy intake and energy expenditure change over time in response to the environment."[13] A key innovation of their hypothetical model of the energy balance system was the focus on energy flux, a "recent concept" defined as the rate of energy flow, or number of calories moving through a system.[14] In this research, it was postulated that individuals with a high rate of energy flow through the body are better able to maintain the energy balance needed to achieve a stable weight. The model suggested that strategies to combat obesity should center on raising energy expenditure by increasing physical activity. This energy flux research, mentioned briefly in the previous chapter, was conducted by Blair and Hand at USC with Coca-Cola support.

Writing for ILSI-CHP a decade earlier, Hill had offered a couple commonsense arguments for the exercise-first solution: we need a variety of approaches, he suggested, and we need to act now. The EBAL research reiterated those points and added new and more sophisticated ones. One emphasized that exercise improves individual metabolic health and otherwise alters human biology in ways that facilitate weight loss and its maintenance.[15]

TABLE 4.1. Activities of the ILSI North America Committee on Energy Balance and Active Lifestyle (EBAL), 2011–15

Date	Event Name	Event Type and Collaborating Organization(s)	City and State or Country
SCIENTIFIC ACTIVITIES			
1. May 2011	The Role of Energy Balance in Health and Wellness, Consensus Conference	Conference in collaboration with the ASN and ACSM	Chicago, IL
2. Oct. 2011	Energy Balance: A New Conceptual Framework and Actions Needed	Session at 11th European Nutrition Conference, Federation of European Nutrition Societies (FENS)	Madrid, Spain
3. Dec. 2011	1-day workshop on energy balance	ILSI-Japan	Tokyo, Japan
4. Dec. 2011	1-day workshop on energy balance	ILSI-China	Beijing, PR China
5. Apr. 2012	Energy Balance: A New Paradigm	Special Conference at Experimental Biology, in partnership with the ASN	San Diego, CA
6. Oct. 2012	Energy Balance at a Crossroads: Translating Science into Action	Expert Panel Meeting in partnership with the ACSM, AND, IFIC Foundation, and the US Department of Agriculture (USDA)	Washington, DC
7. Jan. 2013	Looking at Health Consequences of Physical Inactivity	Session at Annual ILSI Meeting	Miami, FL
8. Feb. 2013	International Symposium on Energy Balance and Physical Activity	Workshop cosponsored with ILSI-Mexico	(n.a., probably Mexico City, Mexico)
9. Sept. 2013	Energy Balance and Active Living	Session at 20th International Congress of Nutrition, International Union of Nutritional Sciences, in partnership with ILSI Europe	Granada, Spain
10. May 2014	The Science of Energy Balance: A Model for Weight Management Intervention	Session at Annual Meeting of the ACSM	Orlando, FL
11. Oct. 2014	The Science of Energy Balance: What We Know and Don't Know	Session at the AND Food and Nutrition Conference and Expo	Atlanta, GA

(*continued*)

TABLE 4.1. (*continued*)

Date	Event Name	Event Type and Collaborating Organization(s)	City and State or Country
12. May 2015	Sedentary Behavior, Physical Inactivity, and Health	Session at Annual Meeting of the ACSM	San Diego, CA
13. Nov. 2015	Physical Activity and Energy Balance: Public Health Priorities for the Americas	Session at the 17th Congress of the Latin American Society for Nutrition (SLAN), cosponsored with ILSI's North Andean branch	Punta Cana, Dominican Republic
WEBINARS			
14. Dec. 2012	New Science on Energy Balance: Exploring All Sides of the Equation	Webinar, in partnership with the ACSM and IFIC Foundation	Washington, DC
15. Aug. 2014	Energy Balance at the Crossroads: Translating the Science into Action	Webcast, in partnership with the ACSM, AND, and IFIC Foundation	Washington, DC

Sources: ILSI North America, *ILSI North America 2010–15 Annual Reports*; Institute for the Advancement of Food and Nutrition Sciences (IAFNS), "Past Events."

Notes: Includes only events with complete or nearly complete information. List of abbreviations appears in the frontmatter.

Another stressed that increasing activity (or energy expenditure) was essential to moving people into a postulated "regulated zone" of high energy flux, where it is easier to achieve energy balance and maintain weight stability. Such arguments based on the newly refined model gave the exercise-first solution added scientific heft.

DIETARY TOKENISM BORDERING ON REJECTIONISM

Dietary approaches, historically small but important parts of soda-defense science, were now increasingly rejected outright. The two webinars sponsored by the committee did cover dietary strategies, arguing in now-familiar terms that: "calories count, but there are no good and bad calories," or "balance calories to manage weight" (i.e., eat what you want, then exercise it off).[16] Such all-calories-equal messages are familiar from *The Step Diet Book* and AOM program, which conveyed the happy news that there's no need to cut out junk food, since simply reducing portion size or adding steps can close the energy gap. These webinar messages were also company friendly, but to the unwary they may have sounded quite reasonable and scientifically grounded.

In contrast, many of the scientific papers presented at EBAL events, some based on Coke-funded research, openly rejected dieting as a frontline strategy in the fight against obesity. The arguments against dietary restriction were many and varied, and in cases a little strained. According to one, widespread calorie restriction alone, having failed so far, is unlikely to work going forward.[17] According to another, calorie restriction at a level needed to make a dent in obesity is too difficult and less feasible than increasing activity. In other words, people can't handle it; it's easier for the average person to move more than to eat less. In a third argument, "drastic" dieting by sedentary individuals brings problems of its own, from compensatory changes like increased appetite to problems such as reduced bone density and impaired immune function. Many of the authors gave lip service to the role of diet, suggesting dietary efforts should continue, yet they said nothing about what those efforts should entail. Diet tokenism did the trick of conveying the impression of balance, without offering much substance.

Not surprisingly, none of the committee abstracts or articles I reviewed mentioned soda taxes or marketing restrictions on junk food, policy options being advocated by a growing number of respected authorities.[18] As noted earlier, taxing soda seems to have been a taboo topic among scientists working with the soda industry. That taboo would certainly have been in force at EBAL, a committee dominated by its member companies.

DISCLOSING, DENYING, DISPLAYING: DOING ETHICS IN THE MID-2010S

In the late 2000s and early 2010s, industry funding remained common in nutritional research, yet the ethics of taking corporate money was generating ever more controversy.[19] Especially in the United States, where the debate was out in the open, it was crucial for ILSI to stay on its toes. One response to the critiques, described in the last chapter, was the development of a set of ethical principles to guide corporate funding of research. ILSI-NA took the lead, and Team Coke members Hill and Allison participated in the effort. As the criticisms moved closer to home, the scientists involved with the North America branch continued to protect the food industry, and their ties with it, by chiding the critics and arguing, sometimes in aggressive ways, for partnership, not adversarial relations.[20]

Following changing expectations in the field, EBAL authors began routinely disclosing their industry funders in their articles. To the critics, this would appear a step in the right direction, but merely naming the funders revealed nothing about the myriad ways in which the support or the sponsor might have shaped the research process itself. Another problem with disclosure is that it tends to be followed by declarations that the funding had no impact on the findings. This practice of disclosing-then-denying was routinely used in publications on the Coke-funded research that informed the EBAL work. In a 2013 report on the big Energy Balance Study funded by Coke, for example, Hand, Blair, and eleven other members of the USC research team wrote: "Funding . . . was provided through an unrestricted grant from The Coca-Cola Company. The sponsor played no role in the study design, collection, analysis and interpretation of data, or preparation and submission of this manuscript."[21] This is the very project whose research design (or protocol) we traced in chapter 3. There we found that Coke technical/scientific staff were centrally involved in the project's development. Here they say Coke had no role whatsoever. While growing numbers of obesity researchers were documenting the impact of corporate funding on research findings, Coke's grantees were declaring no influence, as though the company were offering a free multimillion-dollar lunch.

Far from being concerned about the appearance of industry influence, Blair and Hand at South Carolina seemed to view the generous funding from Coca-Cola as a point of pride. A 2014 PowerPoint presentation of the results of the energy flux study concluded with a prominent yellow box disclosing the funder, followed on the next and final slide by a large color photo of the entire twenty-six-person Energy Balance Research Team in matching shirts and slacks.[22] Rather than cloaking the funding source, Hand (the author here) put it on display. This eye-catching mode of disclosure suggests the USC research team truly believed that their science remained untouched by their sponsor. In another context, Blair would depict his work as pushing good science, and his attitude toward the Coke-funded research must surely have been the same. Blair may also have felt protected from criticism by his status in the field. Around the same time, a colleague at another institution described the name Steve Blair as "a massive brand" that would add prestige to any project associated with it. He was also a giant figure in the American College of Sports Medicine, the apparent audience for the 2014 talk. Colleagues in that friendly association may have been reluctant to question his or his team's ethics.

Another interpretation suggests some fancier ethical footwork. The bright display of the company's name on the penultimate slide could also have been a dare. That yellow box appears on a slide that lists the names of all thirty-nine Energy Balance staff members as well as eleven project consultants from around the United States and the UK. This visual association of Coke's name with the eleven consultants, many well-known figures in the field (four were members of Team Coke), implies that the consultants had no problem with the corporate funding. The ethical message seems to be: if these prominent scientists have no qualms, why should anyone? A final possibility—and these interpretations are not mutually exclusive—is that the success Blair and Hand had enjoyed so far may have left them overly confident, oblivious to the risks they faced in a climate increasingly hostile to industry influence on science. Whatever the research team's objective in posting the yellow box and photo, outside observers are likely to see friction between the public performance of no influence and the private reality, documented in chapter 3, in which Blair worked closely with Applebaum for years, not only on the research protocol but also on promoting his findings around the world. For the scientists involved, any such friction had to remain below the level of consciousness. Perhaps the scientists buried the inadmissible thought by publicly declaring and performing its opposite.

In chapter 1, I argued that the way ILSI and its branches were organized and operated made them remarkably successful vehicles for the creation of corporate sciences like soda science. This account of ILSI's North America branch and its Energy Balance and Active Lifestyle Committee may show ILSI performing at or near its peak. Not only did the committee have sophisticated mechanisms for creating industry-friendly science and keeping any corporate influence from view; its lead scientific advisers, incentivized to believe their ideas were untouched by company hands, were able to state with absolute conviction that conflicts of interest and other ethical problems associated with industry-funded science simply did not exist. ILSI could be uncommonly good at what it did. Precisely for that reason, the way the organization worked and what it did were exceptionally hard for critics or the merely curious to uncover, making its effects unusually insidious. Meantime, behind the scenes, little known to the wider public health community, the key actors in the ILSI-NA effort to advance soda science were preparing a new offensive, this one grander in scale—and riskier in approach—than anything attempted so far.

The Global Energy Balance Network

In the early 2010s, as the energy balance science of obesity was being developed into a more robust, biological science through ILSI-NA, three key figures in that initiative—Applebaum, Blair, and Hill—hit on a bold new idea. Building on the momentum created by the success of the EBAL committee and Applebaum's project of cultivating allies for Coke, in 2014–15 the trio would form a strategic alliance to gain advantage in the obesity wars by building a network of like-minded researchers around the world. The network would place their brand of energy balance ideas, long deemed secondary to more mainstream views, at the center of the understanding and management of obesity *on a global scale*. To build the new organization, the two scientists would form an academic-industry partnership with Coca-Cola, which would quietly underwrite and oversee the network. The idea had especially strong appeal for Applebaum, who, as the representative of the funder, would play a leading role from the beginning. Creating a vehicle to spotlight corporate science and spread it around the world would allow her to build on her ties to Team Coke to do something big and important for Coca-Cola. That something would be called the Global Energy Balance Network (GEBN). This new three-way partnership represented the pinnacle of soda-defense science's development and spread. If it succeeded, it could potentially revamp the handling of obesity worldwide.

Differences in style and substance separated the two scientists, however. Hill was a quasi-corporate scientist who recognized the importance of diet even in an activity-first science. He was also a cautious political actor who had evolved subtle and sophisticated ways to handle the ethical and funding challenges of making a corporate science of obesity. After working with ILSI for many years, he was a believer in the value of Malaspina's ILSI model of multicompany sponsorship of corporate science. Blair was the opposite in many ways. Blair was a virtual corporate scientist, an outspoken advocate for exercise, and a brash political actor whose combative stance, blunt language, and seemingly cavalier attitude toward the appearance of conflicts of interest would invite attention and controversy. After years of collaborating, he was close to Applebaum.

While these differences in political style and views on diet were real, the two researchers belonged on a single continuum of proximity to the corporate world, with Blair falling quite near the corporate end and Hill lying on the corporate side but a bit closer to the scientific mainstream. The evidence suggests that both were complicit with industry, both bent science to industry needs, and both were ethically conflicted in the sense used here. Thrown together by their shared affinity for the energy balance model (a model that could accommodate multitudes of methods and views), the lure of major funding from Coke, and the gung-ho leadership of Coke's top science officer, the two would try to work around, ignore, or conceal their differences to make this grand idea a reality. The result was an organization that would look strong from the outside, but appear weak and wobbly from the inside.

The political climate in the early 2010s was increasingly hostile, posing direct threats to both soda science and soda companies such as Coca-Cola. The need to strengthen and spread the science was pressing. ILSI was designed to solve the delicate problems of corporate science, and it had advanced the soda science of obesity over many years without detection. Now we observe Coke under Applebaum taking a giant step beyond its grants program to set up a highly visible academic-industry partnership sponsored by a single company. This was something new. Even if the political climate were more friendly, it would be a highly risky venture. With no layers of administrative separation between the company and the scientists, the new entity would need to find ways to conceal the corruption of the knowledge, solve the problems of diet and ethics, and keep the breathtaking extent of Coke's involvement secret. The GEBN's founders would try valiantly to solve each problem as it arose, but the responses were often clumsy. So rocky was their road to success that in the end one of the granddaddies of soda-defense science would have to come out of retirement and step in to attempt a rescue.

ESCALATING THREATS TO INDUSTRY AND THE BIRTH OF THE GEBN

In the societal wars over obesity industry remained embattled, as public health critics turned up the volume on their calls for soda taxes and other regulatory measures. These demands threatened not just the profits of the industry but also the viability of the soda science that Team Coke members had been building for a decade. As the attacks began hitting closer to home, Coke's chief science officer amplified the rhetoric of warfare, at once conveying the urgency of responding and justifying to senior leaders at Coke a big investment in the new scheme. "Heat is on," she warned members of her academic team in September 2012 after an attack on a scientist promoting an unpopular position in a different field. "Current environ . . . truly is Scientific McCarthyism." Seeking input on how industry could respond, she floated a trial balloon: "would like to convene a small expert group—views?"

As if to validate her worst fears, a year later (in mid-January 2014) the *Sunday Times* of London published a news item titled "Sugar Watchdog Works for Coca-Cola." The paper's reporters had discovered that Ian Macdonald, a leading nutrition/obesity researcher at the University of Nottingham who was chairing a British government panel tasked with proposing limits on sugar consumption, was a paid consultant to Coke (and Mars). The disclosure of this possible conflict of interest—which Macdonald confirmed, saying he advised Coke on issues of diet, obesity, and exercise and added the (very modest) fee to his personal income—was prompting demands for his resignation from the panel. For Applebaum, the Macdonald affair moved the battlefront in the obesity wars closer to Atlanta and brought out the urgency of responding in a more forceful and collective way. Outraged by the assumption that her company was corrupting its advisers—and, worse yet, that corporate advisers should be kept off government advisory boards!—Applebaum quickly emailed the article to Team Coke with the plea: "Gentlemen—need your guidance. Have discussed with our leadership re what's needed. . . . I asked for a bit of time to check with a few trusted advisors on their POVs." Defending Macdonald as an upstanding scientist who is "evidence based to a fault," she warned darkly of "a coalition of interests working together—and there's an orchestrated attack occurring . . . [that portended] storms ahead." Over the next twenty-four hours, as the members of her team shared their disgust at the attacks on industry partnerships and consultancies and bandied about ideas for a response, Applebaum was preoccupied with developments in the UK and what they meant for Coke: "There will be more—It's obvious the intent is to 'out' these scientists as negative/bad/evil. Similar to the FBI's Top 10 list." Now in clear and present danger, Coke and its scientist friends had to do something big and soon to stop the incessant attacks.

It was at precisely this moment—when one of their own was under attack and they were searching for ways to demonstrate collective strength—that the idea of the GEBN took flight. According to Hill, who claimed authorship, the idea was floated as early as 2009. The first instance in which I found it clearly articulated was in an exchange between Applebaum and Hill in September 2012, just over a year before the Macdonald "outing." Always on the lookout for promising ideas, Applebaum had written to Hill, praising him for attracting five thousand people to a webinar and offering more money for more of the same. Ever the entrepreneur, Hill shot off this message to his sponsor: "We need the *Energy Balance Institute*—then we can do this every day" (emphasis added). He added, "We should reconnect soon." A month later (under the subject line "Ready for a stimulus pkg?"), Applebaum told Hill she had "sold the idea" to the company and "will bring in others," clarifying later that any such project must take the form of a collaboration with private industry (rather than an outright grant). Four days after the Macdonald affair roiled the waters in Atlanta, Applebaum sent a message to her Colorado and Carolina teams (Hill and Peters, Blair and Hand) telling them that a Coke consultant would be getting in touch soon to learn their vision for "the EBI Network." Those ideas would form the background for a working group conference call to be followed by the first in-person meeting on the EBI Network in mid-February. With this plan for action, Applebaum announced, "the journey begins!" And indeed, with hands-on assistance from Coca-Cola staff and consultants, that short message to the two teams set off a veritable tsunami of emails, teleconferences, in-person meetings, trips, funding documents, paper exchanges, and other forms of organization-building activity that in a mere ten months resulted in the establishment of the GEBN.

The GEBN's birth on the battleground of corporate science was no accident. The network was envisioned as an instrument of scientific warfare in which the stakes—for Team Coke and its corporate science of obesity— were nothing less than life or death. In a mid-2014 draft proposal shared with her colleagues, Applebaum framed the GEBN as a weapon in an "escalating war between public health and industry" over obesity strategy. The enemies she named were the same "unreasonable . . . extremists" targeted for attack at the CEO dinner: Brownell at Yale and Jacobson of the CSPI, who "focus[ed] on government regulation to limit, tax or ban foods they consider to be unhealthy . . . despite a serious lack of scientific support." Adopting a rhetorical tactic often used in battles between rival sciences, she labeled her enemies nonscientists (hence lacking in reason, data, and so forth), while declaring her friends scientists (and thus reasonable, reliant on evidence, and so on). Stepping up the battle rhetoric, a new network, Applebaum wrote, is needed to "devise, create and implement a multi-year advocacy 'campaign' that serves as a counterforce to one-sided, regulation-driven proponents." The GEBN would "counteract their shrill rhetoric indirectly, but forcibly with reasonable voices and with science-based strategies that make sense." Although the network was depicted as a "fair and safe place for candid discussions," Applebaum urged use of an ILSI-style invitation-only format for some activities, suggesting that the climate would be "fair and safe" only if everyone present was on the right side of the fight—Coke's side.

Her academic partners, and especially Blair, also struck a combative pose, decrying the "enormous confusion" and "imbalance in information" on obesity—too much on diet, too little on activity—that the network would work aggressively to correct.[23] In the last twenty-five to thirty years, he wrote in a 2014 article, scientific progress in the field of physical activity and health has been enormous; at long last, "the physical activity movement [has] come of age."[24] And yet, what was there to show for it? A 2015 Google search turned up some seventy million hits for "diet and obesity," but not even a million for "physical activity and obesity." A similar search for scientific articles in PubMed (a search engine listing references in biomedicine and the life sciences) yielded forty-five thousand on "diet and obesity" but only two thousand on "physical activity and obesity."[25] The field's failure to attend to the role of physical activity, the focus of his life's work, was galling. In a brief conversation with me in late 2013, he complained bitterly about the lack of funding for research on physical activity and obesity at the National Institutes of Health. Such perceptions of unfairness in the allocation of federal funding may have pushed him into the arms of industry and legitimized in his mind accepting large sums of corporate money to correct the (perceived) misunderstandings of scientific opponents.

That sense that the other side had an unfair advantage in the battle between the sciences certainly drove his passion for the new network. A few months after its establishment, the *Lancet* published a set of articles on obesity that scarcely mentioned physical activity. Blair's frustration boiled over in an email to six comrades in arms on Team Coke, four at the company, and a few others: "I don't know if you say [*sic*] The Lancet last week . . . but it had a major focus on obesity. . . . I am happy that this issue appeared because it gives me a lot of ammo for my [upcoming] talk . . . on energy balance (HOW CAN THE LANCET IGNORE PA [physical activity]?????????????????)." This fury and indignation at being attacked and, worse yet, *overlooked*, may have led Applebaum and Blair to push aggressively forward with their cause at a time when others might have hesitated.

A GRAND PLAN: SODA SCIENCE ON A GLOBAL SCALE

Applebaum and Blair may have felt embattled and bitter, but they were also stoked up by their recent success in spreading EIM around the world. The trips together and the fun of fighting common enemies with Team Coke seem to have given them confidence that together, and with the financial backing of that powerful company, they could achieve anything. At the end of 2014, less than a year after the vision of a network had materialized, the organizers proudly announced the December 7 launch of the Global Energy Balance Network (GEBN). The GEBN was a "public-private, not-for-profit network of experts" dedicated to using the science of energy balance to address the obesity epidemic.[26] Hill and Blair would serve as president and vice president, respectively, while John Peters and Gregory Hand (in 2014, at the University of Colorado and West Virginia University) would be co-vice presidents for North America. By early 2015, the GEBN had lined up vice presidents on seven continents—including one from China, listed as the Asia representative—along with 150 scholars, practitioners, and other experts in eighteen countries. This was an impressive accomplishment, given that recruitment been underway for only a few months. It was also a big win Applebaum could report to her bosses.

The goals for the new organization were nothing if not ambitious: advancing the science, educating the media and the public, training new scientists, developing innovative obesity solutions, and serving as the voice of science in public policy dialogues. The network's primary resource, the website gebn.org, which launched on December 7, would be the leading resource for journalists, scientists/academics, health professionals, and other opinion influencers worldwide seeking evidence-based information on energy balance and its role in active, healthy lifestyles and the prevention of diseases associated with inactivity, poor nutrition, and obesity. It would feature a directory of experts, online forums and meetings, information for journalists, and a library of resources in multiple languages. And the launch was just the beginning. Over the course of 2015, the organizers promised, they would develop the work plan, fill out the membership, line up additional corporate funders, and convene a "Think-Do Tank" Summit to identify promising practical solutions. There was something in it for everyone, and the excitement among organizers and members was palpable in their emails to each other.

DIETARY DISCORD

The GEBN may have aimed high, but it had two sources of vulnerability that left it open to possible public reproach. One concerned the science, the other the funding arrangements. The first was the disagreement among the principals about the role of dietary restraint in addressing obesity. This issue had to be handled delicately because of widespread suspicion that the soda industry was trying to downplay the role of diet in favor of activity. In earlier chapters I've spelled out Hill's views, which many obesity specialists I have spoken to over the years considered fairly moderate. Over many years, Hill had offered a series of finely tuned solutions to the diet problem in soda science—how to handle junk food and sugary drinks—that he had finessed with great rhetorical skill. After he started working as ILSI's adviser, he almost always quietly downplayed the issue of diet in favor of physical activity. Yet for reasons that were both scientific (energy balance had two sides) and political (concern about the perception of conflicts of interest), in the end he always left room for some role for better diets in addressing the obesity crisis. That insistence on including dietary solutions marked his approach to the GEBN as well. His studied position, shared with a couple of close associates, was that "increasing physical activity *cannot fly as the only strategy* to reduce obesity" (emphasis added).

Blair's ideas were more extreme and controversial. As we saw in chapter 3, the South Carolina exercise specialist was known for his position that inactivity was the major public health problem of the day. In a long conversation in his DC office, one prominent obesity specialist told me confidentially that many experts were uncomfortable with Blair's view that obesity was unimportant if you are fit. Even Hill and Peters had distanced themselves from the argument, associating it with Blair in all but name. From *The Step Diet Book*: "A few scientists have argued that obesity, per se, is not a health risk. Rather, they hold low physical fitness accountable for all the health risks of obesity. Few obesity experts agree with this notion, since we have a great deal of information that links obesity with poor health."[27]

In the years leading up to the creation of the GEBN, Blair had been preoccupied with pressing his argument that inactivity is the greatest threat to public health. In a 2007 article reviewing the empirical support for the contention, he wrote: "Our inactive lifestyles are the major cause of obesity, rather than the food/drink we consume; although portion sizes have increased and there are more palatable, energy-dense foods, the predominant cause of obesity is the reduction in our energy expenditure."[28] Other factors explored in chapter 3 (his training in exercise science and closeness to Applebaum) may have biased him toward activity as well. Such views, though, if openly expressed, might have played into public suspicions that any scientific project Coke touched would be biased toward activity, endangering the whole network. As we saw in chapter 3, Blair's actions suggest he was not especially concerned about the perception of industry influence because, like many of his colleagues, he seemed to feel he was immune to it.[29] Though Applebaum was not supposed to influence the science, the reams of emails she sent to GEBN collaborators leave little doubt about her preferences: promote exercise whenever and wherever possible, say next to nothing about diet. And after years of working closely with Applebaum, Blair may have felt they were on the same page and he had her full support to do whatever he wanted.

As the GEBN was coming together in late 2014 and early 2015, the view presented to the outside was one of unity around the shared agenda of promoting energy balance for obesity. Behind the scenes, though, the organizers' views on the importance of diet remained far apart. These differences, which were almost never openly discussed (at least over email), came out in an urgent email conversation within the group in May 2015 about how it should respond to questions from a *New York Times* reporter. In this tense exchange, excerpted below, we see how Hill and Hand, evidently worried that Blair might express his dismissive views of diet publicly, struggled to convince him of the wisdom of leaving some role for diet, or at least of stating that the real issue is diet-activity balance. We see too how, in the end, they failed to change his mind. Blair would not let go of his views, even for the sake of promoting group consensus on this vital issue. Remarkably, six months after the network was established, its principals still could not agree on the scientific basics! That disunity would leave the network unable to speak as one and vulnerable to a rogue act on the part of one or more members. Hoping to reach a group consensus on diet and activity, James Hill opened the conversation with the voice of science:

> Gentlemen, I will be talking with the NY time [*sic*] reporter next week.... Here is my best guess about how he will spin the story: I think he will say that Coke funded GEBN to create a group of people who talk about the importance of physical activity and downplay the importance of food. He will say that Coke is funding us to take the blame off food.
>
> I believe we have to be true to the roots of energy balance—that both diet and physical activity are important. And that energy balance is very complex and it is not possible to point to the problem as either food or physical activity. I believe he is looking for us to say that physical activity is much more important than diet and I think that this would make us look very much like a tool of industry.
>
> thoughts?

Greg Hand concurred, hinting at the discomfort he felt over the issue:

> Tell the truth—physical activity ISN'T more important than diet. The energy balance framework suggests that you can't really separate the two. How can one side of a scale be "more important" for the scale's balance than the other side of the scale? OK, I feel better now. ;-)

Steven Blair joined in moments later, reminding the others of his well-known position but saying he had publicly protected the group in his own interview with the *Times*:

> Greg makes good points. I don't think [I] told him [the *Times* reporter] that PA is more important [than] diet for EB (although all of you know that is my belief!). I did tell him that energy expenditure [is] too often overlooked and needs more attention.

Hand replied, now gingerly disagreeing with his former mentor:

> Can't argue about the focus having been primarily on eating. While I would agree that expenditure is the foundation on which eating decisions need to be made ("if you don't burn it, you didn't earn it!"), I can't agree that it is more important.

Blair then restated Hand's position in an attempt to make the disagreement disappear. Here he admits that he is pushing activity for essentially political reasons:

> I think Greg means that PA is not more important from a physiological perspective and I agree with that. However, [from] an epidemiological/public health perspective, I do believe that inactivity is a more important cause of the epidemic!

Hand, seeking a middle ground, suggested another formulation he hoped everyone could agree to:

> Logic would suggest that the relationship of intake:expenditure is the issue rather than intake or expenditure independently.

Blair, unwilling to back down and evidently aware others would not change their views either, tried to end the discussion on a light note:

> Greg is going to continue his failed task of educating me ☺.

Hill, clearly chagrined, closed the conversation with no comment on substance:

> Well he gets [an] A for effort.[30]

A few months after this email conversation, the dangers of moving ahead without a shared stance and with one outspoken member holding industry-identical views became patently clear. In late summer 2015 the GEBN posted a video on its website in which Blair, speaking as the network's vice president, abandoned all restraint and pushed the scientific claims made in the name of energy balance in even more industry-friendly directions. In a tirade against critics of junk food and sugary beverages, Blair announced the new organization, fuming: "Most of the focus in the popular media and in the science press is, 'Oh, they're eating too much, eating too much, eating too much'—blaming fast food, blaming sugary drinks, and so on. And," he continued, "there's really virtually no compelling evidence that that, in fact, is the cause,"[31] a claim that drew widespread derision because it was so patently untrue. He would soon apologize, posting a statement on the GEBN website saying that his dismissal of diet as a cause of obesity had done a disservice to the work of the GEBN, and had come from an excess of passion about the issues, which he now regrets.[32] What he really believes, he said this time, is that both diet and activity are important to obesity. The video was taken down. Still, the damage was done. Whatever Blair's reasons for making such politically reckless statements—frustration at the neglect of his views? a secret promise of protection from Applebaum? failure to understand the delicacy of the issue?—in this new phase of soda-defense science, the solution to the diet problem associated with the GEBN would border on diet rejectionism, a position with few adherents in the scientific community.

Perils of One-Company Sponsorship

The network's second source of potential vulnerability was its reliance on a single funder. Worse still, that funder had been named the number one culprit in rising obesity rates. The warnings about the network's ties to Coca-Cola began arriving early. In the summer of 2014, as the project was coming together, the GEBN team surveyed potential network participants to gauge reactions to their plans. One question asked: "What would make you hesitate to join?" Of the twenty-four respondents, ten expressed concern about corporate bias, and several of those mentioned problems associated with a single funder:

> Hesitation [to join] could lie in the single sponsorship of an entity like TCCC [the Coca-Cola Company].
> [I would hesitate] if it seemed a marketing strategy and ploy for a potential sponsor (e.g., Coke) to gain "credibility" and avoid attention to products that could be contributing to obesity.
> I worry that the term Energy Balance will be code to critics who will see this as an industry effort. It must have wide financial support and members from all sectors to succeed.

The organizers were aware of the risks yet proceeded anyway. Evidently, they (or some of them) believed any risks could be managed. The network's organizers tried to deflect the suspicion that Coke was biasing their science in two ways: by creating a counternarrative stressing the company's hands-off role, and by devoting sustained efforts to securing additional funders. Both proved difficult to carry off.

COKE INVOLVEMENT: PRIVATE AND PUBLIC TRUTHS

The energy balance network was Applebaum's baby, an organic outgrowth of ideas bandied back and forth over many years with Blair, Hill, and others on Team Coke. No doubt as a result of her hard work, Coke poured millions into the network. In the spring of 2014, it provided $507,000 to the University of South Carolina to develop the GEBN website. The company also awarded the University of Colorado $1 million in targeted funding to establish the network. For work going forward, Coca-Cola quietly endowed the network to the tune of $20 million, producing an annual budget of $1 million.

The funding was just the beginning of Coke's involvement. The emails paint a picture of Applebaum keeping a finger in the pie at all times. In the formal division of labor established in early 2014, she was placed in charge of "high-level organization and governance," "logos and legal issues," and "a one-pager à la ILSI." Informally, she was involved in virtually every aspect of the network's development, from crafting the mission statement to creating the website, shaping the direction of the research, recruiting members, veiling the funding, and much more.[33] She also held the purse strings, so that any activity that required funding would necessarily pass through her office. With superb organizational skills and experience making things happen, Coke's staff were constantly circulating work plans with lists of tasks to do and deadlines to reach, keeping everything on track and moving forward. What the email record says is that Applebaum's drive, determination, and energy and her staff's superb administrative skills were vital to the rapid creation of the GEBN.

Yet none of these facts of Coke's involvement could become publicly known. To allay concerns about corporate influence, the organizers put together a public narrative emphasizing Coke's (putatively) negligible role in the GEBN. One of the fullest versions of the story was laid out in a Q&A document created for GEBN associates who might have to field questions about it. Some of these points will be familiar from Coca-Cola's clean-company narratives we encountered earlier. This sample of Q&As is typical:

Q: Why did Coca-Cola provide this funding?
A: Coca-Cola is interested in better understanding the science of energy balance. For more information about Coca-Cola's position, please contact the company.
Q: Isn't this just the latest in a long line of "front groups" that Coca-Cola is involved with?
A: Not at all. . . . Coca-Cola provided an unrestricted educational gift to support GEBN's start-up, but it is *not* involved in the governance or management of the organization nor in any of its activities.
Q: Why is Coca-Cola the only private company at the table?

A : GEBN is in the early stages of formation and we are actively seeking additional support. . . .

Q : As a scientist, how can you deny the pivotal role of soda and other SSBs
   [sugar-sweetened beverages] as a culprit in America's obesity epidemic?

A : GEBN is not focused on the potential contribution of any one food, beverage, or lifestyle behavior to obesity.

Despite the brow-raising quality of some of these answers (Coke funded
the group but doesn't care how it's run?), the principals used them in communications with potential members and sponsors. In one email, for example, Hill told a prospective network member that because Coke's initial
funding was paid in the form of an unrestricted gift, "they are not involved
with the direction of GEBN at all." How does one make sense of answers
like this that have so little correspondence to the facts laid out in the emails?
Did the principals really believe the answers they gave? Whatever their
mind-sets, the unconvincing quality of many of the recommended replies
hints at the real difficulties they faced trying to create an appealing public
narrative that fit the facts.

THE LIFE-OR-DEATH STRUGGLE TO FIND FUNDERS

The second and far superior solution to the single-funder problem was to
bring more on board. Hill had long maintained that having robust financial
backers was essential to the network's credibility. The results of the mid-
2014 survey underscored the need to find more sponsors. The pressure was
unrelenting. No sooner was the website launched than site administrators
began to get inquiries about the unnamed funder(s). The group's plan was
to find five additional founding sponsors who were willing to provide $1
million over five years. They found few ready to commit. Partly to offset
the limited industry support, the organizers also made major efforts to locate noncorporate sponsoring agencies willing to associate their name with
the GEBN and provide small sums of $5,000 to $10,000. Door after door
remained shut. Blair's invitation to an acquaintance at the National Collegiate Athletic Association (NCAA) to make a verbal commitment before
a planned media launch produced this chilly reply: "I have discussed this
with NCAA Senior Leadership. Although I think the GEBN is a very worthwhile project, it does not match our requirements for NCAA sponsorship."
Throughout the fall of 2014 and spring of 2015, efforts to find additional
sponsors continued at a high level.

A couple of months after the official launch, the group received an un-
welcome tweet from Yoni Freedhoff, MD, an outspoken obesity expert

and industry critic based at the University of Ottawa: "@StevenNBlair. Hi Steven! Hope you're well! I've sent 2 emails to @gebnetwk, inquiring about funding & membership. Radio silence from them." The dreaded "outing" had occurred. Without consulting the group, Steven Blair had quickly replied to Freedhoff from his individual Twitter account, deepening the turmoil within the group. The latest development threatened to destroy everything they had worked for. The network's Chicago-based consultant spelled out the implications, along with a detailed plan to save the organization:

> Freedhoff is an articulate and well know[n] antagonist that uses social media effectively. He is well connected to Michele Simon and they both will be all over this issue. We should be transparent. . . . We need to determine who is responsible for speaking on behalf of GEBN. . . . The way Steve engaged on Twitter is going to make him look disingenuous because Steve could have replied on behalf of GEBN and he did not. . . . We need to inform Coca[-]Cola asap so that they can prepare.
>
> We have lost our ability to launch properly and now need to regroup. Since we're going to be "outed" we need to take control of the story and put one out there. . . . This also means we need to redouble efforts to secure additional sponsors. The development is unfortunate and we need to move quickly. GEBN should have a "stand-by" statement and a tough Q&A document. We will also need to alert the . . . executive committee because this may cause people who have joined . . . to reconsider.[34]

Freedhoff's tweet had brought out another insurmountable problem the GEBN faced as it sought to get up and running: it could not post its major funder on its website without risking a huge blowback from "the other side," the moral crusaders like Freedhoff who would certainly tweet or blog about any news of Coke funding a program on healthy lifestyles. Blair's go-it-alone response, though undoubtedly well-intended, created yet another kind of difficulty for the group as a whole. Seeking to save the network, the group quickly began tackling these projects.

With the GEBN at risk of collapsing, there was a dire need to identify more funders to demonstrate corporate support beyond Coke. As things were looking increasingly desperate, Hill turned to his former mentor and coconspirator in an earlier phase of the obesity-science wars, Alex Malaspina. Retired from Coke in 1998 and from the ILSI presidency in 2001, in late 2014 Malaspina was waxing nostalgic about the good old days at ILSI-CHP, especially the help John Peters had provided serving as president for a few years. "Of all the people in industry," Malaspina wrote Peters in early 2015, "for me you have been the most impressive with your warm heart, diplomacy and high acumen. . . . You were so helpful to me at the CHP in Atlanta, when in addition to your heavy load at P&G you agreed to lead CHP." Far from retiring into leisure, Malaspina wrote, he was "helping Coke a little with some issues," and evidently they included saving the company from the obesity "extremists" in public health. To that end, he launched a personal campaign to help save the global network by getting the company's leaders on board.

In early November 2014, Malaspina had accompanied senior leadership at Coke—Clyde Tuggle (senior vice president and chief public affairs and communications officer, [2008–17]) and Ed Hays (senior vice president and chief technical officer [2015–])—to Colorado to see the great work being done by Hill and Peters, which he considered "the only 'game in town' effort to make a difference [on obesity]." Hill was keen to have them visit as well, seeing it as an opportunity to bolster the case he was making in a series of private emails to these executives aimed at persuading the company to fund a very major research project that apparently was to be separate from the GEBN. (I say more about this in the coda that follows.) Assuring Peters that "our friendship will endure, and I will try and help you and Jim as best I can," Malaspina pressed the GEBN cause with Tuggle and Hays. As the first and longest-serving president of ILSI, Malaspina deeply believed that the success of industry-funded science was contingent on multicompany sponsorship. That was the ILSI model, and he sought to use any powers of persuasion he possessed to move the GEBN in that direction. Hoping Tuggle could connect Hill to high-level corporate executives, especially at companies in other industries, Malaspina arranged a five-way phone meeting at the end of 2014. That discussion was followed by an in-person meeting in Atlanta in March 2015. Though the internal politics of these conversations at Coke remains a black box, there can be little doubt this was an extraordinary intervention by the retired VP on behalf of corporate science. The personal involvement of two very senior leaders suggests the obesity threat remained deeply concerning to the company.

Malaspina also tried to save the GEBN by recruiting corporate sponsors through ILSI, where he still had considerable sway. No sooner did executive director Suzanne Harris approve the use of ILSI to promote the GEBN (in April 2015) than the former ILSI president began contacting the branches in pursuit of his vision of the network as "a global project with great support from hundreds of the ILSI member companies." Hoping to mobilize the ILSI empire he had created to bolster the GEBN, he personally sent enthusiastic emails to branches all over the world, including the one in China, with Peters following up with information on the new network.

∴

Through Malaspina's interventions with the Coke leadership and the ILSI network, as spring turned into summer, efforts to attract corporate funding continued apace. The stakes were inordinately high. With the big investment of company funds in the GEBN website, well over one hundred founding members of the GEBN signed up and ready to contribute, and all the promises made to the scientific community and the leadership at Coke, the reputations of everyone—the four scientists and the chief science officer—were on the line. With fund-raising initiatives developing on several fronts, in the summer of 2015 the prospects for weathering the storm were better than decent. But the era of hope would be short-lived, as we shall see in the coda that comes next.

Reckoning

The Collapse of Soda Science in Its Home Country

For a decade and a half, Coke, ILSI, and a handful of university scientists, working together with a shared vision, built a powerful domain of soda science. But the verdict of history is that product-defense sciences—whether they defended talc, glyphosate (a.k.a. Roundup), or diesel exhaust—are short-lived. Eventually the shoddiness of the science—the corruption at its core—comes to light, and the science has to be scrapped.[1] When the covert collaborations behind the Global Energy Balance Network were revealed in the *New York Times* in late 2015, the science that Coke, ILSI, and their academic partners had spent all those years building together quickly collapsed, spelling the demise of soda-defense science as an active project in its home country.

In the last four chapters I unearthed the knowledge practices the soda scientists developed to demonstrate their integrity and bury their secrets. The sudden exposure of the biggest secret behind the GEBN—the massive influence of Coke—gives us a rare opportunity to see what kinds of practices quasi-corporate scientists turn to when their clandestine worlds fall apart. The first part of the story in this reckoning (the *Times*'s exposé and Coke's response) is fairly well known, at least in the obesity field. I include it because it is an essential component of our account of the life and death of soda science. The second part, the academic scientists' response, is likely to be less familiar.

This unexpected eruption of the soda-science project gives us a chance to reflect on the viability of the two major vehicles for creating corporate science, the GEBN and ILSI, in the reckoning's third part. For while the fledgling partnership between academia and a single company, which had not even started functioning, quickly fell apart once the criticisms became public, the scientific nonprofit, which had done much more over many years to create industry-tainted science, remained untouched. There was a reckoning for Coke and its GEBN partners, but no reckoning for ILSI, which was able to continue functioning with a few changes in practice. This is ironic because ILSI, with its well-honed strategies for burying secrets, posed an equal if not greater threat to the public health understanding of obesity.

Secrets Exposed, Coca-Cola Abandons Soda Science

Despite the massive investment by Coke and the mighty efforts of the scientific principals, the GEBN would not last long. Even before it became fully operational, the secret that had been uncovered by Freedhoff, sender of the "radio silence" tweet, was passed on to a reporter at the *New York Times*, Anahad O'Connor, who, after months of investigating, published a startling front-page exposé on August 9, 2015.

Titled "Coca-Cola Funds Scientists Who Shift Blame for Obesity away from Bad Diets," and illustrated with large photos of the principals—Hill, Blair, and Hand—the article described how the company was "backing a new 'science-based' solution to the obesity crisis: To maintain a healthy weight, get more exercise and worry less about cutting calories." Because this initial unmasking was so influential, it bears quoting at length:

> Health experts say this message is misleading and part of an effort by Coke to deflect criticism about the role sugary drinks have played in the spread of obesity and Type 2 diabetes. They contend that the company is using the new group to convince the public that physical activity can offset a bad diet despite evidence that exercise has only minimal impact on weight compared with what people consume. . . .
>
> Most public health experts say that energy balance is an important concept, because weight gain for most people is about calories in vs. calories out. But the experts say research makes it clear that one side of the equation has a far greater effect. . . . Adding exercise to a diet program helps. . . . But for weight loss, you're going to get much more impact with diet changes.[2]

The response from vocal segments of the scientific community—especially longtime critics of the food industry—was harsh. In a letter to the editor published in the *Times* four days later, Michael Jacobson, president of the CSPI, and Walter Willett, chairman of the Nutrition Department at Harvard's T. H. Chan School of Public Health, excoriated the GEBN for "peddling scientific nonsense," noting that the scientific report accompanying the latest dietary guidelines "provides compelling evidence for the causal link between sugary drinks and disease, as well as the need for exercise. Unfortunately, Coca-Cola and its academic helpers won't accept the well-documented evidence that sugary drinks are a major contributor to obesity, heart disease, and diabetes."[3] The letter was also signed by thirty-four other experts in public health, medicine, and nutrition.

That first unearthing of deeply buried secrets was followed by a series of revelatory news articles over the next few months.[4] With each article, more hidden information about the GEBN was exposed to the light of day, including, most damagingly, the extraordinary extent of Coca-Cola funding and hands-on involvement in the network's construction.

Coke's initial strategy for sponsoring product-defense science had been to remain a distant partner. In the mid-2000s, it did ethics by maintaining a hands-off stance, funding a handful of ethically safe in-house projects (the Beverage Institute, the active healthy living programs) and outsourcing the science to its nonprofit ILSI. When it hired a chief scientific officer in 2004, however, this distanced stance was replaced by increasingly ambitious projects to fund science directly and on its own, rather than in collaboration with other companies. That approach had now failed—and spectacularly.

Caught with its dirty secrets broadcast for all the world to see, Coca-Cola quickly retreated, seeking damage control. Ten days after the *Times* article appeared, Muhtar Kent, the company's chairman and CEO, placed an editorial in the *Wall Street Journal*. In "Coca-Cola: We'll Do Better," Kent publicly apologized for the debacle, expressing contrition for two unacceptable lapses—corrupting the science (poor ethics) and deceiving the public (poor public relations)—that clashed with Coke's self-image as an ethical company:

> At Coca-Cola, the way we have engaged the public health and scientific communities to tackle the global obesity epidemic . . . is not working.
>
> Our company has been accused of shifting the debate to suggest that physical activity is the only solution to the obesity crisis. There also have been reports accusing us of deceiving the public about our support of scientific research.
>
> We have read and reflected on the recent news stories and opinions. . . . The characterization of our company does not reflect our intent or our values.
>
> I am disappointed that some actions we have taken to fund scientific research and health and well-being programs have served only to create more confusion and mistrust. I know our company can do a better job engaging both the public-health and scientific communities—and we will. . . .
>
> In the future we will act with even more transparency as we refocus our investments and our efforts on well-being.[5]

With the company's reputation badly tarnished, corporate leaders appear to have decided that the best way to protect its interests was to quickly drop the science, demonstrate transparency, change directions in research funding, and install guardrails to protect against a repeat scandal. Facing widespread condemnation from the scientific community and public alike, Coca-Cola abandoned not just the GEBN (which was dissolved in late 2015), but the entire soda-defense science project it had been actively fostering since that secret meeting of CEOs sixteen years earlier. Following up on Kent's promise, over the next few months the company established an online transparency list of researchers and groups it funded, updating it regularly; it let Applebaum go; and it adopted new guidelines for sponsoring "well-being scientific research."[6] In a move intended to prevent company corruption of scientific research, Coke would no longer pay the full cost of any studies funded either directly or through a third party (such as ILSI or a trade association). It would sponsor research on health only if a non-Coke entity bore 50 percent or more of the research costs.

In February of the following year, Coca-Cola shut down the last two mainstays of its activity-first strategy to fight obesity: the Beverage Institute and the active healthy living programs. The company's efforts to fight obesity would of course continue; being "part of the solution" was now central to its corporate social responsibility mission. In this new phase, however, the company would rely not on science, but on product formulation and packaging: investing in sugar alternatives and selling more low- and no-sugar products in smaller containers. Its reputation was seriously damaged, however, and the company's efforts to distance itself from this history as late as 2021 (described in the book's conclusion) suggests it still has not fully recovered from the exercise-science scandal.

The Academic Scientists Respond: Corruption Unseen

Over the fifteen years they were involved in creating soda science and the GEBN, the academic scientists had performed ethics and concealed secrets by claiming corporate funding had no effect on their scientific judgment and by promoting partnerships with industry as beneficial, while saying little about the risks. With the publication of that *Times* article exposing the connections between the GEBN and Coca-Cola, they were now caught in the glare of media scrutiny. With nowhere to hide—after all, the academics were the frontline science makers, whose presumed credibility had created the ethical grounds for the soda-defense project to develop—they doubled down on these well-honed practices.

In that first exposé in the *Times*, Hill assured the reporter that Coke leaders are "not running the show, we're running the show."[7] Blair declared that Coke had no control over the group's work or message, adding that the academics saw no problem with corporate support because they had been transparent about it. Hand explained: "As long as everybody is disclosing their potential conflicts and they're being managed appropriately, that's the best that you can do." Half a year later, according to the UCSFL emails, Hill complained to a writer at the *Reader's Digest* that "the coverage to date of my professional relationship with Coke has been steeped in innuendo, and misleading at best. . . . My focus and the focus of the former Global Energy Balance Network has been that both diet and physical activity . . . have to be addressed in order to reduce obesity. . . . My role in working with corporate entities follows a strict code of conduct . . . and adheres to professional standards, as well as my personal standards." In spring 2017, in response to criticisms of another Coke-funded project published in a major medical journal—a series of training sessions for journalists held between 2012 and 2014—Peters and Hill insisted that they accepted Coke funding "with the understanding that the drink-maker would have absolutely no say in the program content. Funding by Coke or any other private business has never influenced our research or educational activities."[8]

A year after the GEBN folded, David Allison, a Coke insider three times over (EBAL adviser, Team Coke and GEBN member), went further, calling any judgment of research by its funding source a form of ad hominem reasoning, and any ad hominem attacks on other scientists uncivil and unscientific.[9] In other words, he declared, any critique of industry-funded research was illegitimate according to the norms of science and polite society.

Doing ethics is not the same as being ethical, however. In the last four chapters I've presented substantial evidence that the two lead scientists quietly if subconsciously distorted the science of obesity to favor physical activity and then concealed the distortion. The important question for us now is not whether the scientists acted in a corrupt manner, but why they didn't see or acknowledge it. Marion Nestle's *Unsavory Truth* makes the case that industry funding is simply the widely accepted norm in nutrition research. Everyone does it, the reasoning goes, so surely it's OK.[10] There are practical reasons as well. The most obvious is self-protection. Acknowledging corporate influence would likely bring the end of their careers as they knew them. To concede they'd been influenced by their funders would be to state publicly that they had violated cardinal norms of their profession and to lose the respect they had enjoyed as upstanding scientists. The costs would have been unthinkable.

Several features of the world they worked in, uncovered in the last few chapters, may have allowed them to reject the charge of corporate influence with consciences clear. At the most basic level, these responses of "no ethical problem" are clearly products of incentivized reasoning. The phenomenal benefits the scientists enjoyed from working closely with Coke were a major inducement to stay the course and find ways to put troublesome questions of ethics aside. These incentives gained power from being embedded in a social project of cultivation in which the scientists came to be enmeshed in an engaging relationship that pulled them into the company's orbit, slowly transforming them into virtual corporate scientists who perhaps even came to see the world through Coke's eyes. As their professional activities, careers, and fame came to be dependent on regular access to industry funding and industry-created opportunities, the scientists' investment in the corporate way of doing science seems to have blinded them to the possibility that those arrangements were skewing their thinking.

Another factor that may have shaped the scientists' ethical calculus was that the impetus for new academic-industry collaborations often came from the scientists, rather than the company, contradicting the conventional narrative about corporate science, which sees companies co-opting scientists. As we've seen repeatedly, Hill was a nonstop go-getter, who actively cultivated the company's executives in pursuit of support for his pet projects. While the GEBN was coming together during 2014, Hill was engaged in a side hustle to secure funds for a separate, very large-scale research project that he pitched as a major game changer. In email conversations meant to be completely private, he approached no fewer than five Coke executives to push his ideas.[11] Although this project did not come to fruition, the communications surrounding it provide an extraordinary record of the shameless molding of science to corporate needs.

In these pitches to the executives, Hill explicitly tailored his scientific ideas to meet the specific needs of the company:

> Arne [Arstrup, of the University of Copenhagen] and I have worked to pull together our ideas. We think the major focus should be on research showing that the framework of energy balance should be the focus of most of your research. We have given you ideas here [in an attached document, "Research Ideas for Coke, 6-3-14"]. We have also given you ideas for research projects that might be very specific to coke [sic] interests.

> When we last spoke I told you I wanted to submit a research proposal on physical activity. Here is my concept. I think it could provide a strong rationale for why a company selling sugar water SHOULD focus on promoting physical activity. This would be a very large and expensive study but

could be a game changer. We need this study to be done. It could be done at our site only or perhaps as a multicenter study with Arne and Steve Blair.

In these private communications, Hill presented himself as an ardent defender of the company and as uniquely qualified to rescue both Coke and "the world" from the obesity crisis:

> I am committing the remainder of my career to trying to guide the field in a different direction. . . . I do hope we can work together to make a difference. The world and the Coca-Cola Company both benefit if we succeed. I realize this is a bold undertaking but I am at a place in my career that I think I can bring together a very powerful coalition of the right people [to] think differently and to change things. We have the passion, the innovative ideas, and the right approach. We need substantial resources. . . . No one else is thinking this way and we can succeed. It is not fair that Coca-Cola is signaled [sic] out as the #1 villain, but that is the situation and makes this your issue whether you like it or not. I want to help your company avoid the image of being a problem in people's lives and back to being a company that brings important and fun things to them. No other groups . . . are thinking the way we are and we can succeed.

How did this kind of language fit with Hill's insistence that his science was immune from corporate influence? There are few hints in the emails. Perhaps he reasoned that because he was taking the initiative in creating these Coke-friendly proposals and marketing them to the company, the company was not influencing him; instead, he was influencing the company. That may be true, but the company's needs had already shaped his proposals. And the effect on the science was the same; whoever was taking the lead, the science was being molded to advance corporate interests.

The larger context of conflict between Team Coke and its adversaries in public health may also have shaped their responses to questions of corruption. From where Team Coke stood, the war on their ideas and life's work was very real. Their daily reality was, in Applebaum's words, a David versus Goliath battle between the "one-sided," "regulation-obsessed," "agenda-driven" NIH-supported "extremists" in public health, and the more "balanced," "evidence-based" (soda) scientists who understood that activity was at least as important as dietary restraint and that close partnerships with industry, not antagonizing regulations, were essential to stopping the obesity epidemic. The members of Team Coke and their sponsor routinely framed their responses to ongoing developments in these terms. If the media uncovered a Coke consultant serving on a government nutrition committee—remember

the Macdonald affair in the UK—the immediate reaction was moral outrage and angry denial. Over time the discourse of combat began to have
effects, dividing the world of obesity science into friends and enemies, us
versus "the other side." The occasional jabs in the emails about the "unreasonable" Kellys, Michaels, Marions, and Yonis of the world (Brownell,
Jacobson, Nestle, and Freedhoff, in that order) worked to "other" these scientists as people whose ethical critiques deserved little consideration. Instead of weighing the validity of the critiques, the members of Team Coke,
hunkered down in their trenches, dismissed them as yet another barrage
coming their way and responded in the only way the narrative allowed: by
returning fire. In the grip of the discourse of endless conflict, Team Coke
may have had difficulty seeing that their "enemies" might just have a point.

None of this would have had the power it did to shape perceptions if
each of the scientists had been operating on his own. Brought together first
by their shared interest in exercise science, and then, with more frequency,
by Applebaum's schemes to groom them as virtual corporate scientists for
Coke, the members of Team Coke interacted frequently over email and in
person, forming a relatively closed community of discourse in which they
plotted the next "defensive and offensive" actions in the campaign against
hostile forces. Team Coke offered not just the fun of promoting shared
agendas and playing (serious) war games together; it offered safety in numbers. *If all these other esteemed scientists are declaring no conflict of interests,
it's fine for me to do it too. Should things get unpleasant, they will have my
back.* The ongoing support of those influential colleagues may have made
the members of Team Coke confident they could weather whatever storms
would blow their way. Their bland initial response to the *Times* article and
their continued inability to acknowledge corporate influence years later
suggest they may have been living in an alternative world in which corporate support for science posed no real ethical problem.

Whatever their claims of neutrality, the scientists caught in this scandal
would face the reckoning of their institutions and their field. Soon after the
downfall of the GEBN, the academic scientists found themselves targets
of criticism and even disciplinary action for perceived ethical violations.
While the email archive contains messages of support from like-minded
colleagues, as a group they suffered public shaming and loss of professional
position. In 2016 Hill stepped down from his position as executive director
of the Anschutz Health and Wellness Center at the University of Colorado.
Hand was removed from his position as dean of West Virginia University's School of Public Health.[12] The setbacks seem to have been temporary,
though. Both Hill and Hand later relocated and assumed leadership positions at their new institutions. In 2018, Hill was appointed director of the

Nutrition Obesity Research Center and chairman of the Department of Nutrition Sciences in the School of Health Professions at the University of Alabama, Birmingham. In 2020 Hand became dean of the College of Health Professions, Wichita State University in Kansas. In 2017, Hill and Peters won awards from the American Society for Nutrition, a close affiliate of ILSI. Hill joined the class of ASN fellows, the organization's highest honor, while Peters won the McCormick Science Institute Research Award.[13] In the end, both Colorado scientists and perhaps Hand as well seem to have recovered with few ill effects. In 2016 Blair retired, assuming the title distinguished professor emeritus.

Whether quietly concurring that the GEBN was problematic or simply performing contrition, the University of Colorado returned the $1 million awarded for its establishment, calling it a distraction.[14] The University of South Carolina did not return the roughly $500,000 it had received, presumably because the funds had been spent on the website. With the (at least short-term) sanctioning of Team Coke's leaders, the GEBN, that ambitious experiment to create a one-company academic-industry partnership to promote soda science, met its death. While Coke took the ethics rebuke very seriously, in the academy the death of the GEBN seems to have settled nothing, leaving the rival views on the ethics of industry funding intact.

ILSI: Safe for Now

ILSI, the most important vehicle for making soda science, largely escaped scrutiny during this first wave of journalistic reporting and scholarly critique. Because the media and public health critiques focused, laser-like, on Coca-Cola and the scientists involved with the GEBN, the longer history of the global network—including the work of ILSI's CHP in laying the foundations of soda science and creating the Take 10! program, and ILSI North America's vital role in strengthening energy balance science—remained publicly hidden.

ILSI was a sophisticated and highly effective vehicle for making soda science, and the success of the CHP and ILSI-NA is testament to that organizational prowess. Yet appearances were sometimes deceptive. Its ability to camouflage industry's role in science making made the organization hard to crack; without extensive probing it was next to impossible to know how deeply industry's influence ran in any of its scientific activities.

ILSI was also a supple organization capable of quickly reinventing itself or rearranging its parts in the face of threat. The initial focus of the *Times* exposé on the GEBN gave the North America branch time to make strategic changes in its EBAL committee. But instead of reviewing the workings of the committee to look for evidence of possible ethical violations, the branch moved to distance itself from the soda-science project and ward off any perception of ethical or scientific wrongdoing. ILSI-NA quickly replaced Hill and Jakicic as advisers to the EBAL committee. In 2016 the branch changed the group's name to the Committee on Balancing Food and Activity for Health and in 2018 quietly dissolved it.[15] ILSI-Global turned its attention from obesity to other issues.

Two Wins for the Soda Industry

Coke's moves, especially the withdrawal of funding, spelled the end of soda science as an organized project in the United States. But research along these lines continues to flourish. In 2022–23 alone, Hand and Blair issued new findings from the Coke-funded Energy Balance Study, Hill published on the small-change approach, and Edward Archer, one of Blair's protégés, declared guidelines on added sugars unscientific and unnecessary.[16] And other food-industry actors have continued to push the exercise-first narrative. Still, Coke's actions marked the end of the project of actively developing and circulating a science to defend the soda industry. Though its grandest materialization failed to reach fruition, we should not think of soda science as a failed venture, at least from the vantage point of Coke and other soda companies. Indeed, the soda industry realized two big wins from this bold adventure in product-defense science.

The first was the profits and other material and reputational gains enjoyed during the lifetime of soda science. Though the science eventually had to be disavowed, during the years it was being built out the company was making money on its unhealthful products. It was also succeeding to some unknown extent in keeping soda taxes and marketing regulations at bay, reaping potentially substantial benefits for over a decade.

The second win for the soda industry was the deep imprint the science left in some ILSI branch countries. In the next part of the book we go to China, where we will see how the very same people who promoted the flourishing of soda science in the United States—Malaspina, Applebaum, Hill, and Blair—worked with ILSI's China's branch to establish it as virtually the only way to think about obesity in that country.

∴

Part Two

TAKING SODA-DEFENSE
SCIENCE TO CHINA

∴

[CHAPTER FIVE]

Laying the Groundwork

Facing a dire threat from the rising tide of obesity, in the first decade of this century the world's largest food companies, working through their non-profit, moved quickly to create a corporate science of obesity to protect their profits. That science in defense of soda was made in the United States, but it was made not just to secure the home market. Instead, it was made to travel. With three-fifths of revenues coming from markets outside North America (in 2000)[1] and obesity rising nearly everywhere in the world, the most critical targets for the dissemination of soda science were the indus-try's major markets abroad: the large, rapidly growing countries of the Global South.

Behind that campaign to spread soda science was an imagined empire of corporate science. In this imaginary, ILSI-Global creates industry-friendly science in the United States, transmits it to ILSI branch countries around the world, and succeeds in getting local branches to endorse it and embed it in public policy. In chapter 1, I traced ILSI's success in creating corporate sci-ence to five features of the organization. The key to its success in this global project was the organization's sixth feature, the stratified structure and op-eration of the entire ILSI network. Concentrating power in the global core (the United States), ILSI had a set of mechanisms in place to facilitate the top-down flow of scientific ideas from the center to the branches. How real versus only imaginary the empire of corporate science was at a given time would depend on how local branches responded to campaigns of scientific influence emanating from the center. And those responses would depend on the politics of science and health in the host countries.

With its 1.26 billion customers (in 2000) with money in their pockets and modern lifestyles on their minds, China was a top-priority target for this campaign. Beyond the size of its market, the country's post-1978 po-litical economy was exceptionally, indeed, perhaps uniquely welcoming to the food industry and its scientific agent. With the rise to power of Deng Xiaoping in 1978 (top leader 1978–93), China's post-Mao leaders embarked on a radical new program of reform and opening up through rapid integration into the global economy (*gaige kaifang*). Articulated by Deng and elaborated by successive leaders, the dream of China's ruling Communist Party (CCP) was to transform China into a rich and powerful nation under its firm leadership. The new formula for China's economic growth and global ascent involved three broad policy currents that would be favorable to ILSI's project. The terms for these policy emphases have been part of everyday political discourse throughout the reform era. *Marketization (shichanghua)*—meaning the shift of governance functions from state to market and the growing role of market actors and logics in the economy—led the state to defund health research, making corporate funding for scientific research attractive and necessary. *Globalization (quanqiuhua, guojihua)*—rapid integration of the Chinese economy into the global economy—opened the door to advanced Western science and large foreign firms, seeing both as vital contributors to China's rapid growth and global rise. And *scientization (kexuehua)*—the vigorous development of science through selective borrowing of Western knowledge—fostered an often uncritical enthusiasm for Western science to serve as the basis for an enhanced *scientific policy making.*[2] All this would leave the door wide open to the soda industry's scientific nonprofit, which was keen to supply the very things China's health sector most needed: research funding and the latest science on the diseases of modern life.

China was the apple of Malaspina's eye, and in this part of the book, which charts the step-by-step success of the soda campaign there, we see why. We track ILSI leaders as they laid the groundwork for ILSI to take control of the obesity issue in China (this chapter), transmitted soda science to the country (chapter 6), got it built into local policy (chapter 7), and framed it as ethical in Chinese cultural terms (chapter 8). In this chapter we follow the ILSI-Global leader as he chose the country's top nutritionist as branch head and then approved the unusual location of the branch within the Ministry of Health. We then see how ILSI's China branch quickly assumed control of the obesity issue, marginalizing rivals; defined obesity as a Chinese disease; and helped seat the food industry at the policy table, clearing the way for ILSI-China and its corporate supporters to have a major voice on policy going forward. The story of Big Food's success in China is full of strange happenings unimaginable in the West. But this account of science and policy gone awry should not be dismissed as just some weird, one-off story about China. The success of soda science opens a rare window on some fundamental flaws in the governing system of a giant nation on its way to superpower status. It deserves our closest attention.

After following the bitter battles over obesity science in the United States, readers may be surprised to discover that once we arrive in China, all such open conflict disappears and the making of obesity science seems remarkably harmonious. That apparent placidity should not fool us into thinking that Chinese scientists are less critical or contentious than American researchers. What it means is that in China the open expression of scientific dissent tends to be culturally disapproved and politically risky. In the book's introduction I described the contextual nature of science and we can see that most vividly as we follow soda science in its international travels. Oversimplifying, we can draw a broad contrast between more democratic and more authoritarian contexts. In the more democratic polity of the United States, science is relatively autonomous from the state, and the open expression of disagreement and dissent is common in virtually every sphere of public life, including science. In the more authoritarian society of contemporary China, science is subordinate to the state, and available evidence suggests that scientific fields, especially those involved in policy debates, tend to be structured hierarchically in such a way that only a few state-approved leaders have a public voice.[3] The leader of the scientific field articulates "the truth," and followers publicly accept and help scientifically construct that truth, while dissidents are silenced or even marginalized. Conflict and disagreement certainly exist, but they often take place behind closed doors or are pushed underground where they become very hard to unearth. We will meet some of those dissidents from corporate-science orthodoxy in this chapter and a few others when we turn to ethical challenges in chapter 8. While promoting harmony on the surface, these hierarchical arrangements would leave China's science of obesity without real critics and thus vulnerable to manipulation by external forces.

My primary story in this chapter centers on ILSI's success in laying the groundwork for China's embrace of soda science. But there is a parallel, rather more grievous, story that must be told as well. This second story charts how Chinese scientists, reading the world through the lens of a particular narrative of nation, systematically (mis)perceived the ILSI campaign as something highly beneficial to China. In that narrative, a once oppressed, scientifically backward nation is now, with ILSI's help, on the path to becoming a global leader in the handling of chronic disease epidemics. China's scientists served ILSI enthusiastically, imagining the connections to American scientists were rewarding them with quality science and top-notch policy, when the true effect was the embrace of tainted science and policy that was far from best practice. These two stories—the victory of corporate science and the (mis)perceptions of researchers in the Global South

fostering a certain naïveté about the intentions of the generous foreigners—
run through the rest of the chapters in this book, creating a subterranean
tension that was never resolved.

ILSI's China Dream: Creating a Branch in China

In the late 1970s Malaspina envisioned a worldwide empire of interlinked
ILSIs stretching across the globe. No sooner was ILSI established in Washington, DC, than he began visiting key countries to "plant the ILSI flag
around the world," as his colleague Suzanne Harris put it, conjuring up
the official image of the world dotted with colorful ILSI flags. From the
beginning, ILSI's leaders viewed China as a prized target for recruitment
into the ILSI family. The potentially vast size of its consumer market was
an irresistible lure. "Was China a pet branch of Malaspina?" I asked Harris.
"For sure!" she replied quickly, adding, "because Coke wanted to expand in
China!"—letting slip the truth that the company created ILSI less to make
better science than to sell more Coke!

Yet China had another draw as well. If Malaspina could find the right
Chinese partner, the political setup in which most scientists work in and
for the state could offer ILSI political access unmatched in the world. In
this section, we see how three of the key attributes that made ILSI-Global
so effective in the United States—its expert-recruitment strategy, public-
private duality, and invitation-only practice—were reproduced in the
China branch. We also see how the broad policy directions established
by the reform state informed the location of the new organization in the
bureaucratic landscape and created a strong demand for the kind of services the new branch was set up to perform: bringing research funds and
advanced science to China. Designed to foster rapid economic growth,
those policy currents would leave the country vulnerable to an influence
campaign that pushed corporate science in the name of the best of international science.

COKE'S CENTURY-LONG LOVE AFFAIR WITH CHINA

In the late 1970s, China's new reform leadership threw open the doors to the
world economy that had been closed for thirty years. The open-for-business
policy, combined with China's vast population, one billion and growing,
and a society yearning to join the modern world, made it a must-grow market for foreign companies selling the consumer goods of modern life. In the

early twentieth century, the Coca-Cola Company had built a vast empire in China. When the People's Liberation Army swept into China's cities in 1948–49, however, the company's bottling plants were seized, and consumption of this crass symbol of American imperialism was soon banned.[4] In the 1970s, in the wake of US president Richard Nixon's historic 1972 visit to China that began the thawing of Sino-American relations, the soda companies began to lust over China, as the historian Charles Kraus has put it, with Coca-Cola lobbying for permission to return as early as 1975. The prospect of Coca-Cola's arrival in Communist China was contentious. While advocates saw opportunities to access the company's advanced technology and production techniques, detractors warned of economic problems for Chinese businesses and health problems for Chinese consumers. With the approval of Deng Xiaoping himself in December 1978 the Coca-Cola Company became the first international company to gain permission to operate in China under the country's new reform agenda.

Coke's Chinese partner was the huge state-owned food trader and manufacturer, COFCO (China Oil and Foodstuffs Corporation, known as Zhongliang). Coke helped COFCO build a new state-of-the-art bottling plant, structured as a joint venture, in exchange for access to the China market. Although it would be several rocky years before Coke was allowed to build more bottling plants and sell directly to Chinese consumers (under the initial agreement the company could sell only to foreigners), once business began to grow it was unstoppable. Coca-Cola soon emerged as the leading seller of carbonated soft drinks in what one executive dubbed "the world's biggest market-to-be."[5]

The China market took on added significance after the turn of the century. After soaring during the 1970s, 1980s, and 1990s, in 1999 soda consumption in the United States began to level off and decline, as American consumers, newly aware of the health risks of sugary sodas, cut back.[6] With profits in the industrialized West slumping, China and other large emerging markets became vital to the long-term growth of Coke and other soda giants.[7] With its record-breaking economic growth and rising middle class hungry for all things Western—young families flocked to McDonald's not for the strange food but for a taste of the American lifestyle—China was especially important.[8] The company's confidence in the China market seemed only to grow. "We don't see China only as a great growth market," chief executive Muhtar Kent declared in 2011. "We see China as a future market for further innovation that will benefit our business globally."[9] For Malaspina and others at ILSI, the message over the years could not have been clearer: protect the China market.

AN OUTSTANDING BRANCH HEAD

The key to the success of this global science enterprise was finding the right people to head up the international branches, people Malaspina could cultivate and incentivize to serve as quasi-corporate scientists loyal to the ILSI mission. Even before Coca-Cola had inked the deal with its first Chinese bottling partner, Malaspina was on the ground scouting for scientists to groom as future branch heads.[10] In 1978, the same year ILSI-Global was founded, Malaspina visited Beijing and identified Chen Chunming (C. M. Chen for short; 1925–2018) as a promising partner.[11]

Chen was an extraordinary person. I first met her in ILSI's bright, crowded, book-stacked office on the ninth floor of a China CDC building on Nanwei Road in south central Beijing. Still formidable at eighty-eight (this was late 2013), she was articulate and energetic. With a degree in nutrition science from the Department of Agricultural Chemistry at National Central University (1947) and decades of experience as a nutrition researcher in the top health research centers in the country, including the prestigious Chinese Academy of Medical Sciences, Chen was one of the nation's leading nutritional scientists.[12] Her scientific reputation would later rub off on ILSI-China, giving its industry-funded science the authority and credibility it needed to be influential. In 1982, Chen became the director of the health ministry's Bureau of Disease Prevention and Control (1982–84), and in 1983 she was appointed founding head of the Chinese Academy of Preventive Medicine (CAPM), becoming a high-level government official. (In 2002, the academy was renamed the China CDC; when referring to it in the years before 2002, I call it the CAPM/CDC.) She was also active internationally and received a number of awards for her achievements in nutritional science.

This résumé of official positions carries political significance that readers unfamiliar with Chinese history and politics might not appreciate. Unlike in the United States and other Western countries, where science enjoys relative autonomy from the state, in China science is subordinate to the party-state. Since the late 1800s, when China first embraced modern science and technology to defend against the incursions of the Western powers, science has been a tool of the Chinese state, deployed for the ultimately statist ends of modernizing the country—and keeping the state (or ruling party) in power.[13] To highlight the state's dominant role in science, we can say that Chen, like the great majority of China's scientists, was a *state scientist*, one who worked in state-run institutions on projects developed primarily to advance state-defined goals. Chen was also a former high official, and in China, once an official, always an official. That career history in the middle to upper reaches of the health ministry would leave her with political connections and lingering policy clout that would be critical to ILSI's success in translating its science into official policy. These invisible connections—known as *guanxi* in Chinese—made Chen a phenomenal asset to ILSI. By drawing on those ties (*la guanxi*, literally pull or pull on relational strings) she could make things happen. All this made Chen not just a state scientist, but a *state scientist-official* whose first loyalty was to her country and its leaders.

ESTABLISHING ILSI-CHINA: "LEARNING FROM THE WEST"

Malaspina's offer to collaborate was evidently appealing, for soon after his 1978 visit, Chen and her colleague Chen Junshi (MD 1956, J. S. Chen for short; the two are unrelated), a prominent food toxicologist and food-safety expert, began working with ILSI.[14] I also spoke at length with J. S. Chen, who was equally engaged and savvy, as well as generous in taking the time to untangle the complexities of Chinese bureaucratic politics for me. As an academician (*yuanshi*) of the Chinese Academy of Engineering, J. S. Chen belonged to an elite group of scientists and technicians who had the right and responsibility to use their expertise to advise the state on policy matters. For the next fifteen years—from 1979 to 1993—the CAPM/CDC worked with the global ILSI organization on projects of mutual interest.

In 1993—as the frosty political climate following the Tiananmen Square massacre began to thaw after Deng Xiaoping's trip to China's prosperous south (dubbed the "Southern Tour") reaffirmed the marketizing and globalizing directions of economic policy—the time was finally ripe to form an ILSI branch in China.[15] Keen to access the many benefits of a more formal association with ILSI—steady funding, advanced knowledge, global connections—in 1993 Chen Chunming formally left government service to establish ILSI-China. Chen would remain at ILSI until her death, initially as director (1993–2004) and then as senior adviser (2004–18). Chen Junshi served as her deputy before taking over as director in 2004. Just as Hill and other top advisers to ILSI in the United States would become quasi-corporate academic scientists, before long Chen (and J. S. Chen) would transition to *quasi-corporate state scientists*, who served the interests of the Chinese government and the transnational food industry at the same time. In China, then, ILSI and the food industry that supported it would have a political insider—a virtual government official—working on their behalf.

In the early 1990s, when Chen decided to formally leave government service, the health sector she had devoted her life to was in chaos. In the late 1970s and early 1980s, Deng Xiaoping was so preoccupied with growing the economy that he essentially abandoned the health bureaucracy to fend for itself in the market.[16] The reform state effectively defunded health work and began urging researchers and health-care institutions to look to the market for support. The bold experiments in rural health care that had made China a model of low-cost health care during the Mao years—the cooperative medical system, the barefoot doctors—were tossed out or allowed to lapse with hardly a second thought. In the new era of reform, the party's strategy of spurring economic growth and boosting incomes at any cost was its primary source of legitimacy. Even as the worrying health consequences of that approach—including the rapid rise in chronic diseases like obesity—began to mount, unabated economic growth remained the number-one priority. It was not until the SARS crisis of 2003 forced its hand that the central government began to pay attention to the health sector.[17] Bureaucratic realities added to the woes of those championing the health of China's people. As a resource-*consuming* rather than resource-*producing* bureaucracy, the Ministry of Health (MOH) had long been a weak bureaucracy, forced to work especially hard to obtain state funding. And within the health sector, chronic disease, Chen's specialty, was among the lowest priorities, attracting much less attention than the management of infectious disease outbreaks.

All of this—and especially the constant struggle to obtain funding— would propel Chen's decision to quit government work and join ILSI. As she explained in an interview, establishing an ILSI branch would give her access to new and much-needed funding from industry and other international sources. Twenty years later, when I was interviewing in Beijing, a dearth of funds continued to stymie research on critical problems of human nutrition in a society facing an onslaught of Western fast foods and sugary drinks. My informants were unanimous on this point, heaving a collective sigh of frustration at how little had changed since ILSI's founding. "There's virtually no basic research on obesity because there is no funding," J. S. Chen told me in exasperation. "China's NSFC [National (Natural) Science Foundation of China] does not support this, and China has no NIH [National Institutes of Health]. The MOH can't do it alone, and there is no Ministry of Obesity!" In this situation, a CDC researcher added, stating the obvious, "corporate money substitutes for government funding."

Establishing an ILSI branch also responded to the party-state's call for upgrading China's science by absorbing the latest knowledge from the international community, part of the process of scientization. In the early reform decades, China was in the throes of science fever. In the late 1970s China's reform leaders elevated modern science to the first of the "four modernizations" (*sihua*) aimed at rejuvenating the economy. In the early 1980s the state mandated that all policy in the new era be based on science.[18]

"Scientific policy making" (*kexue juece*) became the slogan for the new era. This would represent a profound shift in political reasoning. After the chaos of the Mao years, when policy making was often ideologically driven and data-free, this new mode of decision making was needed to strengthen both the "scientific" (empirically based, rational) and the "democratic" (broad-based, involving all kinds of experts) character of the policy process. The assumption was that bringing scientific experts into the policy process would enhance the quality of party policy.

After decades of isolation, Chinese researchers rushed to take advantage of the new opening to the international scientific community. In those early years of opening up, there was extensive academic exchange in almost every field, as Chinese scholars scrambled to catch up with intellectual developments in their specialties and foreign scholars jumped at the chance to visit China and meet their new colleagues in the global enterprise of science. (Throughout the 1980s, I myself was part of a process of scientific exchange, in population studies, that was exhilarating for both sides.)

Chen repeatedly emphasized the goal of upgrading Chinese science on chronic disease to improve health policy. Setting up an ILSI branch, she told me, would enable her to bring the advanced ideas of Western public health science and practice to China, which, after being closed to the Western world for decades, was widely perceived, at home as well as abroad, as "scientifically backward."[19] "Malaspina was very good to China," Chen told me happily. "He brought advanced Western ideas and contacts to the country." Importing Western science did not mean recognizing, let alone accepting, China's inferiority. Far from it! Chen's—and China's—project of assimilating Western knowledge was driven by a strong sense of national pride. After more than a century of national humiliation at the hands of the Western powers, China's leaders and intellectuals like Chen believed that by learning from the West, China would quickly boost the quality of its science, catch up, and, in time, join the ranks of global scientific powers. That aim of copying to catch up motivated projects of learning from the West across the sciences. Chen's assessment of the gaps in Chinese public health science was of course correct, at least in the early post-Mao decades. Yet it would foster an uncritical, at times almost worshipful, stance toward Western science that, in the case of obesity science, would not be warranted.

AN AGENT OF CORPORATE
SCIENCE IN THE MINISTRY OF HEALTH

In 1993, Malaspina introduced five companies to jump-start the branch's growth. Building on the substantial donations of these founding firms, Chen began pouring her energies into building up the branch. In the 1990s (1994–99), the number of companies supporting the branch tripled from five to seventeen. In the next fifteen years (2000–2015) it more than doubled again, to thirty-seven.[20] The vast majority were well-known transnational corporations in the food, beverage, and restaurant industries. By 1999, the year ILSI took up the obesity issue, the list included many of the world's best-known megacorporations, including Ajinomoto, Coca-Cola, Danone, Heinz, International Flavors, Mars, Monsanto, Nestlé, and PepsiCo. Indeed, the membership of ILSI-China looked a lot like the membership of ILSI North America: lots of food companies eager to protect and grow their markets.

In many respects, the branch Chen created with Malaspina's enthusiastic support conformed to ILSI's bylaws on branch organization and looked and functioned like an ordinary ILSI branch. Its official mission—bridging government, academia, and industry to provide the most current scientific information for policy decisions in nutrition and food safety—paralleled that of ILSI-Global.[21] Like the larger organization, ILSI-China fulfilled its mission by organizing conferences, task forces, and research activities, inviting experts from the local CAPM/CDC and universities, as well as a few foreign specialists, to lend their expertise. Chen also followed ILSI's invitation-only rule on participation in ILSI activities. In a conversation, she proudly described how, when organizing conferences, ILSI-China did not just issue calls for abstracts, but rather hand selected most of the invitees, presenters and experts in the audience alike. This practice, which to her marked ILSI as a highly selective organization that brought only the best of the best nutritional science to China, actually operated rather differently. Just as it had in the United States, in China the invitation-only rule would work to create a quasi-closed scientific world in which ILSI could control what science was labeled "the best" and exclude both dissidents from ILSI orthodoxy and critics of corporate influence more generally. This little rule would be crucial to ILSI's success in advancing soda-defense science in China.

The China branch was distinctive in two ways, though. First, in violation of ILSI norms and bylaws, ILSI-China would take the form (if not the name) of a central government–based think tank rather than a nongovernmental organization (NGO). (Branches of foreign NGOs were not permitted at the time it was formed.) ILSI-China was designated an affiliate of the CAPM/CDC, which, recall, Chen had led for the previous ten years. Remarkably, the office of the China branch was physically located in the CAPM/CDC headquarters in Beijing (see figure 5.1).[22] Also unusually, ILSI was assigned to report not to the CAPM/CDC, but to the head of the Bureau of Disease Prevention and Control within the Ministry of Health, who in turn reported to the minister of health. These arrangements gave Chen daily access to of-

FIGURE 5.1. CDC office building in Beijing, home to ILSI-China. Photo by author.

ficials in the CAPM/CDC and a nearly direct line of communication to the top health official in the country. With this location, ILSI-China exhibited the same public-private duality that characterized ILSI-Global. Publicly, it was just another government-based scientific think tank; privately, whether wittingly or unwittingly, it was an outpost of corporate science. An agent of corporate America was now located in China's Ministry of Health. To mark itself as different from a foreign NGO, the branch's founders called the organization "ILSI Focal Point in China" (Guoji shengming kexue xuehui Zhongguo banshichu) and referred to its funders not as members but as "supporting companies." I call it simply ILSI-China.

In a second major violation of ILSI's rules on branches, ILSI-China lacked a board of directors. "We have lots of freedom; we can do anything we like," Chen told me elatedly. Chen was exulting in her freedom from the control of a government bureaucracy known, like many official bureaucracies, for its sluggishness, inefficiency, and complicated politics. And within the organization, decision-making power was concentrated in the hands of the two Chens.[23] With her incorporation into the ILSI system, however, a new form of control was being substituted for the old one. While Chen saw herself as free to do what she liked, we will see that what Malaspina had granted her instead was the freedom to serve ILSI and its corporate project well. Through the powerful incentives provided to ILSI branch leaders and subtle forms of top-down compulsion that pervaded the organization, over the years Chen would become not just a willing quasi-corporate state scientist, but a powerful (if not fully aware) agent for the food industry that ultimately called the shots at ILSI. In the blunt assessment of one knowledgeable Western observer, "ILSI-China was a two-person show, with Malaspina calling the shots."

ILSI-Global and ILSI-China: Hidden Channels of Corporate Influence on Science

If Malaspina called the shots, how did he do it? In chapter 1 I mapped out the distinctive features of ILSI-Global, showing how they facilitated the creation of soda-defense science in ILSI's home country. Here I describe the mechanisms it created to spread that science around the world. The ultimate goal—stated with greatest clarity by the Coke-funded GEBN—was, to the extent possible, getting soda science endorsed as the dominant understanding of obesity and the basis for public policy in countries with swelling markets for the food industry.

ILSI's official rhetoric about its worldwide network implies an egalitarian ethos, depicting the parts of its far-flung empire as "work[ing] together to identify and resolve outstanding scientific questions."[24] The reality is rather less noble. Close study of how ILSI worked in practice reveals that its goal of "harmonized use of science" across the globe was achieved by the imposition of subtle and not-so-subtle controls from the top. The corporate-science empire that Malaspina worked to create functioned hierarchically, with the science—the choice of issues, their framing, the conclusions—decided by US-based experts, then exported around the world through collaborations with local branches and partners (universities, CDCs). ILSI leaders and ILSI member companies based in the United States also had a set of mechanisms they could deploy to get the branches to accept ILSI science and follow the organization's ways more generally. Some were legal, others operational, and still others social in nature. Such devices to quietly benefit ILSI's companies existed at the branch level as well. In China, companies had access to hidden financial channels that allowed them to shape the science, along with an official ethics rule that worked to their advantage. Virtually all these mechanisms, global and local, would be brought into play in getting the energy balance science of obesity created in the United States endorsed in China.

Beyond these strategies aimed at wielding influence in ILSI branch countries, ILSI-Global also worked, undoubtedly through the office of its president, to shape food, and especially sugar, policy in global institutions such as the UN World Health Organization (WHO) and Food and Agriculture Organization (FAO). Because of the secrecy surrounding these illicit activities, I have only one example, but it suggests that, when an opportunity arose, the organization was willing to set aside scientific niceties and resort to what amounts to bribery, providing funding in exchange for promises of favorable scientific outcomes. Let's start there.

HIJACKING SUGAR POLICY AT THE UN

The global sugar industry, and the food industry more generally, have long fought aggressively against governmental restrictions on sugar. The primary tactic has been to undermine scientific evidence linking sugar to a host of diseases, including obesity and type 2 diabetes. In chapter 1 we saw how, in the early to mid-twentieth century, sugar industry scientists mobilized the energy balance framework to argue that sugar was innocent of the charges. At the end of the century, as the obesity epidemic was becoming more visible to the makers of all that tasty junk food laden with sugar and fat, the sugar industry combined forces with ILSI to meddle in the work of the UN. In 2004, *BBC Panorama*, an acclaimed current affairs documentary program, uncovered evidence that the sugar and food lobbies had surreptitiously influenced a key meeting of nutrition experts held in Rome in 1997.[25] At the WHO/FAO Expert Consultation on Carbohydrates, top nutritionists had convened to establish definitive conclusions about the proper amount of carbohydrates, including sugar, in the diet. Governments around the world, including China's, use guidelines such as these when making decisions on health standards, nutrition policy, food labeling, and other matters crucial to health. The BBC discovered that the consultation was supported by $40,000 from ILSI and $20,000 from the World Sugar Research Organization. In exchange for their support, the funders were invited to suggest participants, nominate the chairperson, and comment on the agenda and the background papers for the meeting. The consultation was supposed to be independent; these opportunities granted to industry funders were highly irregular and, to put it mildly, incompatible with the way UN agencies are supposed to operate.

Interviewed by the BBC in 2004, experts who had participated in the consultation were shocked. Jim Mann (University of Otago, Dunedin, New Zealand) said they had no idea where the funding came from or what influence the support had bought. Yet the new evidence made sense of some of the odd things that had happened in Rome. When they arrived for the weeklong meeting, Mann recalled, he and some others were summoned by an official and told in no uncertain terms that it "would be inappropriate for [them] to say anything bad about sugar in relation to human health," and that doing so "would have profound political implications." Whenever sugar was mentioned in the discussions, John Cummings (Ninewells Medical School, Dundee, Scotland) reported that one official, there to observe, tried to block the debate. "Normally these officials sit and listen and just sort of prod you when they think something needs doing but this was quite amazing." The consultation concluded that carbohydrates including sugar should make up no less than 55 percent and no more than 75 percent of daily calories. But when the written report appeared, the upper limit had vanished. Without an upper limit, food companies seeking to compensate for the fat they had removed from their products by increasing sugar could keep adding more and more sugar to improve taste. The scientists interviewed found the whole thing "absolutely appalling" and felt their integrity had been severely compromised. For several years, until the BBC uncovered this history, the industry was able to use the Rome Report, as it came to be called, to argue against limiting how much sugar we eat and how much sugar governments allow in our food. This was a major coup for the industry engineered in part by ILSI.

ILSI-GLOBAL: MEANS OF TOP-DOWN CONTROL AND CORPORATE INFLUENCE

ILSI was also involved on a more routine basis in efforts to shape national policies through the hierarchical workings of its international network. Key mechanisms of top-down influence that I uncovered in my research are listed here.

MECHANISMS OF TOP-DOWN INFLUENCE WITHIN ILSI

A. ILSI-Global Mechanisms of Top-Down Influence over the Branches
  1. Direct controls from the top
    a. Legal mechanisms: Bylaw provisions on ILSI-Global and ILSI branch relations
  2. Operational mechanisms

    a. Expert recommendations by DC-based leaders, in context of ILSI's invitation-only policy

      Allow DC executives to shape what counts as "good science" in the branches

      May result in international conferences packed with ILSI- and industry-friendly speakers

      Create a quasi-closed world of ILSI science with few if any dissidents

    b. Intra-ILSI transfers of funds, from US-based ILSI entities to ILSI branches, to support favored programs

    c. Leader visits to ILSI branches, often used to advocate for preferred programs

  3. Social mechanisms

    a. Annual meetings, facilitate networking and sharing of leader priorities

    b. Personal cultivation of branch heads by ILSI-Global leaders

B. ILSI-China Mechanisms of Influence over Chinese Science and Policy

  1. Location within China's health bureaucracy

    a. ILSI-China is affiliated with and located within the CAPM/CDC

    b. ILSI-China reports to the MOH

    c. C. M. Chen, well-connected former high MOH and CAPM official, operates as de facto policy maker

  2. Funding arrangements

    a. Corporate funding via ILSI to support preferred public health interventions; funding is both direct and through a Healthy Lifestyle Fund

    b. Corporate funding via ILSI to support scientific meetings promoting the health benefits of their products

    c. "Special donations" by companies to support large-scale events, with expectation of something in return[26]

ILSI-Global's bylaws gave the larger organization broad oversight over the branches. In exchange for a charter and certain benefits, branch heads were responsible for a sweeping set of obligations: implementing all orders of ILSI's board, coordinating branch activities with ILSI activities, reviewing their finances with ILSI's officers, and controlling the dissemination of information in accordance with ILSI policies. If fully enforced, these legal provisions would have left branch heads little choice but to comply with requests from the top.

ILSI-Global actors (ILSI executives as well as member companies) also had informal or operational channels they could use to steer how scientific issues were treated in the branches. Most were invisible to the untrained eye. Three were most important in the dissemination of soda-defense science. Tellingly, none was mentioned in any of the ILSI documents I examined; instead, I chanced upon them while studying how things worked on the ground.[27] First, ILSI leaders in Washington recommended experts from the advanced countries (and mostly the United States) to share "the latest international science" at branch meetings. As we have seen, under the invitation-only practice, presenting at ILSI events was by invitation, giving ILSI leaders a powerful way to determine who would count as "the global experts" in branches around the world. In China, the American soda scientists we met in part 1 would be introduced as the best obesity scientists in the world. Because the branches were dependent on ILSI-Global for support, these *expert recommendations* may have been hard to rebuff.

In a second channel of influence, DC-based leaders transferred funds from US-based ILSI entities to support preferred programs and interventions like Take 10! *Intra-ILSI financial transfers* were routine parts of doing business at ILSI, and the amounts could be substantial. In a third mechanism, *leader visits*, ILSI executives based in DC occasionally visited a branch locality and used the occasion to promote particular scientific ideas or public health interventions. Alex Malaspina would use this to great effect, as would Rhona Applebaum. These three channels served as powerful pathways for the transmission of the preferences of ILSI-Global leaders and companies to the branches. All three would be used to steer the way ILSI-China handled the obesity question.

Nurturing social bonds among far-flung experts was an important part of how ILSI achieved the "harmonized use of science" across its branches. "Alex likes meeting and greeting people," his colleague Suzanne Harris confided. "He has a wonderful network around the world." The main vehicle for fostering shared goals and collaborations was the annual ILSI meeting, held every January in a sunny location not too far from DC (think Bermuda or Miami). The meetings brought together all parts of the extended ILSI-Global family—company representatives, ILSI trustees and officials, participating scientists, and branch heads—for several days of meetings, receptions, and just plain fun.[28] With attendance at many sessions designated by invitation only, the meetings provided an important vehicle for top-down communication about what mattered most to the organization and its leaders. The delivery of annual reports by the branches, often in the form of posters displaying their main accomplishments, provided opportunities for branch heads to demonstrate their enthusiasm for ILSI projects and success in meeting organizational goals. Working together and with cumulative effect, these varied mechanisms gave ILSI leaders and companies based in the United States powerful means to shape behavior in the branches.

ILSI-CHINA: FUNDING MECHANISMS AND THE WORK OF THE "NO-COMMERCE RULE"

As a producer of corporate—and thus corrupted—science, ILSI-China, like ILSI-Global, was ethically compromised. And like headquarters, the branch worked hard to conceal that fact. The branch's public narrative insisted that supporting companies gained no commercial benefit from their association with the branch. To ensure that companies were not advantaged and, more generally, that its science was (or appeared to be) protected from corporate bias, ILSI-China followed a strict ethics rule prohibiting companies from promoting their corporate logo at meetings or using their ILSI connection in advertising. "Companies may not do commercial work at ILSI meetings," Chen declared, stressing: "This policy is very important for the ILSI organization." Let's call it the *no-commerce rule*. Others attributed it to the Chinese state. "The [party-state] Center has very strict regulations on this," a top government scientist advised me. Undoubtedly the rule was vital to both. So central was it that every time the subject of business involvement in ILSI-China came up, my informants immediately mentioned the rule, as though that was the beginning and end of the matter.

This simple rule struck me as rather—indeed, comically—inadequate to the task of constraining giant (and paying) companies like Danone, Nestlé, and PepsiCo from using ILSI-China to promote their commercial agendas. Yet virtually all my expert-informants, ILSI leaders and researchers alike, were adamant that it fully protected ILSI's science. Perhaps, like the 50 percent rule followed by ILSI-Global, the no-commerce rule at ILSI-China, while appearing to ensure scientific neutrality and ethical behavior, was actually designed to protect people (and the ILSI branch) from allegations of unethical conduct. It did that by enabling plausible deniability ("we have a rule") and being open to flexible interpretation. My interviews with ILSI-China affiliates suggest that the point of the no-commerce rule was not to gain substantive compliance from the companies; not one person I spoke to even mentioned such efforts. Instead, the point was to provide an official norm, a defensible code within which ILSI's work could go on and people could readily deny wrongdoing through flexible interpretation of the rule. By supposedly guaranteeing the neutrality of its (corporate) science, this little rule anchored everything at ILSI, fully legitimizing its project and allowing everyone to engage with no worries. As Chen insisted, this policy was vitally important for the ILSI organization, though not necessarily in the way she meant.

Privately, however, the China branch had a number of informal, behind-the-scenes channels by which companies could covertly sway the science. An awkward conversation with Chen Chunming suggests that the most consequential had to do with funding arrangements. Not surprisingly, ILSI-China does not disclose details about its funding or operating budget. Yet Chen was willing to answer my questions about how things worked generally. Each year, she explained, ILSI-China's leaders set standard levels of support, and companies chose how much to give. ILSI leaders then asked the companies to recommend activities. "Those who provide more money have more say," Chen added. But then, as though realizing her remark may have seemed out of step with the bylaws, and growing uncomfortable with my questions about ILSI's connections to its supporting companies, she backtracked and offered two different accounts of the use of company fees, both circumscribing company power within ILSI. In the first, she and J. S. Chen made the final decisions on which activities to sponsor, making sure that no company had greater weight than the others. In the second, company contributions were pooled before being disbursed, so that no company would have outsized influence. A different, *special donations channel* (my term) operated indirectly. When ILSI needed additional funds for specific projects or major conferences, it asked companies for special donations. Often, it asked the bigger companies to help in this way. My research suggests that those providing moneys for these projects received something in return. In this way, the larger supporting companies could have wielded extra influence in China. To see how all this worked in practice, we turn now to how ILSI-China took charge of the obesity issue and turned it to corporate advantage.

ILSI Takes Ownership of the Obesity Issue, Marginalizing Industry Critics

Along with the dramatic rise in incomes and living standards, China's post-1978 reforms ushered in an epidemic of unhealthy lifestyles and, in turn, the chronic diseases of modern life. As public health scholars have lamented, China's traditional plant-based diet of coarse cereals and vegetables has gradually been replaced by an American-style diet with growing consumption of animal-based foods, refined grains, and highly processed, high-sugar, and high-fat foods.[29] Lifestyles have turned more sedentary as transportation has been motorized, household chores have become less physically taxing, and production processes have been mechanized and automated. While traditional mind-body practices such as qigong, taiji, and yoga continued to be popular with the fifty-plus set as forms of self-cultivation and alternative healing, during the years of interest to us there was no widespread culture of exercise that viewed strenuous activities like running or lifting weights as socially appropriate and beneficial to health.[30]

This dramatic transformation in lifestyles is an only-too-familiar story, told of nations around the world. What made China different is the speed at which everything changed. In the blink of an eye, fast-food outlets began popping up all over the big cities. KFC, which opened the first Western fast-food outlet in China on November 12, 1987, was followed by McDonald's (1990), Pizza Hut (1990), Starbucks (1999), and Burger King (2005) (see figure 5.2). By 2020 there were more than eighteen thousand of these oases of Western junk-food culture in Chinese cities. At the same time, neighborhood dumpling and noodle stalls were being shut down in the name of urban renewal, while traditional wet markets with their fresh vegetables, fish, and meat (see figure 5.3) were being replaced by supermarkets selling produce packaged in Styrofoam and plastic wrap. A better recipe for diet-related chronic disease can hardly be imagined.

FIGURE 5.2. McDonald's on Wangfujing Street, central Beijing (open since 1992). Photo by author.

FIGURE 5.3. Traditional Sunday wet market, Gu Lou (Drum Tower), Beijing.
Photo by author.

In a long conversation over lunch, one of the country's top child nutrition experts reflected on how, in less than two decades, a little joking about
chubby kids had morphed into a national angst about childhood obesity.
In the mid-1980s, he recalled, child health specialists began to notice a few
heavy kids on the streets. It was a "fun new thing" that was completely non-
sensical. In a land long haunted by malnutrition and the threat of famine,
how could there be so many fat Chinese children? By the late 1980s, with the
growing popularity of Western fast food and sugary drinks, the trend had
become "explosive," and health specialists grew worried. Around the time
JAMA published its alarming report on rising obesity in the United States,
China was seeing a tripling of overweight and obesity among adults, from
5.5 percent in 1982 to 20.0 percent in 1992 (using Chinese criteria; see table
5.1). Those numbers would continue to climb, reaching 29.9 percent in 2002
and 50.7 percent in the years 2015–19. Silently but swiftly, the epidemic of
overweight and obesity had arrived in China. (The Chinese BMI cutoffs are
slightly lower than those used internationally because Chinese people tend
to have higher percentages of body fat and higher rates of cardiovascular
risk factors and all-cause mortality than white people at given BMI levels.)
    The critical question is who would manage it? In identifying a director
for the new Beijing-based branch, Malaspina had made a brilliant choice.

TABLE 5.1. Trends in Overweight and Obesity among Chinese Adults, 1982–2019

Year	Percent Overweight	Percent with Obesity	Percent Overweight and with Obesity
1982	5.4	.1	5.5
1992	16.4	3.6	20.0
2002	22.8	7.1	29.9
2010–12	30.1	11.9	42.0
2015–19	34.3	16.4	50.7

Sources: Pan Xiong-fei, Wang Limin, and Pan An, "Epidemiology and Determinants of Obesity in China," 376. Data are from the China National Nutrition Surveys, which cover nine selected provinces.

Notes: Classification of overweight and obesity is based on Chinese criteria (BMI cutoffs of 24 for overweight and 28 for obesity). Statistics are for adults aged eighteen and over.

In Chen Chunming he had tapped a well-connected, politically savvy, state-backed scientist-official whom he could cultivate into a quasi-corporate scientist loyal to the ILSI mission. In the crucial early years of China's engagement with the obesity issue (roughly 1999–2003)—with official public health bodies overburdened and short of resources, a ruling party favoring market solutions, and a well-funded corporate-science organization eager to sponsor a program of obesity research and practice—the issue would be captured by ILSI. Though obesity was a new issue of health governance in China, China was not a blank slate. Quite the contrary. A group of researchers had been working on nutrition, including obesity, for at least a decade. Yet their story would be swept away into a hidden history that has been erased from the triumphalist narrative of ILSI's successful management of the national obesity epidemic.

THE TRANSFORMATION OF A PUBLIC OFFICIAL AND THE MARGINALIZATION OF INDUSTRY CRITICS

In the early 1990s, nutritionists began to grow concerned about the thickening bodies on the streets of China's cities. Height and weight data from two national surveys had shown increases in unhealthy weights since the earliest days of reform, the early 1980s. The question of who would manage it took on growing urgency. In the 1990s, the answer seemed obvious. It was not ILSI. ILSI's China branch was not formed until 1993, and it had no demonstrable interest in obesity until 1999, when Malaspina at ILSI-Global asked all branches to place obesity on their agendas.

From the late 1980s, the major player in research on the growing epidemic of diet-related chronic disease was instead a team of nutritionists working with Barry M. Popkin, the renowned global nutrition expert at the University of North Carolina, Chapel Hill (UNC). Popkin had developed the concept of the *nutrition transition*, the dramatic shifts in dietary intake (increased consumption of high-fat, high-added-sugar foods) and physical activity (declines) that emerge with urbanization and integration into the global economy, and the rise in debilitating diet-related chronic diseases that follows.[31] Using data from the newly launched China Health and Nutrition Survey, a collaboration between UNC and the Institute for Nutrition and Health at the CAPM/CDC, this pioneering group of researchers published dozens of important papers on the rapid shifts in Chinese diet and activity patterns and the corresponding rise in obesity and related diseases. By 2000, researchers and students associated with the UNC group had published some 125 articles on related topics.[32]

Some members of the Popkin group, working with the Chinese Nutrition Society (Zhongguo yingyang xuehui), were involved in preparing the 1997 Dietary Guidelines for Chinese Residents. For the first time ever, the guidelines on healthy diets included a principle on managing body weight: "Balance food intake with physical activity to maintain a healthy body weight" (*shiliang yu tili huodong yao pingheng, baochi shiyi tizhong*). Although unified standards for overweight and obese had not yet been devised, this was a major step forward in recognizing excess weight as a health risk.[33] Yet these researchers in the Popkin group lacked the financial resources and political clout needed to bring the scientific community together to define it as a disease threat and—perhaps more importantly—to get China's government to pay attention. With a powerful, state-backed leader, funding from industry, and a strong nudge from ILSI-Global, ILSI-China had what it took to essentially seize control of the issue in the late 1990s and early 2000s, and then lead the effort to turn obesity into a nationwide concern—*on ILSI's terms.*

For many years, Chen had been a committed public health official who had done important work that included the development and management of a national food and nutrition surveillance system. Several of my European and American expert-informants who had worked with her when she led the Academy of Preventive Medicine (1983–93) relayed to me how, after she transitioned to ILSI branch head, her work came to be more closely aligned with the interests of ILSI. In Chen they saw a subtle shift from the public health official looking after the interests of the Chinese people to the agent for the global food industry who would pay lip service to public health needs but whose primary concern was looking after the needs of the industry. As head of an ILSI branch, she would now have the resources to pursue that new agenda and quietly exclude those who saw things differently.

PLACING OBESITY ON THE AGENDA:
ILSI-CHINA TAKES CONTROL

In 1999, around the time the CEOs met in Minneapolis, ILSI-Global took the giant step of asking all branches to put obesity on their agenda. The time to create an industry-friendly science of obesity for China had now arrived, and Malaspina made every effort to ensure things went according to plan. Just as he had nurtured James Hill, Malaspina personally cultivated the head of the China branch, providing generous incentives to gain her cooperation. Most relevant to the obesity issue, in 1999 he gave her a seat on the board of trustees of his Center for Health Promotion, which created the model Take 10! program.[34] The two were also reportedly close on a personal level. Malaspina "had the greatest respect for Mme. Chen," Harris at ILSI-Global told me. "He was close to her until he formally retired" in 2001. As her health began to decline in the mid-2010s, Harris added, "he still called her on occasion." "I liked that lady so much," he divulged to a couple colleagues at the time. "She did make the ILSI Focal Point a Star." Malaspina's affection for and personal investments in Chen would pay off for the global organization.

In 2000, Malaspina returned to Beijing to make sure the Asia branches of the organization understood the issue's importance and how they should address it. In a lecture delivered to about three hundred ILSI-affiliated participants from thirty-six countries, he named obesity a top-priority nutrition concern in the region, using the CHP's signature activity program Take 10! to illustrate ILSI's approach to the issue.[35] In this lucid example of the leader-visit mechanism of top-down influence within the ILSI network, Malaspina sent an unmistakable message to all Asia branches that activity was the preferred solution to this new ILSI concern.

In the late 1990s and early 2000s, China's health ministry was facing problems far more pressing than diet-related chronic disease. As my informants explained, to a state concerned above all about economic growth and social stability, "slow-developing diseases" (the Chinese term is *manxing bing* [slow-nature disease]) were invisible epidemics. Compared to food safety, which produced one public scandal after another—the sale of ham soaked in pesticides, for example, or counterfeit baby formula or pickled vegetables sprayed with pesticide—rising levels of chronic disease lacked urgency and so commanded little attention. That gap in state governance left a hole for ILSI and its supporting companies to fill.

No sooner did ILSI-Global ask the branches to work on the condition than ILSI-China made it a major focus of its work. Chen Chunming took charge of the obesity issue, while J. S. Chen handled food-safety questions. "Different branches have taken the obesity issue to different levels," Harris told me with an amused shrug. "Chen Chunming just grabbed it [the Take 10! model program] and went off." According to her Chinese colleagues, Chen tackled the obesity issue with remarkable focus and speed. During 1999–2003, ILSI-China organized some of the country's top public health researchers into a task force, the Working Group on Obesity in China. Under Chen's leadership, the group gathered all the data available pertaining to obesity and chronic disease, arrayed them in charts showing disease prevalence at each BMI point (18, 19, and so forth), and on that basis created China-specific Body Mass Index (BMI) cutoffs. This was vitally important work: for the first time ever, this research defined obesity and overweight as Chinese diseases, to be managed by the government and its health agencies. All that was accomplished in a mere eight months.

There was something peculiar about the process, though. These official activities of health governance had all been generously funded by a foreign company; this helps explain Chen's focus and speed. The funder was the Swiss pharmaceutical giant Roche, one of ILSI-China's supporting companies. Roche had sky-high hopes for its obesity drug orlistat (brand name Xenical). Shared by other Big Pharma firms in China, the hope was that defining obesity as a bona fide disease would create a big market for its new drug, as growing numbers of Chinese would be diagnosed as having obesity and thus being in need of pharmaceutical treatment. Chen, acting as de facto policy maker, then oversaw the preparation of the official guidelines for its prevention and control. Issued in 2003 and published later in an attractive grape-colored booklet decorated with figures of athletes and a doctor-patient consultation superimposed on a BMI chart (figure 5.4), this document would be China's very first policy on obesity. In defining the new disease and preparing the guidelines for its management, ILSI had operated as a de facto division of the Chinese government and used industry money to do official work.

Big Pharma had made a bad bet on China. By the time the guidelines were issued, the promise of big profits from anti-obesity drugs had begun to fade. Hospital-based nutritionists I talked with a decade later explained where pharma had gone wrong. Despite the public health warnings about obesity, they lamented, their patients did not see excess pounds as a disease or even a health problem. If they visited the doctor it was for a "real medical problem," not obesity.[36] The appeal of market solutions in these decades of go-go economic growth also made the work of health professionals difficult. Their patients with chronic conditions tied to obesity (diabetes, fatty liver disease, and so on) got their nutritional advice from advertisements,

FIGURE 5.4. *Guidelines for the Prevention and Control of Overweight and Obesity among Chinese Adults* (2003) (cover)

not doctors. While the doctors labored to stop the spread of diet-related disease, their patients chose the popular *baojian shipin* (health foods) as treatments and, as one nutritionist bemoaned, couldn't distinguish *baojian pin* (health products) from *yaopin* (drug products) in any case. After all, both ended in *pin* (Chinese for item).

In the mid-2010s fatness seemed to be a nonissue, culturally and politically. More people than I care to remember told me I was wasting my time studying obesity. Unlike in the United States, where fatness has long been deemed a moral flaw, larger body size did not have negative meanings in China.[37] Quite the contrary, fatness has long had cultural cachet. In children, chubbiness has been seen as a sign of good health, while in adults, at least men, a bigger body with excess body fat was clear evidence of prosperity and success—not unhealthy overconsumption. To be sure, decades of growing consumerism and the ubiquitous ads associating thin with beautiful have fostered the ideal of the paper-thin, childlike, young female body.[38] My informants who had visited the United States were shocked at the massive bodies they had seen on the streets of San Francisco or New York, but they did not view those individuals as morally deficient, just unfortunate. Nor was there the kind of political discourse reflected in that image of the wide-bodied Statue of Liberty (see again figure 0.3), in which "irresponsible individuals" who "refused to lose weight" were said to be harming the American nation. A talk I gave on America's efforts to manage the rapid rise in obesity brought this home to me. Some in the audience found it extremely funny that the US government was devoting so much attention to something as trivial as body weight. The Chinese government, they assured me, had more important matters to attend to, like economic development. Concern would grow, but slowly.

As for Roche, the company clearly had misread Chinese culture. By around 2004, Roche's interest had moved on to other products, and the company stopped supporting ILSI-China.[39] Yet obesity was beginning to attract the attention of another industry—Big Food—and with ILSI support, around 2004 it would step up to offer solutions.

THE FIGHT OVER BMI CUTOFFS AND THE DEPARTURE OF DISSIDENTS

Having introduced the obesity issue and shown exceptional leadership in getting the BMI cutoffs finalized, ILSI-China under Chen Chunming came to be the undisputed center of gravity on the condition. To gain and retain power she had to quietly silence rivals. The major point of contention in those early years was the decision on BMI cutoffs. In delineating the cutoffs for overweight and obesity, the ILSI-led group used data on chronic disease risk rather than following methods used internationally.[40] In adopting China-specific cut points, China bucked the WHO, which did not agree with its adoption of idiosyncratic standards. In a marathon interview in London in 2016, Philip (W. P. T.) James (MD 1960s; 1938–2023)—a top

British obesity expert who was advising the WHO and helping China gather data for the cutoff research—reported that Chen was adamant about using her criteria. To him, that approach was more than "a bit embarrassing" because it was "not quite international." Even with the contributions of many researchers, the data they had assembled were "embarrassingly weak," he said, cringing. Some Chinese experts also disapproved of the methods and felt the work was done in a somewhat slapdash manner, potentially compromising quality. Even Chen Chunming, who oversaw the process and generally emphasized interpersonal harmony, disclosed that there were big debates and disagreements over the cutoffs.

One centrally placed Chinese participant described what happened this way: "At first maybe there were many different opinions, but in the end we got all the data and they showed us the curve . . . and almost all agreed [on the use of Chinese cutoffs]." Eventually, with Chen "ramming her ideas through" (in the view of one dissenter), the controversies died down, and ILSI's methods prevailed. Some of those who disagreed most strongly distanced themselves from the ILSI work and sought professional refuge elsewhere. Although those struggles took place a decade before my interviews, it was clear even in late 2013 that differences of viewpoint and hard feelings persisted. While I was not able to learn the details of these highly sensitive matters, the evidence suggests that consensus at ILSI was achieved at least in part by the departure or exclusion of those with differing views. In this early phase of obesity science making, then, we can see the emergence of a hierarchical pattern of scientific decision making, in which power to make key decisions was concentrated not only in ILSI-China, but also in the hands of its powerful leader. To my knowledge, in the 2000s and early 2010s, the small number of Chinese who obtained short- or, in a few cases, longer-term training abroad in obesity research, unless they agreed with Chen, would not participate in the conversations about obesity policy. This pattern of reaching accord by ensuring that only friends of ILSI had a voice would persist and have serious implications for China's policy.

Over the next decade-plus, this node of corporate science would become the major sponsor of obesity activities in China—more important even than the country's Ministry of Health, according to many expert-informants. From all indications, the ministry was only too happy to let ILSI under Chen take care of the issue. The arrangement of course had huge benefits for the organization; over time, as obesity management became a major focus of ILSI-China's work, Chen's power and influence would continue to grow. The larger significance of this history of corporate encroachment on a public health field needs to be underscored. What this means is that ILSI-China placed obesity on the public health agenda, took charge of the issue, and over the next ten years would heavily color its solution simply because ILSI-Global and its member companies had mandated and funded the work, and no governmental entity had the incentive or the financial and political means to do so. And ILSI's—if not Chen's—ultimate concern in taking on the issue was to protect corporate profits in the name of promoting health. (Chen's concern may have been to promote ILSI's agenda and continue receiving financial support, while making every effort possible *within those constraints* to improve the public's health.) To be sure, the public health need was real and the scientific effort serious. At that very time there was a parallel movement on the global level to define obesity as a disease and specify treatments.[41] But the role of the corporate sector in the creation and management of this new disease has received virtually no attention. In China and perhaps other countries as well, corporate money and agendas would dominate the handling of the obesity issue from the outset.

Marketization of Health Policy: Bringing Industry to the Policy Table

ILSI's aspirations went beyond creating soda science to embedding it in official policy. Following a strategy of all-out marketization and globalization, the country had long been pursuing a path of market-oriented health governance into which the ILSI model of industry-funded science fit perfectly. With the obesity guidelines now in place, in the mid-2000s Chen would take the governance of chronic disease further in the direction of market management by inviting food companies to sit at the policy table. In this section we see how the three parties involved in that decision—ILSI-China, China's MOH, and the WHO—quickly reached consensus and how, in their enthusiasm for including market actors, they considered the benefits but failed to weigh the risks.

Following the mantra *copy and catch up*, China's public health experts regularly monitored the publications of the WHO, the most authoritative institution in the field of global health, for ideas they should adopt. In 2004, the WHO issued an important document on chronic diseases called *Global Strategy on Diet, Physical Activity and Health*.[42] The Global Strategy recognized the growing burden of obesity and other noncommunicable (or chronic) diseases as causes of death and disease around the world and sought to place them high on the world's health agenda. Focusing on two of the main risk factors for the major chronic diseases of modern life (cardiovascular disease, type 2 diabetes, some cancers), the Global Strategy offered recommendations for, one, improving diets and, two, boosting activity levels. For the first time, the WHO assigned private companies responsibilities in the fight against chronic disease. While member states were assigned by far the most duties (fourteen in all), the private sector had two small but important parts to play, the first as "responsible employers," the second as "advocates for healthy lifestyles." This was big, and ILSI's China branch took note.

In the ensuing years China, with its probusiness culture and state enthusiasm for market solutions, would take the corporate role in chronic disease governance to a new and remarkable level. This represented most fundamentally a confluence of interests among the government, ILSI, and the food industry, but Chen seems to have played a critical role bringing the parties together around this goal. Stressing the need to carry out the WHO recommendations, in 2004–5 Chen began actively mobilizing her supporting companies to become partners in the effort. At a pair of roundtable meetings in 2005, ILSI-China's director offered the organization's "technical assistance" in developing the Global Strategy for China.[43] All the major speakers, including the top ministry official in charge of chronic diseases (the director of the Noncommunicable Disease [NCD] Division of the Bureau of Disease Control), endorsed an industry role in addressing these crises by selling nutritious food and teaching the public the fundamentals of nutrition—precisely the roles specified by the Global Strategy. Reflecting the government's hands-off approach to health in the new market economy, the NCD director announced that the health ministry would take a "market orientation" and cooperate with industry on health-promotion projects that were not for profit. Companies were not just encouraged; they were actively mobilized to contribute "under the overall leadership of the government." With this, the Chinese government became a willing and vocal partner in opening the process of making and implementing health policy to (mostly foreign) companies.

To an American such as myself, that lack of skepticism or wariness toward Big Food seems disturbingly naïve. But this inclusive attitude, in addition to according with the state's market-oriented policy preferences, had deep cultural roots. During my time in Beijing, I kept hearing the phrase "all social forces should work for the common good [as defined by the government]." Citing the Chinese expressions *jianrong bingbao* and *qiutong cunyi* (all-encompassing, seeking common ground while preserving difference), a thoughtful nutrition researcher explained the thinking to me:

> It is like a natural wisdom that we should employ all the resources to solve social problem[s], especially when China is backward regarding science. Hence, they included the food industry as an important actor.... Chinese

people's mindset is like, it is clever to utilize every resource we could think of. It is silly to refuse some actors out of the door. It seems that no one in China doubts this idea.

[In China's] one-party political structure . . . the top government naturally thinks that all sectors should work for the target of Beijing, which is public health. Asking the food industry to work together [is] like an order from the top officials. In a collective communist society, rarely [would] people consider including the food industry to be problematic. (Emails from Wang Yixi, May 3 and 8, 2023)

But perhaps the "natural" and "clever" embrace of the food industry was a bit shortsighted. All parties were enthusiastic about the benefits, but no party seems to have expressed concern about the risks, including self-serving corporate behavior such as putting profits over health. And while the companies were to contribute under governmental leadership, no mechanism was set up to ensure state control over industry. Evidently, ILSI-China's no-commerce rule was supposed to do the work of keeping industry in line. The new arrangements for policy making left China unprotected from corporate attempts to benefit their bottom lines at the country's expense.

Company representatives responded enthusiastically, at least according to accounts in the ILSI-China newsletters. Political calculation played a role. In addition to demonstrating corporate social responsibility for (supposedly) contributing to the public's health, here was an opportunity to present oneself as a corporate good citizen, one that actively supported the Chinese government by answering its call to contribute to this important public health project.[44] In a risky environment in which Western companies could easily be targeted in anticorruption campaigns, staying in the good graces of the state made good corporate sense. Joining public health campaigns would also provide opportunities to present their often sugar-, fat-, and salt-rich products as healthful.

Constantly referencing the country's policy of decentralizing services to the market, Chen Chunming was a tireless promoter of corporate participation. And of course, she was right; the question, though, was how big the company role would be. By stressing the corporate responsibility theme in meeting after meeting, over time she worked to normalize the idea, making it the only approach that was thinkable. And by creating concrete mechanisms within ILSI that enabled their participation, Chen would help insert firms in the food industry directly into the nation's core strategy to combat obesity and chronic diseases more generally.[45] In Chen's hands, ILSI would become a powerful vehicle for corporatizing health science and policy in China.

··

With obesity officially defined as a disease, it was time to bring the condition under governmental control. What was needed now was a science, a set of concepts to explain how obesity levels had risen so rapidly and what should be done to stop the epidemic from getting worse.

CHAPTER SIX

Getting Soda Science Endorsed in China

With the essential groundwork laid, the first step in making Big Food's dreams come true in China was to get soda science endorsed as the dominant science of obesity. In this chapter, we follow the leaders of ILSI-Global and ILSI-China as they orchestrated the direct transfer of that science from the United States to China. They did it in remarkably straightforward fashion: by flying Hill and, a few years later, Blair to Beijing on multiple occasions and presenting their ideas as the best obesity science in the world. ILSI and its corporate partners were strikingly successful in getting their *exercise-first (soda-tax-never)* science accepted as authoritative. Over the years that soda science was actively promoted, China came to rely ever more on exercise solutions, while dietary ones declined proportionately. At the same time, the terms that undergirded the new corporate science— energy balance, energy gap, exercise as medicine—came to be integrated into China's scientific discourse, becoming the key concepts structuring China's thinking on obesity and how best to manage it. The evidence suggests that the idea of taxing soda was scarcely mentioned.

While energy balance science protected soda by diverting attention to physical activity, in this branch of the ILSI network the science also had a noteworthy—or perhaps notorious—dietary component. We visit some fifteen of these "nutritional" conferences where Mars, Danone, Yum!, Coke, and other ultraprocessed-food companies promoted their foods as healthful and repeated the boilerplate corporate line that all foods can be part of a healthy diet. This trip to these spaces where science literally was for sale reminds us that although ILSI introduced soda science to China, it would not have taken root had the country not already endorsed the marketization of everything, science included. Everywhere one looked, market actors and logics permeated and often dominated the field of action.

ILSI also had mechanisms that enabled member companies to invest directly in projects to fight obesity in China. Largely hidden from sight, these arrangements created a tangle of invisible ties that connected Coke to ILSI-

Global to ILSI-China and ultimately to the Chinese government. A close look at Coke's investments suggests that food-industry influence pervaded the field of obesity work, and that influence was largely hidden from sight by chains of connections too complex to fully unravel.

Foreign Experts Bring "the Best International Science" to China

In the mid-2000s, ILSI-China began organizing scientific meetings to present "the latest information" on strategies to control obesity. Obesity was a huge focus of the branch's activities. Between 2004 and 2015, it organized some sixty scientific meetings—an average of five per year—aimed at bringing the most up-to-date knowledge on the condition to China.[1] For major activities, ILSI-China routinely sought cosponsorship with the leading governmental and intergovernmental health bodies in the country (China's MOH and CDC, as well as UNICEF and WHO). Those larger events, because they brought together many of the most influential figures in the obesity and chronic disease community in the country, were sites not only of scientific discussion, but also of deliberation on policy solutions. In the last chapter, I described the various mechanisms ILSI-Global devised to help ensure the acceptance of its preferred understandings of obesity in branch countries. By studying the concrete activities the China branch organized to address the issue, we can see the remarkable maneuvers by which ILSI-Global and its most powerful member companies were able to use the ILSI setup to promote soda science as the latest and best obesity science in the world.

The newsletters the branch published twice a year provide a record of all these meetings, along with important information on agendas, speakers, and much more. Because ILSI-China was the lead organization on obesity in the country, the activities profiled in its newsletters represent the great majority of all obesity-related activities in China. The newsletters allow us to distinguish between events focusing primarily on nutrition and events focusing primarily on physical activity as the main solution to the obesity problem. By tracking changes over time in the proportion of obesity-related events focusing on activity versus diet, we can gauge how successful ILSI-Global was in its campaign to spread its exercise-first soda science to China. Because the larger conferences that included health officials had the weight of quasi-official policy-making events, the newsletters also allow us to see the germinating of official policy ideas through the ILSI process.

In this and the following section I mine the newsletters from 1999 (when obesity landed on ILSI-China's agenda) to 2015 (when Coke headquarters abandoned the exercise-first approach) to see how successful Coke and ILSI-Global were in spreading corporate-scientific orthodoxy on obesity to China. We will see that, as industry gained a place at the policy table in the mid- to late 2000s, the emphasis of China's obesity activities shifted markedly toward physical activity. That shift was accompanied by a growing presence of foreign experts with ties to Coca-Cola and ILSI, with our friends from America James Hill and Steven Blair invited most often. These two findings—which I pursued on a hunch, and never expected would be so robust—leave no room for doubt that ILSI and Coke played a critical role in pushing China's choice of obesity solutions toward exercise.

A HISTORIC SHIFT FROM NUTRITION TO PHYSICAL ACTIVITY

Let's begin by looking at overall shifts in China's approach to obesity between 1999 and 2015. If ILSI-Global and its industry members influenced China's work, we would expect to find an increase in the share of events focusing on activity after 2004, when Coke and other food giants began energetically promoting exercise for obesity. The results, shown in table 6.1, reveal a striking upward shift in the proportion of activities focusing on exercise that started in the mid- to late 2000s.

In 1999–2003, no ILSI-sponsored obesity activities focused on physi-

TABLE 6.1. ILSI-China-Sponsored Obesity Activities, by Emphasis, 1999–2015

	Number of Activities	Nutrition Total Number (and Percent of Total)	Physical Activity Total Number (and Percent of Total)	Both Total Number (and Percent of Total)	Neither Total Number (and Percent of Total)
1999–2003	12	5 (41.7)	0 (0)	0 (0)	7 (58.3)
2004–9	30	12 (40)	11 (36.7)	5 (16.7)	2 (6.7)
2010–15	30	7 (23.3)	18 (60)	4 (13.3)	1 (3.3)

Sources: ILSI Focal Point in China newsletters, supplemented by interviews.

Notes: Includes all activities on obesity, or obesity and other chronic diseases, sponsored or cosponsored by ILSI-China. For the handful of activities that are the focus of more than one newsletter column (because they generate more than one conference or intervention), the table counts the activity, not the news column(s). Activities dealing with neither nutrition nor physical activity generally focused on the definition of obesity or other measurement issues.

cal activity. (Most dealt with measurement issues, not prevention strategies.) Some prevention activities focused on nutrition, but none centered on physical activity. Between 2004 and 2009, roughly a third (37 percent) of ILSI-China's obesity activities focused on physical activity. Between 2010 and 2015, the proportion nearly doubled (to 60 percent), while obesity programs focused on nutrition sank to around one in four or five. Dietary approaches were not neglected, but they gradually lost out to solutions emphasizing activity. There was no discussion in the newsletters of soda taxes or advertising restrictions, suggesting these possibilities may not have been even raised.

This shift toward exercise is pronounced, but did those engaged in China's obesity work perceive that a switch was underway? In my conversations in Beijing, two key informants spontaneously mentioned a growing emphasis on physical activity in China's obesity work. "After solving the cutoff question," J. S. Chen (ILSI director since 2004) remarked enthusiastically, "we decided to promote exercise." The head of the chronic disease section of the Beijing CDC made a similar report: "Over 50 percent of the work of the Beijing CDC staff involves encouraging citizens to engage in physical activity. *Basically, this is our anti-obesity work*" (emphasis added). At the time, I scribbled down their words without grasping their significance. But now, having studied the newsletters covering all the years from 1999 to 2015, I realize these authorities on China's obesity work were conveying something real and important. To be clear, I have no evidence that anyone working with ILSI-China knew that the balance of obesity activities was shifting *away from diet*, a troubling finding, especially when connected to the corporate project to promote exercise; what they said was that work on exercise was *increasing*. They did not connect the dots because there was no reason to do so. If China was moving toward exercise solutions, how did that happen?

A GLOBAL HIERARCHY OF SCIENCE

Virtually all the experts I talked to saw global science as hierarchically ordered, with the most advanced science being produced by the West. That was true of both Chinese and Westerners. That notion of the proper flow of knowledge made ILSI's campaign to transfer corporate science from the United States to the Global South seem natural and unquestionable. These cultural assumptions never got articulated, but they help us understand how ILSI had such an easy time of it.

Among the many mechanisms ILSI-Global had developed for promoting its preferred scientific ideas in the branches, none was so vital as the expert recommendation. In China, the ideas of those handpicked foreign scientists invited to share "the best of international science" came marked as authoritative, and they were often promoted as key concepts to inform Chinese policy. Widespread Chinese perceptions of the backwardness of the nation's public health research reinforced the view that Western (primarily American) experts were the bearers of authoritative knowledge that China needed. In a conversation, William (Bill) C. Hsiao, a Harvard University health economist, described the "blind emulation" of US public health strategies by his Chinese colleagues, adding that the fellowships many receive to get training in the United States socialize them even further to deem America the global standard. In interviews in China, obesity researchers relayed how they routinely took US research as the standard, consulting dozens, even hundreds, of American-written papers before adapting their methods to the China context. One directed my attention to over three hundred pdfs on his computer to make the point. In a similar vein, Chen described how her organization structured its conferences as opportunities for Chinese researchers to "learn the advanced knowledge of the West." Foreign speakers generally occupied the most prominent slots on the program.

Chinese scientists who looked up to the West also felt tremendous pride in their own country. In this strategy of emulating the West—which was encouraged by Deng Xiaoping and pursued in many academic fields in the early reform decades—there was a strong nationalistic agenda at work. I mentioned this in the last chapter, but it deserves underscoring because it helps us understand why China's researchers accepted ILSI's science so enthusiastically. Chen and her colleagues energetically embraced Western knowledge because they deeply believed that, by "learning from the West"—absorbing its advanced knowledge, sending Chinese to the West for training, and so forth—their proud country would soon catch up and gain acceptance as a respected member of the global public health community.[2] In embracing Western science as their own, they thought they were doing something for China. American and European speakers at ILSI-China conferences often reinforced this notion of a global hierarchy of scientific worth by urging their Chinese colleagues to learn "best practice" from the industrialized West. China's public health science of obesity was indeed less developed than that of Euro-America, making the transfer of knowledge appropriate. At the same time, this widespread narrative of backwardness, and the intense desire to absorb Western learning, also worked to boost the interests of corporate science by encouraging an unquestioned acceptance, even veneration, of the knowledge produced by ILSI-Global in the United States.

Attitudes toward Western science and Western ways more generally have shifted since around 2010, and especially since Xi Jinping became top leader in late 2012. China has become more authoritarian at home, with power increasingly concentrated in the hands of the party and Xi himself. Moving away from Deng's project of learning from the world, Xi has adopted a more assertive posture emphasizing the superiority of China's statist model, the country's growing dominance in high-tech fields, and China's active leadership role in the world.[3] These shifts were just getting underway in the mid-2010s. In the (low-tech, minimally state-supported) community of specialists I studied, the general zeitgeist continued to reflect the globalist longings of Deng Xiaoping to throw open the doors to foreign investment, multiply cultural and educational interactions with the West, and learn from the outside world.

WHICH EXPERTS WERE INVITED TO PRESENT "THE BEST SCIENCE"?

Who then was invited to share the most up-to-date international science with these large audiences of Chinese experts? The newsletters allow us to answer this question with great precision. And the answer is *the very same people who made soda science in the United States.* The key actors were American scientists with ties to the food industry, and in particular, the Coca-Cola Company.

Let's start by considering all the major activities ILSI-China sponsored between 2004 and 2015 that emphasized exercise for obesity. These are listed in table 6.2. I discuss the first set of events—conferences and workshops—in this section, returning to the others later in the chapter. Between 2004 and 2015, ILSI-China sponsored nine conferences and workshops on obesity or obesity and other chronic diseases chiefly emphasizing physical activity. The most important were the six international conferences, each with over two hundred attendees, including several from abroad. Remarkably, most of the speakers at those six international conferences had ties to ILSI or the Coca-Cola Company, if not both. Of the eighteen foreign experts invited to the six conferences, thirteen were based in the United States, making America the go-to country for obesity expertise.

The two most prominent experts at these conferences, Hill and Blair, received repeat invitations, and over time their views gained enormous traction. The ideas of James Hill were introduced to Chinese researchers early on. In a long conversation in Denver in late 2015, he told me he had first met Chen Chunming at WHO headquarters in Geneva in 1997, and was invited to attend the very first obesity meeting in China in April 2000. (Their paths

TABLE 6.2. ILSI-China-Sponsored Obesity Activities with Physical Activity Focus, 2004–15

Conferences and Workshops	
1. Dec. 2004	International Conference on the Health Benefits of Physical Activity
2. Mar. 2006	Joint Meeting of the Working Groups on Obesity and Physical Activity in China
3. Nov. 2006	Conference on the Control of Obesity and Related Diseases in China: Maintaining Healthy Weight—a Priority in Chronic Disease Control and Prevention
4. June 2007	2007 International Beverage Forum on Sport and Health
5. June 2009	Working Groups on Obesity and Physical Activity in China Symposium
6. Nov. 2010	Conference on Physical Activity and Health: Exercise Is Medicine
7. Dec. 2011	2011 Conference on Obesity Control and Prevention in China: Energy Balance and Active Lifestyles
8. Dec. 2013	2013 Conference on Obesity Control and Prevention in China: Appropriate Technology and Tools for Weight Control
9. Nov. 2014	2014 Conference on Physical Activity and Health: Exercise Is Medicine

Other Scientific Activities	
10. Apr. 2005	ILSI-China forms Working Group on Physical Activity in China
11. July 2011	MOH issues Physical Activity Guidelines for Chinese Adults (Trial), followed by Advocacy Conference in Sept.
* 12. June 2012	Launch of Exercise Is Medicine (EIM) in China, numerous training courses 2012–15
* 13. 2011–14	Training Fellowship Program, Coca-Cola Beverages (China) and ILSI-China Scholarship Program, with Winner Report Meetings
*14. Apr. 2014	ILSI-China forms EIM China Working Group

Public Health Interventions	
* 15. 2004–15	Happy 10 Minutes, from trial to inclusion in national campaign
* 16. 2007–9	Community-based Physical Activity Promotion Project
* 17. 2007–15	Healthy Lifestyle for All Action, develops into a national campaign

Sources: ILSI Focal Point in China semiannual newsletters, supplemented by interviews.

Notes: Includes major activities on obesity, or obesity and chronic diseases, sponsored or cosponsored by ILSI-China and focusing on physical activity. Does not include activities with major emphasis on both nutrition and physical activity. Each activity is listed only once, even though it may have been the subject of more than one newsletter column.

* Coca-Cola heavily involved.

could also have crossed at ILSI-CHP, where Hill was an adviser and Chen was a trustee, or at an annual ILSI-Global meeting, which Chen would have attended as a branch head.) Invited back to China virtually every year thereafter, Hill reported, he visited the country six or seven times in that fifteen-year period, becoming something of a fixture at ILSI-China gatherings on obesity. In 2010, Steven Blair was recommended as a prominent expert by Rhona Applebaum at Coke. "Rhona introduced me to people here," Blair told me in a casual conversation in late 2013, adding, "She's been funding my research for a long time." At the time I was astonished by his casual revelation of the close tie to Coke. Now, of course, I understand it as a product of Applebaum's strategy of cultivating exercise scientists for Team Coke. Through repeat invitations and praiseful remarks by Chen, the ideas of these two top experts—in particular, energy balance, energy gap, and the neglected importance of physical activity—would be presented to China's public health community as the best international science on obesity.

Of the remaining eleven US-based obesity specialists, seven were exercise scientists, including "father of aerobics" Kenneth H. Cooper. Five had known funding, advisory, or professional ties to Coke, ILSI, or both, and two more were employees of Coca-Cola or its Beverage Institute. Only two were nutrition specialists. Of thirteen American experts, then, nine were exercise specialists and nine had known ties to Coke, ILSI, or both. The numbers speak for themselves. These were no ordinary scientific gatherings; instead, they were showcases for Coke's and ILSI's industry-friendly ideas camouflaged in the language of "the best international science." They were, in short, exercises in the global dissemination of corporate science. A product of the expert-recommendation mechanism, this practice of packing international conferences with ILSI- and Coke-friendly speakers can be added to the list of ILSI mechanisms for spreading soda science around the world.

All this was hidden from sight, however. Except for the unusual practice of company representatives giving talks, the ILSI branch conferences appeared to be ordinary scientific gatherings. They were held in big hotels scattered around the capital—from the Beijing Guangxi Hotel in the southeast to the Beijing Capital Hotel in the city center—whose entryways were flanked with posters and exhibition boards advertising related products. The featured speakers were presented not as "ILSI experts" or "corporate scientists," but as the best scientists internationally. For example, at the 2011 obesity conference, the four foreign speakers (all American) were introduced as "renowned specialists and professors" (assuming the newsletter description is accurate). Two were members of Team Coke (Allison and Pratt), a third worked with Coke's Beverage Institute (Buyckx), and the fourth was based in Beijing with the US CDC (Michael Engelgau).[4] Three-quarters of these "renowned [foreign] specialists" had close ties to Coca-Cola. Unless they were familiar with the inner workings and finances of ILSI, no one attending the meetings would have had any idea how tilted the balance of top-billed speakers and scientific viewpoints was toward industry. Nor would they have seen the hidden corporate logic that shaped everything.

One person who would have grasped these corporate logics was Barry Popkin, the American who had done more than any foreign scientist to advance the understanding of the nutrition transition and its consequences for population health in China. Popkin's name was conspicuously absent from the lists of invited speakers. In fact, there is no evidence in the newsletters that he was invited to speak at any of the many conferences organized by ILSI on obesity in China. (He confirmed that by email on December 17, 2021.) ILSI's invitation-only rule permitted exclusions as well as inclusions, and it is easy to see why Popkin was not welcome. He was a fierce advocate of restrictions on sugar-sweetened beverages and an independent scientist with decades of experience who was unlikely to kowtow to ILSI's industry-friendly line on obesity policy.[5] In his view, change in diet was the main driver of unhealthy weights in China, and a focus on minor extra physical activity was completely misplaced. He was not the only one doubting the exercise-first mantra. A number of his highly respected Chinese colleagues concurred—to little avail. His absence—and that of other leading experts critical of the food industry—is a pointed reminder that ILSI's science actually was not the best international science, but just one science, and a corporate one at that.

Energy Balance Concepts: Transferring US-Made Corporate Science to China

If experts with ties to Coke and ILSI were the dominant voices at China's obesity conferences, what difference did it make to Chinese understandings of the issue? We've seen that exercise approaches came to prevail over dietary ones, but we can take things further and look for specific concepts that made their way from the United States to China. In this section we trace the main concepts promoted by Hill and Blair and how they were taken up. We will see how energy balance, energy gap, and the other ideas that were advanced by scientists associated with ILSI's Center for Health Promotion and ILSI North America were introduced as authoritative and quickly endorsed in China. These ideas from America changed the scientific discourse on obesity in China, producing new framings of the problem and its solution. The influence of ILSI-Global can also be seen in the organizing themes of the conferences. During the 2010s the themes of ILSI-China meetings were virtually identical to those of meetings being held contemporaneously by ILSI entities in the United States These patterns say that ILSI-Global's campaign was a phenomenal success in China. Its expert-recommendation practice was a big part of that success. Let's start by shadowing Hill as he comes to be anointed the world's number-one expert on obesity.

"SMALL STEPS," "ENERGY GAP," AND "EXERCISE-FIRST" IN THE WILD WEST OF CHINA

Hill was invited to China in the earliest years of its efforts to find public health solutions to the obesity epidemic. As a result perhaps of that early visibility and his stature at ILSI North America, the ideas he championed had a profound influence on the scientific discourse on obesity in that country. Hill himself felt his hosts had been highly receptive to his ideas. In our 2015 discussion, he exclaimed: "China is better able [than the United States] to get the exercise component. China is a like a wild west! There's more hope there." The ideas on exercise were not his alone, of course. The WHO's Global Strategy also promoted physical activity (and energy balance) as part of the solution to the epidemics of chronic disease.[6] Yet Hill put a distinctive stamp on these ideas and, as we saw in part 1, was instrumental in placing them at the core of the larger complex of ideas that made up the corporate science of obesity.

That receptivity Hill sensed may have reflected the Chinese government's growing realization that rapid urbanization combined with a hyper-competitive market culture had left the new generation of single children no time for play and thus woefully unfit. In the mid-1990s the government established the first national fitness program to promote activity and improve fitness in the general public. That was followed in 2007 by the Sunshine Sports Project to increase physical activity among young people. Although the number of measures fostering activity increased only slowly in the years before 2015, there was an awareness in public health circles that activity levels had seriously declined among all age groups, and that was harmful to the population's health.[7] A concrete plan to promote exercise for health would have been enthusiastically received.

To see how Hill's ideas took hold in China, we need to do a deep dive into the ILSI-China newsletters. Hill was first mentioned in the newsletters in 2006, when his name appeared on three separate occasions. The first was a March meeting of the two working groups that would spearhead the development of China's anti-obesity work, one on obesity, the other on physical activity. At that meeting, Chen Chunming, the gatekeeper between China and the United States, introduced "the concept of energy gap . . . put forward by Professor James Hill" and his notion that "small changes [in diet and activity] can bring large impacts" as "the latest development of the international research [on] obesity."[8] With this, she declared the Colorado scientist's ideas the best, or at least the latest and most promising approach to obesity in the world. At the same gathering, Chen presented new research demonstrating the usefulness of the "energy gap theory" in the Chinese context, suggesting that that theory, which was being advanced by Hill and his associates at the time, was being actively considered for use, if not already on its way to being translated into Chinese practice. Remarkably, Chen seemed to be repeating Hill's published words verbatim and passing them along to her Chinese colleagues as the best ideas, without elaborating on their advantages over other approaches or their possible disadvantages in the Chinese setting.

On the second occasion, at a mid-November meeting, Chen introduced that same research on the energy gap in China, explaining that if people ate just one bite less or walked thirty to forty minutes a day, they could avoid excess energy accumulation.[9] These were precisely the messages Hill and his colleagues were conveying to Americans around the same time in *The Step Diet Book* and the America on the Move (AOM) campaign. And like those messages, the ones Chen was promoting in China were aimed at creating scientifically minded, self-responsible, health-preoccupied citizen-consumers who would internalize the new advice, learn basic energy balance concepts, and calculate their daily energy in and out to achieve balance and thus a healthy weight. Finally, at a major international conference on Chinese obesity in late November, Hill contended that his model of small behavioral change is "a practical and effective measure in the fight against obesity."[10] The Chinese were highly amenable to Hill's suggestions. Before long a euphonious Chinese phrase for energy balance—*chidong pingheng* (literally, eating-moving-in-equilibrium)—entered the discourse. In the ensuing years, the approaches Hill was advancing would be discussed regularly at ILSI events, suggesting they were now part of official thinking.[11]

While Hill continued to be a regular visitor, from 2010 Steven Blair became an increasingly prominent voice at Chinese conferences. Predictably, in China he vigorously promoted physical activity and the Exercise Is Medicine (EIM) program he had helped develop in the United States and spread around the globe in partnership with Applebaum. He popularized the notion that exercise should be routinely prescribed as a medical treatment, the key theme of EIM, and helped design and develop China's EIM program, which launched in 2012. The idea of physicians prescribing exercise to their patients proved popular in China (or at least at ILSI-China). From the early 2010s, columns in the ILSI-China newsletter began filling with talk of how to train clinicians in *tiyi* or *tiyi ronghe fazhan* (literally, movement-medicine or the integrated development of exercise and medicine). Blair too played a critical role in China, validating and strengthening China's already solid commitment to the role of physical activity in chronic disease prevention, and institutionalizing it in the local EIM program.

PARALLEL PROGRAMMING: COPYCAT SCIENCE

Turning to the organizing themes of China's conferences, we find a pattern of copycat science in which ILSI North America seems to have essentially dictated meeting topics to the China branch. That pattern was especially flagrant between 2011 and 2015, the years the Energy Balance and Active Lifestyles (EBAL) committee was actively promoting soda science. During those years, the China meetings mirrored developments unfolding at the EBAL committee, and they brought people closely involved with EBAL to China to spread "the gospel," as Applebaum might put it with tongue in cheek.[12]

China's 2011 obesity conference centered on the theme "Energy Balance and Active Lifestyles"—the very name of the ILSI-NA committee established in 2010.[13] Indeed, ILSI's North America branch cosponsored the meeting, illuminating yet another mechanism by which US-based ILSI entities swayed science in the branches. Three people from the EBAL committee, all affiliated with Coca-Cola (which, readers will recall, dominated the committee) were present—the chair, Maxime Buyckx of Coke's Beverage Institute, and two scientific advisers, David Allison and Michael Pratt—and they took the opportunity to promote its core concepts and encourage their translation into active lifestyle programs in China. The 2013 conference featured the principals of the soon-to-be-formed Global Energy Balance Network: Hill, Blair, and Applebaum. Because of its significance, I discuss this event separately just below. The 2014 conference on physical activity and health was organized around the theme of EIM and featured presentations by Blair and Xuemei Sui, the junior scholar at USC who had eagerly done research for Applebaum, and like Blair was a major Coca-Cola grantee.[14] Although the GEBN scandal that broke in August 2015 may have kept Hill and Blair away from the obesity conference in Beijing in November of that year, Hill helped organize the meeting. Most likely at Hill's suggestion, one of the network's most enthusiastic vice presidents—Wendy Brown from the University of Queensland in Australia—did attend and, predictably, advocated the use of energy gap notions to guide anti-obesity strategy.[15] And the meeting's theme was, once again, energy balance. This dual process—of top-down export of soda science by researchers working with ILSI entities in the United States, and bottom-up mimicry and assimilation by a branch seeking to demonstrate advanced understanding and devotion to ILSI goals—was the key to ILSI's success in getting soda science endorsed globally.

The GEBN Founders Deliver Their Message in China

In December 2013 ILSI-China hosted a major international conference on obesity control and prevention in China, inviting three Coke-affiliated scientists to be the featured speakers. Those speakers, Applebaum, Hill, and Blair, were just then beginning to put the GEBN together behind the scenes. Late 2013 was a critical time in the global consolidation of the energy balance approach by the food industry. In May of that year, Coca-Cola chairman and CEO Muhtar Kent, reiterating Coke's commitment to being "part of the solution [to] . . . today's most challenging health issue," announced four worldwide business commitments "to further contribute to healthier, happier, and more active communities." The third was to "help get people moving by supporting physical activity programs in every country where we do business."[16] Coke was now pushing exercise from the very top.

As chief science and health officer, Applebaum certainly played a part in the formulation of those commitments. Now, perhaps feeling empowered by her boss's support, Applebaum and her chief partners were together in Beijing delivering Coke's (and ILSI-Global's) message to leading scientists and decision makers in this vital market. Careful dissection of this key event shows us how rich and powerful member companies of ILSI and its China branch were able to mobilize the funding, participation rules, and other hidden channels within the global organization to get their preferred scientific positions presented as the unquestionable scientific truth. A close reading of their presentations shows how the GEBN scientists further contributed to the effort through the strategic deployment of a rhetoric of science.

At Chen Chunming's invitation I attended the conference, where I was able to observe the goings-on and chat informally with many participants, including all the foreign speakers. The 175-page bound conference program contained the program and photocopies of the PowerPoint presentations of every talk, giving me an exceptional written record of what was discussed.[17] To capture what was actually said, I typed the oral presentations and con-

FIGURE 6.1. Conference participants exercising during wellness break.
Photo by author.

ference discussions into my computer in as verbatim a fashion as possible. From my perch in the audience, I had a rare up-close-and-personal look at how ILSI conferences worked in practice. Especially noteworthy were the structured fifteen-minute "wellness breaks," guided by a Pilates coach, that put the exercise-first mantra into action (see figure 6.1).

The gathering, attended by over two hundred people, brought together these Coke-affiliated speakers with leading representatives of all parties concerned with obesity in China: researchers, government health officials, clinicians, and, of course, food companies. As with most major events, ILSI-China partnered with leading Chinese health policy bodies, making the conference a quasi-official activity. The conference program lists the following organizations as supporters of the conference:

Supporter
     Bureau of Disease Prevention and Control, National Health and
          Family Planning Commission
Sponsors
     Chinese Center for Disease Control and Prevention
     ILSI Focal Point in China
     China Physical Fitness Surveillance Center

Chronic Disease Control and Prevention Society, Chinese Prevention Medicine Association

Organizers

ILSI Focal Point in China

Executive Office of the China Healthy Lifestyle for All Initiative

Special Thanks

Coca-Cola Beverages (Shanghai) Ltd.

Nestlé (China) Ltd.

Herbalife (China) Health Products Ltd.

COFCO Corporation[18]

The conference was also typical of large-scale ILSI events in soliciting corporate funding. In a marvelous illustration of the workings of the special funding mechanism of corporate influence, four companies were listed as conference supporters. Coke was named first, suggesting it had provided the most funds. The fourth invited speaker from abroad, David Heber of the University of California, Los Angeles, was a nutritionist and chairman of the Nutrition Institute and Nutrition Advisory Board of Herbalife. Herbalife was a supporting company of ILSI-China and one of the funders of the event.[19]

SPREADING SODA SCIENCE: SCIENTIFIC RHETORIC HELPS THE CAUSE

Reflecting Coke's financial contribution to the meeting (or so we can assume), Hill and Blair gave the lead presentations. (The honor may also have stemmed from their status as "old friends" [*lao pengyou*] of ILSI-China.) Each was allocated forty minutes, twice the time offered the majority of speakers (fifteen to twenty minutes). Applebaum spoke a little later in the first morning for thirty minutes. In their talks, the GEBN organizers offered many sensible ideas about obesity prevention. At the same time, they reiterated the arguments for physical activity developed by the EBAL committee, weighting their presentations toward industry-friendly views. These views were then delivered directly into the heart of China's obesity science, policy, and health-care community. Though few listeners would have noticed, each speaker made skillful use of the rhetoric of science to advance their version of corporate-science orthodoxy. I've argued that corporate science, as a corrupted form of knowledge, tends to be packaged in a strong rhetoric of science, assuring listeners that the (actually problematic) science is of the highest caliber. These talks provide a superb opportunity to see that rhetoric at work.

In his talk, Hill managed to serve industry interests with almost every point. He struck the main themes of soda-defense science early, mapping out the "energy balance system" and the vital role of physical activity in improving metabolism and, in turn, realizing weight loss, maintaining weight loss, and preventing weight gain.[20] He then raised a set of pressing questions at the center of the wider public debates on obesity. But instead of engaging the debates, he simply shelved the questions as unanswerable. On a slide titled "Where's the Data?" he listed a host of issues that were contentious in the field but anathema to the soda industry: sodas and obesity, marketing and obesity, fast food and obesity, vending and obesity, physical activity/inactivity and obesity. Despite the growing body of research on subjects like soda and obesity, his answers at the bottom of the slide were this: "We'll never get the data—too complex" and "Problem is too great to wait for the data." Placing topics that were taboo to the food industry outside the field of science certainly served the industry well. Here and elsewhere in his talk, Hill skillfully used the rhetoric of scientific rigor (e.g., "data too complex") to promote his corporate science of obesity. Regulatory measures that the industry loathed—restrictions on soda, marketing, fast food, and vending—were spurned as "haphazard tactics," suggesting, oddly, that a piecemeal approach targeting specific causes of obesity had little use.

How then should we tackle obesity? Hill's answer was: "Move More and Eat Smarter." Although the slogan implies he was giving equal weight to diet and exercise, if we thumb through the full set of PowerPoint slides we discover an overwhelming emphasis on exercise. Astonishingly, his suggestions for smart eating occupied just one slide out of forty-nine, taking diet tokenism to an extreme. Once again, Hill plugged his dictum that small behavioral changes can prevent weight gain, reassuring his audience that "China can do this successfully." The conclusion identified *individual motivation* as "the missing piece" of the puzzle, again absolving governments and corporations of major responsibility.

If Hill served the soda industry by taking governmental regulation off the agenda, Blair obliged it even more by asserting that obesity was not much of a problem. To change the subject from obesity to inactivity, Blair deployed a rhetoric of scientific fact. In a hard-hitting, data-packed presentation full of bar graphs, pie charts, and line graphs (these "ocularly powerful" symbols of science occupied twenty-six of fifty-two slides), Blair contested the findings of the many mainstream obesity researchers who advocated dietary change.[21] After dismissing common understandings of energy balance, he presented the results of his and Hand's Coke-funded Energy Balance Study. Citing extensive data demonstrating that "overweight is good for you, and class 1 obesity [BMI 30–34.9] is not so bad," he ended with his signature claim that the major public health problem was not obesity but physical inactivity. In the conference wrap-up session the next day, he laid out the implications for policy: "[It] will be controversial [but] let's forget about weight. . . . Focus on healthy living. . . . If you're active and fit, your weight doesn't matter very much. . . . The US . . . ha[s] proven that focusing only on weight loss . . . does not work. Let's drop [the] focus on weight, let's try something else."

Declaring that "well-being starts with energy balance," Applebaum delivered a hearty endorsement of Coca-Cola's so-called science-based approach to obesity.[22] In her hands, the language and facts of science, delivered in bold proclamations, became powerful tools to defend her company's reputation and market its products. "We collaborate with folks who are fact-based and credible," she declared. "It's not our science, it's theirs," a statement belied by the long history of Coke's involvement in shaping the "facts" of energy balance science. That science-based approach—based on several Coke-funded research projects, including Blair's Energy Balance Study—involved three elements. The first was "Education"—in energy balance concepts, among other things. The second was "Physical Activity," meaning the active healthy living programs Coke had sponsored globally since 2004. The third was "Variety," a pointed reminder that more than 25 percent of Coke's products were no- or low-calorie options, implying the company had already addressed the obesity question. While noting in her oral remarks that Coke focuses on "a sensible, balanced diet and regular physical activity," her emphasis was almost exclusively on activity. Only one of Applebaum's forty-three slides dealt with the dietary side of the equation, and that was the slide highlighting the variety of beverages the company offered! Like Hill, Applebaum ended by placing responsibility for weight management in the hands of individual consumers, whose "motivation" was "the next frontier" to be explored by Coke and friends.

BREAK-TIME CHATS

As I sat watching Applebaum run through her colorful slides, I was astonished to see her actively promoting her company's portfolio of beverages. ILSI-China leaders had impressed on me the importance of their ethics rule forbidding companies from plugging their products at ILSI events. Curious about the power struggles between ILSI's China branch and the powerful companies supporting it, I caught up with her on a break and asked if ILSI-China's leaders had reviewed her presentation in advance. Surprised at my question, she gave me a smile of recognition and replied, "Oh, yes. They said the first version was too commercial." Yet the version I had heard was still rather commercial. What this told me was that it was impossible for ILSI-China to say no to the vice president and chief science and health officer of the Coca-Cola Company, especially at an event the company had helped underwrite. The ethics rule the branch had instituted was exposed as but a normative statement designed not to stop corporations from interfering in the science, but to provide ethical cover and plausible deniability. As I stood there in the hallway outside the meeting room, the larger truth suddenly became clear to me. Despite the heroic job ILSI-China had done performing ethics and concealing its inner dynamics, the balance of power in the organization was overwhelmingly weighted toward rich, powerful member companies.

With these three talks, soda science was portrayed to China's public health community as not just the best thinking, but the only approach that was thinkable. Heber, the nutritionist, spoke after the other three and used his forty minutes to advocate a diet with lots of color (red, purple, green).[23] In a brief exchange during a break, he told me he "didn't necessarily agree" with the other academic speakers, before letting me know he had no interest in engaging further. This little cameo—a potentially illuminating conversation quickly aborted—offers a momentary glimpse into the ways the ILSI setup, especially the rules on participation by invitation only, worked to create a closed world of ILSI science, in which dissidents from orthodoxy dare not express their views.

"PHYSICAL EXERCISE IS A MUST": A CHINESE ENDORSEMENT

Given the consistency in message conveyed by these three prominently placed speakers and the dearth of alternative perspectives, Chinese researchers attending the conference might well have concluded that in the United States there was a consensus that promoting physical activity is the most important strategy for combating obesity.[24] They may also have come away feeling that China's model of multisector health science and policy making, with industry as a major partner, was the standard everywhere, unquestioned and unproblematic. As he had done so often in the United States, Hill, in China, championed the partnership model, saying in his oral presentation: "I'm pleased to hear the talk about partnerships. That's the only way forward; we must bring the private sector in." Blair directed his praise at ILSI, telling the audience: "ILSI came from corporate America. We need more of this—NGOs with money to support scholarly research." With these two prominent "best scientists" from "the world's most advanced country" urging industry participation in science and policy making and, as far as I can tell, expressing no concern about risks such as industry distortion of science and policy, members of the audience would have no reason to question the correctness of China's corporate-heavy approach to solving the chronic disease problem. If they quietly harbored doubts, it's unlikely they would have shared them at this ILSI-run event.

ILSI-China's big obesity conferences had a history of being places where national policy directions were established or affirmed, and this would be no exception. In the conference's concluding session, Chen Chunming, the unquestioned leader of China's obesity field, highlighted Blair's points, contending that his approach should become a national priority:

> Our meeting is a success. . . . We've learned lots of strategies from other countries, especially the US. . . . [One] strategy is *smart eating*. . . . We shall educate people and raise awareness. . . . Also, the importance of physical activity. . . . Look at Dr. Blair, he seems to be overweight, but he says he has lots of muscle. I've known him for years, he's very healthy. BMI is not important. This shall be our goal: become fit. *Physical exercise is a must.* This is another thing we learned from this meeting. (emphasis added)

Chen's comments suggest that Blair's views and those of Hill ("smart eating") had influenced Chinese thinking at the highest level.[25] By this time, the language of soda science had been normalized as the best and virtually only way to think about obesity in China. The campaign to get soda science endorsed in China had succeeded beyond all expectation.

Corporate Investments in Activity Programs: A Tangle of Invisible Ties

Coca-Cola and its corporate peers also advanced their interests in China by investing in concrete exercise programs. Working through hidden funding mechanisms within ILSI, during the years 2004–15 the company supported three important programs promoting exercise for obesity. Because Chen worked closely with decision makers in the China CDC and MOH, all three programs gained official support, and two were integrated into the ministry's nationwide anti-obesity campaign (described more fully in the next chapter). These corporate investments through ILSI—which are listed in table 6.2—represented a distinct pathway by which Chinese chronic disease policy came to be subject to corporate influence. These exercise programs were certainly beneficial in some ways. Even if their impact on obesity was minimal or unclear, they most likely contributed to the general health of those who participated. What needs emphasizing, though, is their outsized impact on China's obesity work. Because other approaches to obesity control failed to enjoy similar private funding, these investments by the food industry worked to further skew China's approach toward activity. Through the power of its financial investments, Coke left a major mark on China's obesity work.

By tracing the pathways followed by these three key investments, we can see the intricate tangle of visible and invisible ties that linked Coca-Cola in Atlanta through ILSI-Global and its quasi-corporate academic partners in Washington, DC, to ILSI-China and eventually the Chinese government in Beijing. Whether constructed to obfuscate those pathways or not, this multilevel tangle of ties, which is almost impossible to unravel, certainly helped keep ILSI's secrets secret. Coca-Cola clearly intended these investments in the health of the local population to burnish its reputation as a virtuous company, and conversations in Beijing suggest the effort was a genuine success. Reflecting China's probusiness political culture, my expert-informants viewed the programs initiated by the giant food companies not only as unobjectionable, but as positively beneficial to the country's development. Far from being questioned, Coke's and ILSI's corporate-science project had become thoroughly normalized, accepted as business as usual in the fight against chronic disease.

COKE-FUNDED, ACTIVITY-THEMED PROGRAMS

In 2004, ILSI-China and the China CDC introduced Happy 10 Minutes, a Chinese variant of the ILSI-CHP's Take 10! program. Coca-Cola's connection to Happy 10 Minutes remains hidden, but a collection of random facts about it—all the facts I was able to gather—provides clues to the ingenious ways ILSI was able to bury its corporate secrets, especially when operating transnationally. In 1999, readers may recall, Malaspina invited Chen Chunming to serve on the board of the Atlanta-based ILSI-CHP. In February and May 2004, the CHP arranged an intra-ILSI transfer of just over $53,000 to the China CDC. These funds (or some portion of them) were earmarked for a Chinese program based on Take 10! Chen asked the China CDC to develop it, and in September of that year, the CDC, with technical and funding support from ILSI-China, introduced Happy 10 into schools in Beijing.[26] The following year, the CDC launched a national campaign aimed at expanding the program to multiple cities. Coke's China subsidiary supported the launch ceremony.[27] In the mid- to late 2000s, Happy 10 became a standard feature of the ministry's Healthy Lifestyle for All campaign to fight obesity. Perhaps reflecting Chen's effort to show her enthusiasm for Malaspina's pet program, Happy 10 was one of the most visible and enduring components of ILSI-China's work on obesity, featured regularly at ILSI conferences on obesity between 2004 and 2015. Though the extent of Coca-Cola Global's involvement is unknown, the Atlanta-based company claimed credit for Happy 10 in its sustainability reports (dubbing it "Happy Playtime"), suggesting the company may have been a shadow funder from the beginning.[28] By following the money and motivations back to their source in Atlanta, we can see how something as sweet and innocuous as a little ten-minute program encouraging Chinese schoolkids to move was deeply penetrated by global capital, which had its own interests in mind.

This is not to say Happy 10 was bad. Quite the contrary: the program addressed a very particular need. The major obstacle to getting schoolkids to be more active, a Chinese insider explained, is the tiny size of the classrooms. In too many cases rooms built for fifty children must squeeze in seventy. The CDC created exercises specifically for such miniature spaces, allowing children in schools that authorize Happy 10 to get twenty minutes of movement a day, up from ten. Surely the doubling of activity time was beneficial for their health generally, even if it didn't make much of a dent in the childhood obesity epidemic.

After 2004, when Applebaum joined Coca-Cola and began investing in science, the company's direct investments in activity programs in China multiplied. In 2011 the soda company funded a three-year Coke/ILSI-China scholarship program enabling young professionals to obtain training in physical activity and health in the United States.[29] Blair's home institution, the University of South Carolina, served as a training center and in 2012–13 received $67,525 for the work.[30] A second center for training was the US CDC, where Michael Pratt, another member of Team Coke we met in chapter 3, served as host. In the 2010s, the major Coke-backed project in China was Exercise Is Medicine. The EIM-China Program, launched in 2012 by ILSI-China and the American College of Sports Medicine (ACSM), held numerous training events for clinicians.[31] When I was interviewing in China in 2013, EIM was ILSI-China's top-priority activity on obesity, emphasized by several of the experts I talked to. Through its investments in these two programs, the company—its money, its ideas, its associates and grantees in academia—came to be a major player in China's chronic disease work. And in turn, China's public health field came to be tangled up in webs of global capital that no one could unravel and few could even see.

This is just a handful of the many programs Coke funded, directly or indirectly, to boost enthusiasm for activity solutions to obesity. With the government welcoming industry contributions to public health efforts, many of these were adopted as official health-promotion activities, producing a nice halo effect for the company. The researchers I talked to brought up Coke's public health projects on their own, applauding what the company was doing for China. Two experts told me about Coke's annual Health Incentive Plan (Jiankang zhili jihua), carried out in cooperation with the MOH's Bureau of Disease Control. In Shanghai, Coke launched the Beat the Street program, an international walking competition that encouraged children to walk to school.[32] Shanghai, the lead city for walking to school in China, was in second place globally, lagging New York but beating out London and Liverpool. Coke was far from the only food company involved in China's health campaigns; Nestlé, Mars, the retailer Carrefour, and many others joined in as well. Yet Coke was one of the most visible, generous, and committed to the effort. Indeed, the company's use of the motif of "health" to promote its good name—and its business—went far beyond the obesity campaign. For example, Coke had a longtime collaboration with the China Academy of Chinese Medical Sciences, a research center under the Ministry of Public Health and the most respected center of traditional Chinese medical research and practice in the country. In 2007 the company established the Coca-Cola Research Center for Chinese Medicine, becoming the first international company to open a research center within the academy. The aim was "to bring . . . Chinese medicine to the world through packaged beverages."[33]

Briefing me on Coke's activities, a well-known Chinese nutritionist described its message as "balancing eating and moving," which he wryly translated as "Eat more, then exercise more." Aside from this little wink about an underlying profit motive, my informants seemed genuinely impressed and grateful to Coke for bringing its clever ideas to China. No one mentioned possible risks. The state's receptive attitude toward rich (and cooperative) foreign firms was thus mirrored in the public health community. The experts involved with China's chronic disease work had no reason to even mull over possible industry influence on their science because, they assured me, ILSI's no-commerce rule had taken care of all that.

THE DIFFERENCE ONE COMPANY MADE

Coca-Cola's influence on China's obesity work may not have troubled Chinese experts, but it should have. Coke and other widely admired food companies left a large and nonbenign mark on China's approach to the condition. The company's influence can be seen in table 6.2, which lists ILSI-China's major activities promoting exercise for obesity. Of the total of eight "other scientific activities" and "public health interventions" listed, Coca-Cola was deeply involved in six. This simple accounting exercise actually understates Coke's influence, because the company also helped fund at least one and probably quite a few more of the large-scale obesity conferences, as well as other projects that remain hidden from view.

The growing importance of these Coke-funded interventions also helps explain how, between 1999 and 2015, a rising proportion of ILSI-China's obesity activities came to prioritize physical activity over dietary restraint. Indeed, the handful of activities just described that were funded (directly or indirectly) by Coke account for a fifth of all obesity-related activities sponsored by ILSI-China between 1999 and 2015 (fourteen of seventy-two) and nearly half of all events emphasizing physical exercise (fourteen of twenty-nine). If we eliminate all Coke-funded activities emphasizing exercise, the proportion of ILSI-China obesity activities emphasizing exercise rises much more modestly. Instead of being more than twice as common as dietary activities between 2010 and 2015, the proportion of obesity events emphasizing exercise is the very same as that emphasizing diet. The gap between dietary and exercise approaches virtually disappears. In other words, these Coke-funded programs account for *all* the increase in the proportion of ILSI-China activities emphasizing physical exercise. Coke funding for these programs unmistakably skewed ILSI's work on obesity toward physical activity. And yet, because of ILSI's success at invisibilizing corporate influence, without an in-depth analysis such as this the company's impact would remain undetected.

Science for Sale: Nutritional Conferences on Obesity

Though China's obesity conferences increasingly took physical activity as their focus, a subset dealt primarily with nutrition.[34] Between 1999 and 2015, ILSI-China sponsored an assorted collection of scientific gatherings—some fifteen in all—presenting "the most up-to-date scientific information" on the connections between diet, nutrition, and chronic disease, including obesity (the details can be found in table 6.3). While all took up obesity to some extent, only a handful were labeled obesity or chronic disease conferences. The reason is not hard to find. A close look at what transpired shows that, far from genuinely combating obesity by encouraging the consumption of nutritious foods, the vast majority of these "nutritional science meetings" were devoted to advancing the commercial ends of the ultraprocessed-food companies supporting ILSI.

The nutrition conferences defended soda, and highly processed foods more generally, in three main ways: emphasizing the health benefits of soda ingredients, promoting the healthfulness of foods criticized as contributing to obesity, and presenting the food companies as corporate good citizens.

TABLE 6.3. ILSI-Sponsored Activities on Nutrition and Health Dealing with Obesity, 1999–2015

Date	Activity
1. Aug. 1999	Seminar on Sweeteners
2. Oct. 2000	Third Asian Conference on Food Safety and Nutrition
3. May and June 2004	Symposium on the Health Impact of Dietary Fat
4. Sept. 2005	Workshop on Carbohydrate and Human Health
5. May 2006	Symposium on the Importance of Water as a Nutrient *
6. Nov. 2006	First Workshop on Restaurant Foods and Balanced Diet
7. June 2007	International Beverage Forum on Sport and Health *
8. Sept. 2007	Symposium on Snacks and Health
9. Apr. 2008	Workshop on Sweetness and Health
10. Oct. 2008	Symposium on Nutrition and Metabolic Syndrome
11. Oct. 2008	Hydration and Health Symposium *
12. Aug. 2012	MOH, China Health Forum 2012; Subforum on NCD Prevention and Control
13. Oct. 2012	Research Results on Weight Control in Communities Briefing (Weight Watchers)
14. Aug. 2013	MOH, China Health Forum 2013; Subforum on Nutrition Improvement and NCD Prevention and Control: The Role of Enterprises
15. Feb. 2015	ICN2 Seminar (following the Second International Congress on Nutrition, held by FAO and WHO in Rome)

Sources: ILSI-China newsletters, 1999–2015.

Notes: Includes activities sponsored or cosponsored by ILSI-China dealing at least partly with obesity.

* Addresses obesity indirectly.

Although few of the meetings engaged with the ongoing discussions about obesity prevention in China, their prosugar, pro-food-industry messages, conveyed repeatedly over the years, may have cast doubt on arguments that sugary drinks and junk food are problematic. With their barely disguised defense of commercial interests, these meetings deserve a close look because they lay bare, in a way the physical-activity-focused meetings do not, what the ILSI enterprise of "science for the public good" was all about.

DEFENDING THE INGREDIENTS OF SUGARY SODA

A first group of meetings worked to blunt critiques of sugary soda by presenting carbonated soft drinks as a good source of hydration and exonerating sugar as a main cause of obesity. Coca-Cola was especially active in these efforts to emphasize the benefits and deemphasize the dangers of key ingredients in sugar-sweetened carbonated beverages, with Maxime Buyckx of the company's Beverage Institute taking the lead. Two symposia on hydration and health concluded that water—the most important ingredient in soda—is critical to overall health as well as sports performance, and that caffeinated beverages are "fine sources of water."[35] In other words, soda was good for you because it met the body's needs for liquids.

Two workshops on sweeteners concluded (much evidence to the contrary) that sugar consumption was not a factor in the development of obesity or type 2 diabetes, and that sweetness should be "managed not banished."[36] These flagrantly self-serving (and sometimes dubious) health claims by ILSI companies may strike some readers as preposterous, yet such messages were business as usual in many ILSI-China activities, and they had effects. Among other things, these claims might have left participants doubtful or confused about how unhealthful sugary sodas really are, helping keep soda taxes and sugar limits off the agendas at ILSI's meetings on obesity.

MARKETING PRODUCTS AND CORPORATE CITIZENSHIP

A second cluster of meetings introduced the health benefits of specific food products to China's nutrition studies community. While these events were presented as public health activities, they were essentially scientific marketing tools for ILSI's supporting companies, which funded the events. The marketing aim was far from subtle. Over the years a wide range of supporting companies and trade associations—including Mars, PepsiCo, DSM, and the Almond Board of California—used the mechanism of ILSI-China's scientific symposia to hawk their products as healthful for China's people. In many cases, the health claims appear to have been legitimate, but in the materials I read qualifications and evidence of potential risks were not presented.

In a subset of these gatherings, ILSI's supporting companies sponsored workshops aimed at defending certain food products (snacks, restaurant food) or ingredients (dietary fat, carbohydrates) in the wake of mounting evidence of their role in the rise of obesity and other chronic diseases.[37] Danone Biscuit China supported a symposium on the health impact of dietary fat, for example, while Yum!, owner of KFC and Pizza Hut, funded a workshop on restaurant food and the role of Western fast foods in promoting health, nutrition, and balanced diets. In the workshop on snack foods a rep-

resentative from Mars outlined the health benefits of cocoa flavanols, leading the workshop to conclude that healthy ingredients such as chocolate
should therefore be added to snack foods. The newsletters describe how
speakers advocated "balance" and "diversity" in diets. The message—that
any food, no matter how unhealthful, can be part of a healthy diet, as long
as there is "balance"—was a main argument associated with soda science,
and we've encountered it many times before.

A third set of meetings defended Big Food by presenting the companies
as corporate good citizens who contribute to public health by educating
consumers to make smart, scientific choices. "Lots of money for public
health comes from companies," a popular health blogger declared over dinner one night in Beijing, naming Nestlé, Coke, and KFC and its parent Yum!
as among the most active. As part of their corporate social responsibility
work, these companies actively supported the government's campaigns
to fight diet-related diseases by sponsoring colorful displays on nutrition
and healthy lifestyles, putting health halos around the firms and their products.[38] While conveying positive images of the companies, such messages
deflected attention away from the unhealthy foods the companies did sell,
effectively protecting their core products and their corporate reputations
from blame for contributing to the obesity problem.

Of course, the food-themed meetings were not completely dominated
by companies' promotion of their products. Many if not all conveyed standard public health messages on diet as well. In the symposia on sugar, for
example, J. S. Chen exhorted colleagues to urge citizens to cut back on
sugar-rich foods and drinks. Chen Chunming constantly badgered her
supporting companies to be more responsible and active in improving the
nutritional content of their foods. Still, in a now-familiar pattern, these
meetings virtually all served industry needs by emphasizing education
of the public to "choose healthy lifestyles." Chen did not call on the government to regulate the food industry or its marketing practices. Instead,
industry was to regulate and police itself, while consumers were made responsible for their own dietary choices. The many ILSI meetings on physical activity I discussed earlier protected food-industry interests indirectly
by diverting attention from the role of high-fat, high-sugar food in obesity.
These nutrition meetings supported the industry much more directly—
including by literally marketing junk food in the name of science—showing
us the face of corporate science with no filters whatsoever. The very fact
that such meetings were held, and were presented as public health events,
also tells us something important about the larger climate surrounding science in China. What they suggest is that in China, the marketization of

science—the hijacking of science for purely commercial purposes—is fully acceptable practice.

⁘

Working hand in hand with ILSI-Global and Coca-Cola, ILSI-China had served as agent for the corporatization of Chinese science, overseeing both the importation and the endorsement of a corporate science of obesity. Now, leaving their US-based partners behind, the branch leaders would navigate their way through their country's guanxi-based, who-you-know-is-what-matters political system to get that science turned into official policy.

CHAPTER SEVEN

Translating Soda Science
into Chinese Policy

The food industry's campaign to protect sugary drinks was not confined to getting soda science endorsed in major markets. For soda to be well defended, that science had to be built into local policy as well. In this project China presented unique opportunities that lay not just in the country's business-friendly environment, but also in the location of Chinese science within the institutions of the state. In part for this reason, ILSI president Alex Malaspina had a special place in his heart for China and its capable head, Chen Chunming. Malaspina was secretly thrilled to learn that the ILSI branch would be located within the Ministry of Health. "Let them do it their way!" he and Suzanne Harris, ILSI's executive director, declared, apparently ignoring the violation of ILSI's paramount rule of no lobbying or policy advocacy. Such a location would give ILSI direct access to policy makers in the ministry and thus the potential to shape policies favorable to member companies. Malaspina personally selected, cultivated, and incentivized the head of the China branch and, as we shall see in this chapter, was well rewarded. According to Harris, ILSI-Global's leaders were happy with the results of their work in China. And well they might be, even if they had only an inkling of everything ILSI achieved in that country.

This chapter takes us into the tangled inner world of Chinese science politics to see how a local agent of a global corporate-science organization negotiated the convoluted politics of a government trying to marketize, globalize, and scientize all at once. We follow Chen and her colleagues at ILSI-China to see what tactics they used to get soda science built into Chinese policy, how successful they were, and what hidden effects their methods and maneuvers had on Chinese public health institutions more generally.

How central-level policy is made in China is a closely kept secret. The information gathered on chronic disease policy allows us to penetrate the wall of official secrecy and sketch an account of one set of policies that is at once intimate in detail—told in the words of the actors themselves—and transnational in scope—connecting policies adopted in China back to their origins in the United States. We will see that the soda industry and its scientific agent left a deep imprint on China's policies. They not only played a central role in creating China's two main anti-obesity policies; they got the core ideas of soda science built into the country's primary strategy to combat chronic disease and appear to have had a hand in the omission of soda taxes from China's policies. The industry and its scientific agent also left their mark on China's public health institutions, policy-making dynamics, and political logics.

In using a foreign nonprofit to bring Big Food–friendly policy to China, ILSI-China was playing a potentially dangerous game. Chinese policy is supposed to be made by the Chinese government. The Chinese government is usually suspicious of foreign organizations. Under normal circumstances foreign involvement in the creation of China's policies would be scandalous. But Chen seems to have had the political skills and capital to make everything work.[1] Even as it managed to quietly mold policy to industry's needs, ILSI's China branch mobilized the discourses of the state and ILSI-Global to keep the molding out of sight, ensuring the corporate nature of Chinese policy remained hidden from view and the scandal of foreign involvement in policy did not materialize.

How did ILSI-China engineer these extraordinary outcomes? The answer lies in the workings of China's hierarchical, personalistic, guanxi-based political system and the favorable location of the organization and its leader within it. With these advantages, ILSI was able to structure the political field to give it significant control over the who, how, and what of obesity policy making in the country. With such well-connected and politically savvy friends on the inside, Big Food would fare remarkably well in China.

Political Secrets of ILSI's Success in China

What exactly was ILSI-China? So far I've been writing about it as though it were a unique organization, one of a kind. It was not. ILSI's emergence in the early 1990s was part of a larger, politically significant historical process in which China's party-state sought to open up policy making to experts outside the state. The preferred vehicle was the Chinese-style think tank. Though ILSI was not registered as a think tank, it closely fit the broad definition introduced just below, and so I treat it as a member of this class of organizations. I begin with the political fundamentals, laying out the characteristics of the Chinese-style think tank, then unravel the political secrets to ILSI's success in China.

What is a Chinese-style think tank, and why is it so important? In the

early 1980s, the frenzied opening of the country to the global economy trig-
gered a cascade of social and economic problems never encountered in the
Mao decades. With a state bureaucracy fragmented into separate silos and
lacking the expertise needed to manage the new challenges, the state authorized the formation of various kinds of think tanks (*zhiku*) to provide
expert consultation and scientific policy advice to decision makers in the
state.[2] In China's authoritarian system, where power is concentrated in the
party-state, bringing outsiders in to advise the state was a politically sensitive task that required recruiting trustworthy experts—those certain to
uphold the goals and norms of the party-state—and creating mechanisms
to ensure state oversight of the process.

In the United States and other Western countries, think tanks are typically registered as nongovernmental organizations. In China that kind of
arrangement was politically impossible. As Zhu Xufeng, a leading scholar of
Chinese think tanks, explains: "Strictly speaking, there are no purely independent organizations in the governing system dominated by one party."[3]
China's think tanks have instead been affiliated to varying degrees with a
governmental agency. Despite being subject to government oversight, in
Zhu's characterization China's think tanks have been autonomous in important ways. These include being able to decide what research projects to
take on and when. For many think tanks, it has also included the freedom to
accept research tasks from other bodies and to receive grants from foreign
entities. ILSI-China enjoyed all these freedoms. Regardless of the source
of a think tank's projects and funding, a fundamental expectation has been
that it will provide *neutral policy advice*. This expectation has two parts: the
services provided were supposed to be *advisory*, with the actual policy formulation handled by the state bureaucracy; and the advice was supposed to
be *disinterested*. How these norms were to be enforced, however, remained
unclear. As we will soon see, these expectations would pose challenges for
ILSI, which sought to use the vehicle of the think tank to provide industry-
friendly—that is, nonneutral—advice and to move beyond policy advising
to policy shaping and even *making*. How it navigated the contradictions is a
big part of the story told below.

In chapters 5 and 6, we observed Chen in her capacity as a scientist and
leader of a major scientific project aimed at bringing the latest science on
diet-related chronic diseases to China. Turning now to her political status
and activities and digging more deeply into her extraordinary résumé, we
discover that Chen was a giant in the field of chronic disease work, a political heavyweight with the kinds of credentials and connections that would
lead the government to entrust her with the important task of advising it on
chronic disease policy. Her new organization was located within the core

health policy institutions of the government, giving it a place in the official policy process and, equally important, direct access to state policy makers. These two sources of ILSI's power—Chen's status as a powerful leader in a hierarchical political system in which power is personalized, concentrated in the hands of leaders at various levels, and not subject to interrogation by followers; and the branch's status as one of the newly prioritized think tanks established to provide expert advice to policy makers needing to ensure that all policy is made "scientifically"—are vital keys to the organization's success. There is more to the story, though. These arrangements, while beneficial for ILSI and its soda-defense goals, carried hidden risks for the country. Chen and ILSI-China had dual identities and dual loyalties—to the Chinese state and to the corporate masters of ILSI. Success for ILSI would mean jeopardy for China.

WHO WAS CHEN CHUNMING? STATE SCIENTIST-OFFICIAL WITH SUPERLATIVE GUANXI

At the heart of many Chinese think tanks was a politically well-connected individual, often a retired senior government official in the relevant ministry. Such individuals were likely to enjoy the trust of current officials on the basis of long-standing personal ties as well as their experience in making public policy. In this delicate process of allowing outsiders to advise the state, trust was paramount, and, as we have seen, in Chinese politics personal connections (*guanxi*) are the single most important source of it. Guanxi ties tend to involve two things: positive sentiment and an exchange of favors between the parties. Such ties have exploded in the reform years of rapid socioeconomic change. As sociologist Yanjie Bian explains, these carefully cultivated personal connections have helped to "plug . . . the institutional gaps and holes" in still weakly developed political and economic institutions that lack formal means to ensure interpersonal trust.[4] The art of guanxi is most finely honed at the apex of Chinese society. *Red Roulette*, Desmond Shum's gripping account of wheeling and dealing between China's wealthy entrepreneurs and the political leaders to whom they must pander to succeed, describes a system of exchange—lavish dinners for lucrative business deals—in which every interpersonal detail is calculated in advance and nothing is written down. So too in Chen's ILSI, it was interpersonal ties—rather than formal or professional qualifications—that formed the underpinnings of the mutual trust between the two Chens and officials in the health ministry. And in both cases—Shum's businessmen and ILSI's science and policy entrepreneurs—ties to the party-state were the essential key to success.

ILSI-China was a highly unusual health organization, the experts I talked to explained, permitted to form only because of Chen's political status and the guanxi that came with it. "ILSI is a very special model," a former high-level health official confided. "Its leader was previously head of the Chinese Academy of Preventive Medicine and had good working relations with people at the health ministry. Those natural links, combined with Chen's leadership experience, made ILSI-China one of a kind." A well-connected university professor put it more strongly, spelling out what was at stake: "The government would not have approved Chen's application if she weren't formerly an important government official. The government is afraid that [independent social forces] might have other, particularly political, agendas (*mudi*). It's especially concerned about international organizations because of worries about intelligence and security—spying. So it strictly controls the formation of organizations. The government believes ILSI is scientific, and not involved in any other political matters" (loose translation). Without a doubt Chen enjoyed the backing and trust of key officials in the state. In this hierarchical system, as long as she played by the political rules, she would enjoy substantial power and influence that those working under her authority were not supposed to question. Chen's state-backed power was the first secret of ILSI's success in China.

As my expert-informants indicated, Chen's ties to key MOH leaders were crucial to ILSI's influence.[5] Yet there was a twist in this case, for Chen was not only a trusted state scientist-official; from at least the turn of the millennium she was also a well-incentivized quasi-corporate scientist, who could be relied on to make sure the interests of ILSI-Global and of ILSI-China's supporting companies were taken care of in her policy proposals. Remember, there is no assumption that Chen intentionally or even consciously served as agent for the companies. More likely is that, with the ample professional rewards she gained from her association with ILSI, she was heavily incentivized to *not see* that her policy proposals aligned with corporate interests. As a quasi-corporate scientist working with ILSI, she would import an industry-friendly science of obesity into Chinese policy circles, quietly (perhaps unknowingly) violating the think tank norm of neutrality in policy advice. This dual character of ILSI entities—public and private, state and corporate—would allow Chen to pivot between identities when useful and, in the end, to work political wonders for ILSI-Global.

Beyond her status and ties in the health establishment, Chen was known to have connections high up in the central government. According to a Chinese friend and colleague, her father, Chen Fang, served as secretary to Chiang Kai-shek (known also as Jiang Jieshi, head of the KMT or Nationalist Party), before moving to Hong Kong after the Communists took control of the mainland and Chiang and his Nationalist Party fled to Taiwan. We can assume Malaspina was aware of the family's political background, since Chen Chunming sent him a book of her father's calligraphy and bamboo ink painting created during his time in Hong Kong.[6] Chen the daughter remained in China, where she developed collaborative relationships with a range of prominent intellectuals and political figures. Among the most impressive of her connections was the friendship with Yu Ruomu, the wife of Chen Yun, a major architect of the reforms of the 1980s and 1990s and in those years the second most powerful person in China (after Deng Xiaoping).[7] Yu was a well-known expert on and advocate for good nutrition and thus a natural ally.

In interviews with prominent European and American global health experts who had worked with Chen in the 1980s, 1990s, and 2000s, I repeatedly heard that she had extraordinary political connections that she drew on when she needed political backing. When things got tough, the solution was "just a phone call away." Who she was calling, though, no one was quite sure. Each had heard a different rumor. Was it a classmate on the party's Central Committee? Was she a personal adviser to a previous prime minister? Did she have ties through her husband to the president of China? Although I was unable to confirm any of these suspicions, both Chinese and foreign observers relayed stories of her ability to mobilize bureaucracies of the government and organize large-scale collaborative projects to stress that Chen was extraordinarily, even "terribly" powerful, an "unchallengeable" force to contend with in the high politics of Chinese nutrition policy. High-level connections of this kind would most certainly have amplified Chen's power and ILSI's influence within the health field.

On top of these political connections, Chen had a set of personal attributes that marked her as a leader. Those same American and European scientists described her as charming and charismatic, well respected, and tremendously influential. She was also, in their opinion, wily, reserved, and discrete, playing her cards close to the chest. Chen had, in short, all the attributes of a skilled scientist-politician. ILSI-Global had found a brilliant leader to run its China operations. How brilliant? Let me answer with one personal anecdote. In a long, animated conversation in London in 2016, Philip James, a top British expert on obesity (whom we met briefly in chapter 5) described how, one day in the early 2000s, he and Chen were in her Beijing office talking. By then they had worked together for some time, and he had imagined her as a well-connected scientist working for her country. Who should walk in just then but top officials of Unilever, Pepsi, and Nestlé! Only then did he realize, "to my horror," that he—a physician who had spent years fighting the influence of the food industry on the British diet—

had been collaborating with a branch of ILSI, and that his Chinese partner was in all probability "corrupted" (his word). "Industry got to China first," he continued morosely, "to make sure nothing happened on obesity." By which he meant nothing that would make a dent in the profits of Big Food.

WHAT WAS ILSI-CHINA? A THINK TANK LOCATED IN THE MOH

The second secret of ILSI's success in China was the branch's politically felicitous location in the administrative system of the People's Republic of China (PRC). As we saw in chapter 5, the ILSI branch in China was designated an affiliate (of unspecified sort) of the Chinese Academy of Preventive Medicine, the agency Chen had led for a decade before establishing ILSI-China. For policy makers, keeping ILSI within the health sector of the state was an obvious and advantageous move. With this arrangement, the health ministry could make use of Chen's valuable expertise (scientific knowledge, policy insights, long experience conducting research on chronic disease) and obtain much-needed foreign funding for research. Health-sector leaders could also take advantage of ILSI's greater operational flexibility and speed, relative to the administrative organs of the state, to get things done. ILSI's office was physically located in the CAPM/CDC (from 2002, CDC) headquarters in Beijing and made ample use of CDC staff and professionals. With that location, Chen could take advantage of informal interactions with CDC leaders to float and promote policy ideas, which the CDC, a technical unit supervised by the Ministry of Health, could pass on to its parent body, the official policy-making agency in the field of health. Housed at the CDC, Chen was smack dab in the middle of the hustle bustle of making policies and programs to safeguard the health of China's people.

But ILSI's China branch also had close—and much less visible—relations to the Ministry of Health. As noted in chapter 5, ILSI reported to the head of the Bureau of Disease Prevention and Control within the ministry, who in turn reported to the minister of health. Although think tanks are not supposed to make policy, this reporting chain gave Chen as head of ILSI-China a place of some sort in the policy-making process within the health ministry, and thus the potential to influence official policy. In discussions, Chen repeatedly stressed that ILSI's role was merely to provide "evidence" and at most "recommendations." Her strong insistence that ILSI simply supplies evidence suggests she was keenly aware of the dangers of crossing the line into policy creating, and was constantly working to stay on—or to appear to stay on—the right side of it. Because of Chen's background and status as a former high official, during her years with ILSI she worked closely with the

MOH and, as we will see shortly, in practice operated as a de facto ministry official. As one well-placed informant put it, ILSI-China was a "loose organization" (*songsan de zuzhi*) ruled more by people and connections than by formal guidelines. With these arrangements, the ministry had unwittingly placed an agent and advocate for corporate science at the heart of the health policy establishment. The trust inhering in her guanxi ties was supposed to ensure things were done in politically correct ways. Did it? Or would corporate logics overshadow Chinese political ones?

Turning Soda Science into Policy: ILSI as Policy Maker

Chen took advantage of this extraordinary location to translate the new scientific ideas brought to China through ILSI into public policy. We saw in chapters 5 and 6 how, as the lead organization on obesity, ILSI-China took charge of virtually every step involved in addressing the condition, in the process getting obesity defined as a Chinese disease and soda science endorsed as the authoritative approach to understanding and managing it. In the first decade of this century, when obesity was largely a stand-alone policy issue, Chen and her organization created China's two most important policies on the prevention and control of obesity and related diseases. The first was the official guidelines for obesity prevention and control in adults (promulgated in April 2003).[8] By placing obesity on the policy agenda, establishing new bodily norms the public should abide by, and laying out the interventions needed to achieve them, these guidelines constituted the first obesity policy of the PRC. The second was the government's first nationwide campaign to slow the rapid rise in obesity and other chronic diseases. Launched in 2007, after soda science had been embraced in China, the campaign built that science into China's official response to the epidemic. Reflecting ILSI's (*soda-tax-never*) approach, there was no attempt to promote taxes, marketing restrictions, or other measures regulating the soda industry. Quite the contrary, there is evidence that Chen's organization deliberately blocked an effort by others to get soda taxes on the agenda, a subject we return to at the end of the chapter. From around 2012–13, obesity would be gradually merged into the larger class of noncommunicable diseases (NCDs, also known as chronic diseases), and ILSI's influence reportedly waned. In one of its last notable contributions to national policy, ILSI provided some input into the first major policy on NCDs, the 2012–15 National Plan for Chronic Disease Prevention and Treatment. I say more about that in the coda.

Creating these policy documents was a remarkable achievement for a

tiny organization like ILSI-China. How did it do it? In remarkably frank discussions of sensitive policy issues, Chen and a handful of close colleagues
described a highly informal, collaborative process involving the MOH,
CDC, and ILSI-China that was driven and at points dominated by ILSI itself. The MOH essentially outsourced policy making to the trusted head
of ILSI-China, inadvertently allowing ILSI's industry-friendly science to
become the foundation of Chinese policy. In this section, we examine that
process, using Chen's own words and those of some of her key associates to
document ILSI's role as policy initiator and decision maker. Initiator and
decision maker? ILSI-China was not supposed to take over such roles, and
doing so blatantly violated foundational rules established by China's government and the larger ILSI organization. Yet the branch head found ways
to quietly reframe (that is, camouflage) the violations, allowing the policy
work to proceed. To make sense of these dynamics, we begin with a primer
on Chinese policy making.

CREATING CHINA'S OBESITY GUIDELINES

China watchers from afar tend to think of policy making in China as a top-
down, highly secretive process concentrated within the party-state. That
image is essentially correct, but only for a certain class of policies.[9] Guard-
ing its power and prerogatives, China's party-state has long reserved to itself
the making of policies on issues of strategic importance or bearing on the
power or legitimacy of the party. Such policies have been created in top-
level, highly secretive venues. Yet in policy areas dealing with "technical"
issues—where neither of those pertained—during certain times, stakeholders outside the formal policy apparatus, including scientists and engineers,
have been welcome to participate.[10] Health care was such an issue, as were
environmental and science and technology policy.

In the first decade-plus of this century, when Hu Jintao was the top
leader, the initial stages of the policy process on "technical issues" were
opened up to nonstate actors. Experts, NGOs, foreign actors, and other
groups lobbied furiously to get their pet issues on the agenda and submit-
ted proposals and petitions in hopes of getting their views built into state
policy. Later in the same policy cycle, though, the experts were locked out,
as the open door gave way to a closed-door process in which only state bu-
reaucrats were involved in formulating the policy.[11] (After Xi Jinping's rise
to power in late 2012, decision making became much more centralized in
the party, but that happened largely after my story ends.) The making of
chronic disease policy between 1999 and 2012–13 was very different from
this process of opening followed by closing during one policy cycle. Perhaps

because so few stakeholders were concerned about the invisible epidemics of chronic disease, and one entity was heavily subsidized to address these issues, we find a different dynamic in which a low-key, nonstate entity took charge and managed the making of policy from start to finish.

Chen was at the center of the whole process, and she relished relaying how she orchestrated it from beginning to end. In her account, the guidelines were a product of a bottom-up, highly personal dynamic in which she, acting as a de facto official, first persuaded the health ministry that the obesity problem was serious and that China needed official guidelines, adding pointedly (with a coy smile here) that ILSI was prepared to provide them. The MOH then asked ILSI to prepare the guidelines. Happy to comply, ILSI quickly drafted them. Seeking to ensure high-quality scientific work, ILSI sent the draft guidelines to a wide range of experts for comment, revising them seven or eight times. It then presented its recommendations to the government for review. After a few back-and-forths, they were accepted and published as a standard Ministry of Health document, with ILSI's name removed. Chen's success in bringing the guidelines to fruition was facilitated by her dual identity as scientist and official, and by the porous nature of the boundary between policy recommending and policy making in an iterative policy process such as the one just described. The procedures followed a few years later in the creation of the guidelines for children and adolescents were similarly personal and thorough.[12] From the outside it looked like the ministry ran everything, but from the inside ILSI was indisputably, if discreetly, in charge.

INITIATING A NATIONAL CAMPAIGN TO FIGHT CHRONIC DISEASE

A few years later (2006–7), ILSI notched a second notable policy achievement, the adoption of the National Healthy Lifestyle for All Action (Quanmin jiankang shenghuo fangshi xingdong). This was the government's first major public health measure to slow the rising tide of obesity and related diseases, and it was a pathbreaking development. In the 2003 guidelines, China had taken a biomedical approach, defining obesity as an individual disease that could be treated with clinical and preventive medicine measures, including medication. That work was funded largely by Big Pharma, which hoped to sell obesity-reducing drugs. In the healthy lifestyle campaign, China would switch to a public health approach, seeking to manage obesity at the population level. This new phase of obesity research and policy advocacy would be funded in good part by Big Food, which hoped to encourage measures that spared junk food and sugary drinks from the threat of government regulation.

The active involvement of ILSI's American advisers in China's obesity conferences allows us to trace the remarkable process by which specific scientific ideas were not just advocated as solutions to its problems, but also embedded in official policy. By the mid-2000s, Hill's core concepts had been introduced as the best international science on obesity and were circulating within public health circles. The idea of launching an annual campaign to combat diet-related chronic disease was born on November 23, 2006, the last day of one of ILSI's big conferences on Chinese obesity.[13] Hill had helped organize the event and attended as a featured speaker. In his talk he presented his America on the Move campaign as a case study illustrating the "big effects of small behavioral change on maintaining and reducing body weight." His audience needed little convincing. Evidently inspired by the AOM example, in the conference's final session, a vice president of the Chinese CDC suggested that China launch a national campaign, to be structured as a Chinese patriotic health campaign (*aiguo weisheng yundong*).[14] One of Chairman Mao's most famous contributions to global health, these mass movements—launched in the 1950s, 1960s, and 1970s to improve sanitation, eradicate pests, and eliminate infectious diseases—would now be repurposed to improve daily health along lines suggested by corporate science. The lifestyle campaign would address unhealthy diets and physical inactivity, the two major risk factors for obesity and related chronic diseases. The plan was enthusiastically endorsed by those attending.

The policy process was similar to that followed in creating the guidelines. As Chen recounted it, after the conference ILSI made a recommendation to the MOH, and the plan was quickly endorsed. ILSI then helped the ministry work out the details, and in the fall of 2007 the Healthy Lifestyle for All Action was launched.[15] It has been carried out every year since and in 2012 was scaled up into a national campaign. Thus did ILSI's fingerprints end up all over China's nationwide movement for good health.

Were there no dissidents? There were, but they remained publicly silent, voting with their feet. Philip James, the British obesity expert, also helped organize the 2006 conference and spoke at the event. As he narrated it to me ten years later, he was aghast at what transpired. Not only did Chen change the program they had agreed on, but she invited representatives of KFC (Yum! Restaurants) and other food giants to talk at this supposedly scientific event. James was horrified. Aside from rendering some help with publishing a couple years later, that was the end of his involvement in China. Knowingly or not, he was leaving the field to the American scientists associated with ILSI and Coke.

ILSI-China also succeeded in getting its signature intervention—Happy 10 Minutes—incorporated into the official healthy lifestyle campaign. The CDC researcher who created and for years championed Happy 10 described for me an equally informal process eased by personal guanxi ties. After testing its feasibility and effectiveness and publishing the results, those in charge of Happy 10 proposed it to the CDC. The CDC then recommended it to the MOH, which in turn made it a standard feature of the governmental campaign.[16] It was that simple. Though less weighty than creating a national policy or health-promotion campaign, getting a public health intervention adopted for nation-wide promotion, especially one so closely tied to Malaspina and Coca-Cola, ILSI's founder and its founding company, was a small triumph for the branch.

This reconstruction of the policy process suggests that ILSI's lead role in translating soda science into official policy was rooted in its prior success in structuring the larger political field. ILSI conferences were the main sites of reaching agreement on policy ideas; ILSI's invitation-only rule named industry-friendly scientists as featured speakers while excluding or silencing dissenters; and ILSI's leader was able to produce broad consensus on policy questions because she was a de facto government official who reported to the MOH and enjoyed status and connections throughout the health establishment. With Chen seemingly managing virtually every phase of the policy process in China, soda-science ideas moved quickly and easily, without evident obstruction, from Hill to Chen, to the conference participants, to the head of the ministry's Bureau of Disease Control, to the minister of health himself. China's hierarchical, leader-centric system of power relations worked wonders for ILSI-Global and its goal of fostering a "harmonized use of [corporate] science" around the world.

"SCIENCE ADVISING, NOT POLICY MAKING": VEILING POLITICAL REALITY

ILSI's pivotal role in the policy process ran afoul of critical norms, however. As noted earlier, in the Chinese administrative system, think tanks were supposed to provide policy advice but not actually engage in policy making, the prerogative of the party-state. Within the ILSI-Global network, the "Code of Ethics and Organizational Standards of Conduct" (below, simply "Code of Ethics") forbad advocacy, lobbying, and making policy recommendations. These deviations by the China branch would need to be erased, at least in the language, and both the Chinese government and ILSI-Global had discursive resources Chen could use to make that happen. My investigations suggest that, like the ethics rules discussed earlier, these administrative rules were meant not to actually keep scientists from making policy (that is, to achieve their stated objective); remember, in both cases there were no evident means of enforcing the rules. Their aim instead was to ward off criticism and allow plausible deniability (or, more colloquially, provide political cover), allowing experts to engage in policy making whenever it served their needs. By claiming adherence to the rules through discourse while disregarding (even violating) them in practice, ILSI staff effectively hid the political reality from sight. (A second potential problem—the scientific advice think tanks offered was supposed to be neutral—was handled in another way. We see how in the next chapter.) To be clear, I am not suggesting that ILSI leaders consciously set out to hide the deviations from sight. More likely, they saw themselves as simply taking advantage of ambiguities in the political field, including their flexible identities as experts and officials, and opportunities afforded by the political discourse by choosing to ignore or downplay certain inconvenient norms in the pursuit of other worthwhile goals. The effect, though, was to hide the violations from sight.

The arrangements in China clearly violated ILSI-Global's rule of no advocacy. ILSI's "Code of Ethics" stated baldly that ILSI does not influence policy, it merely "provides scientific evidence as an aid in decision making." This formulation proved useful to Chen, allowing her to manage and effectively cloak the contradictions between the code created in Washington and the reality unfolding in Beijing. While wooing corporate sponsors with the promise that ILSI-China "does not just sponsor scientific conferences . . . [it] puts scientific evidence into policy," Chen explained in an interview, she added a crucial caveat. ILSI-China does not make policy; it merely "provides scientific evidence and recommendations for policy decisions" (language from the code of ethics).[17] With this phrasing, Chen was able to exert extraordinary influence over official policy—while publicly alleging she was not.

Since the beginning of the reform era, when China's ruling party declared modern science a vital key to China's global rise, "science" (*kexue*) has been a powerful term of political discourse, commanding widespread belief.[18] In the hands of a pervasive propaganda apparatus, science has been associated with national rejuvenation, technological sophistication, modernity, and, most generally, hope for a brighter future guided by reason. Exploiting its advantage as part of an "international *life sciences* institute," ILSI-China made frequent and strategic use of a rhetoric of science to stress both the authority and the noncommercial character of its policy advice. Sensitive to potential charges that their own industry-funded knowledge bore a commercial imprint, J. S. Chen told me, he and Chen Chunming made a point of including the word "scientific" in the names of their conferences to distinguish them from the many profit-oriented, company-sponsored meetings held in China. From the earliest days of its work on obesity, ILSI-China, denouncing the commercialized and inaccurate information on weight-loss products saturating the media, countered it by promoting ILSI's own "scientific information" on obesity prevention.[19] Activities and materials for the general public—educational pamphlets, health classes, BMI calculators, and so on—invariably stressed the scientific nature of the information.[20] To be sure, ILSI's work *was* more legitimately scientific than much of the health information circulating in China. My expert-informants complained endlessly about companies that hired medical experts to advertise weight-loss products that ended up harming people's health. Yet ILSI's science was science with a difference, and the difference was not disclosed. Because ILSI's leaders were esteemed scientists and science enjoyed utmost respect, bordering on worship, there was little apparent questioning of the label. The political weight of the word "science," which was heralded by the government as the essential key to China's global rise, may have helped shield ILSI's work from scrutiny.

The semantic distinctions in the bureaucratic discourse of the party-state worked similar magic, allowing the two Chens to describe something that was seriously awry—the location of a corporate-science organization inside the Ministry of Health—as administratively unexceptional, of no concern at all. In official communications, Chen Chunming routinely described ILSI-China as a "nongovernmental academic [i.e., scientific] institution" fully compliant with ILSI rules.[21] Yet its physical location within the CDC clearly made it governmental and, given Chen's official background, a de facto policy organization. J. S. Chen unspooled the mystery for me. Technically, he explained, ILSI-China was indeed *non*governmental. ILSI-China was affiliated with the CDC, which was not a government agency, but a *technical unit* (*shiye danwei*) ("more like a university or research institute") under the Ministry of Health. As merely a technical scientific organization, CDC's close ties to another scientific organization (ILSI-China) were fully legitimate. This distinction in Chinese bureaucratic discourse between "government agency" and "professional unit" would allow ILSI to work in unrestricted partnership with the China CDC—while presenting itself as nongovernmental. Here again, the rules worked not to produce a technical/political divide, their purported job, but to allow things to move forward while invisibilizing the political and policy-making reality. This may sound like a distinction without a difference, but in China political labels are everything, and the label technical unit was critical to ILSI's viability.

Healthy Lifestyles for All: A Policy Anchored in Soda Science

The policy process just traced had a profound impact on the content of China's policy. The 2003 obesity guidelines advocated combining "dietary adjustment" with "increased physical activity" to achieve a negative energy balance (that is, to lose weight) but made clear that dietary change was the centerpiece of therapy: "Limiting total energy and fat intake is the primary measure for weight control."[22] Once the ideas associated with Hill and ILSI-Global began to reach China around 2006, however, the discourse began to subtly shift. The primacy of dietary change disappeared, replaced by the primacy of exercise. This shift, never before uncovered, is of utmost political significance, for it connected China's chronic disease policy to a global political economy of sugar and soda dominated by giant food companies.

In this section I trace the imprint ILSI's soda science left on China's main strategy to contain the obesity epidemic, the Healthy Lifestyle for All Action, inaugurated in 2007. We will see that ILSI-Global and its cooperating scientists were remarkably successful in getting their favored ideas embedded in China's measures to fight obesity. Not only did the action embody the concepts that Hill had promoted; it was modeled on his America on the Move (AOM) campaign.

AOM IN THE PRC

Through the policy dynamics just described, the central ideas in soda science became foundational elements of China's response to the epidemics of diet-related diseases. The energy balance framework is commonly used to understand obesity, but it was Hill's version of the framework and related concepts (energy gap, small steps, and so on) that came to predominate in China. Along with those concepts came the ideal of the self-cultivating, scientifically minded citizen-consumer carefully adjusting their daily routines and calculating their energy in and out to reach balance. As the slogans were propagated among ordinary people in health-promotion campaigns, that figure, familiar to many Americans, was presented as the ideal to which China's people also should aspire. Let's see what this looked like at the ground level.

The notion of energy balance—of individuals "maintaining eating and moving in equilibrium" (*chidong pingheng*)—became the main slogan and a central organizing concept of the healthy lifestyle action.[23] The campaign urged healthist citizen-consumers to walk ten thousand steps a day[24] and propagated "small change and big effect"—Hill's pet rallying cry—as the campaign's "happy life concept."[25] Under the small-changes formula, it encouraged Chinese citizens to eat a few bites less and move a few minutes more every day. The action also advocated the catchy phrase "Healthy 121" (it's catchy in Chinese: *jiankang yi'eryi*): 1 for ten thousand steps a day, 2 for eating and moving in balance, and another 1 for a whole lifetime of health. This slogan also became the domain name for the action website:

FIGURE 7.1. National Healthy Lifestyle for All Action, carried out in Gansu in 2014

jiankang121.cn. (The numbers 1-2-1 jump out in the Chinese original: *rixing yiwanbu, chidong* **liang***pingheng, jiankang yibeizi*.) Photos of the lifestyle campaign being carried out in the northwestern region of Gansu in 2014 show banners featuring these very slogans (see figure 7.1).

Of course we should not overstate the similarities between the Chinese and American health-promotion movements. The Chinese campaign was structured as a government-run, nationwide patriotic health campaign, complete with numerical targets and an annual cycle of preparatory and summation meetings. Still, the resemblances between the Healthy Lifestyle for All Action and AOM are striking. The two healthy lifestyle campaigns shared conceptual underpinnings, slogans, and objectives, and both involved a short-term, focused effort every year to mobilize a substantial fraction of the population to improve their lifestyles. Both employed multiple means to veil the impact of corporate funding on the programs introduced in the name of public health. As if these similarities were not enough, the Chinese campaign was even held on the same date every year as AOM in the United States: September 1.[26]

EMPHASIS ON ACTION, SILENCE ON SUGAR

Like AOM, in ways subtle and not so subtle the campaign for healthy life-styles gave special weight to promoting exercise and fitness, both beneficial to health generally though of less use in weight loss.[27] The campaign's "action logo" was an eye-catching lime-green-colored body within a circle, meant to represent "a running and jumping humanoid figure enjoying a happy life" (see figure 7.2).[28] Several of the annual slogans promoted between 2008 and 2014 had movement themes, subtly equating more activity with the achievement of health:

> "I'm acting [or moving], I'm healthy, I'm happy" (*wo xingdong, wo jiankang, wo kuaile*) (2008)
> "Science-based exercise and [i.e., leads to] good health" (*kexue yundong yu jiankang*) (2010)
> "Walk 10,000 steps a day, keep eating and moving in balance, enjoy lifelong health" (2014, the same slogan as "Healthy 121")

On the dietary side, the propaganda for the healthy lifestyle action included tips on healthful eating. Yet only one annual slogan had a dietary theme, and that was "Reduce salt to prevent high blood pressure" (*jianyan yufang gaoxueya*, adopted in 2011). (Three of the slogans urged prevention of high blood pressure.)[29] None of the materials I reviewed on the first phase of the lifestyle campaign (2007–15) mentioned education on reducing the consumption of free sugars, a growing concern at the WHO since 1989.[30] That was not part of the AOM program.

By around 2008–9, the notions of energy balance, energy gap, and small changes that Hill and others had developed in the United States had been

FIGURE 7.2. Healthy Lifestyle for All Action logo

baked into China's official response to obesity as the best, indeed, the only thinkable approach to obesity management. These constructs, which emphasized exercise at the expense of diet, would soon form the conceptual groundwork for a growing body of Chinese policies on the prevention and control of diet-related chronic diseases developed between 2016 and 2020. (That story is told in the coda to this part.) Once again, the importance of a handful of American researchers affiliated with ILSI should not be exaggerated; Chinese policy was influenced by other sources, including the US CDC and the WHO. According to several informants, standard practice when creating new chronic disease policy called for first studying how various advanced countries had approached the issue and adopting useful ideas where appropriate. Still, given our ability to historically trace the ideas of Hill and his collaborators from their entry into China to their promotion to key audiences, to their new home in Chinese policy, it is hard to escape the conclusion that their imprint was pronounced. The energy balance framing at the heart of soda science was embedded in everyday public health practice as well. Asked how obesity work was progressing, the head of the chronic disease section of the Beijing CDC replied in frustration: "The biggest problem is that eating and moving are not in equilibrium. [People] eat too much meat and oil and do too little exercise. People do not understand the dangers . . . and so remain unwilling to change. The problem of fatness is just too big [*tai dale*]!"

This close look at ILSI-China's most important policy contribution suggests that ILSI-Global, working through Hill and Chen Chunming, was remarkably successful in getting its soda science translated into official policy. In a conversation Chen proudly pointed out the cleverness of the secretive arrangements, describing ILSI-China's policy achievements as win-win situations for the MOH and ILSI-China. I mentioned earlier that guanxi ties entailed an exchange of favors, and it was here, in this conversation, that Chen articulated what the favor exchange involved. In that everyone-wins scenario, the MOH got what it believed was internationally cutting-edge policy on a new and unfamiliar issue, while ILSI got to quietly influence health policy. "The Bureau of Disease Control has good things to say about ILSI," Chen reported. "We're happy that we can do good things for the people, the government accepts them, and ILSI does not claim credit." In China, ILSI-Global's dream of soda-defense science becoming the basis for official policy had come true. With no one keeping track of all the borrowings, ILSI's leaders in Washington probably had no idea how successful they had been in China.

EDUCATING THE MEDIA AND THE PUBLIC: PIVOTING
BETWEEN PRIVATE AND PUBLIC

ILSI-China had a dual identity—as scientific agent of the food industry and as government-based think tank serving China's state and society—and could rhetorically pivot from one to the other as circumstances dictated. Committed to the broader mission of public health and seeking to help the government carry out its public education role, Chen and her colleagues sought to ensure that ILSI-China's latest findings on obesity reached the Chinese public. With a public that found chubby kids cute and ranked losing body fat far below buying a car or a house—or so one informant assured me—ILSI's efforts to educate ordinary people addressed a real need.

In mid-2001, ILSI, working with a companion organization, the Think-Tank Research Center for Health Development, created the first of many "media salons" aimed at providing reporters for major newspapers and magazines the most up-to-date scientific information on obesity.[31] ILSI and its partners also engaged the public directly. Starting in 2007 ILSI partnered with the Office of the Healthy Lifestyle for All Action to organize exhibits and conduct community educational sessions. At those events, citizens could learn scientific approaches to body weight (such as the small-changes and energy-in-balance formulas) and pick up information packets and BMI calculators that would enable them to make those changes in their daily lives. In these activities ILSI was operating as a virtual governmental agent. Though its intent was to bring scientific knowledge to the public, one of the effects was to extend ILSI (and industry) influence beyond policy and policy makers to the public at large.

The little tools distributed at these public events show the pivot strategy at work. The nifty little round BMI calculators, produced in bright lime green on heavy, laminated paper, carried the message that ILSI-China and the ThinkTank Research Center had designed the tool in accordance with the MOH obesity guidelines (which, recall, ILSI had developed) (see figure 7.3; the message, on the back, isn't shown here). The calculators and other educational materials carrying the ILSI name (posters, flyers) positioned ILSI as a close collaborator with the public sector, not an agent of the food industry that ultimately funded almost all of ILSI's scientific work. Whenever a new set of obesity-related guidelines was issued or a new strategy for obesity prevention was adopted, ILSI and its partners would organize media events and public health campaigns to educate the public. From ILSI's point of view, the aim of presenting itself as an official implementing agency was undoubtedly to assure the public that its message was officially approved. Once again, though, from the outside observer's point of view,

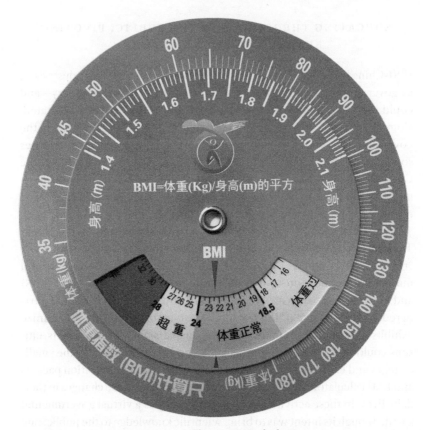

FIGURE 7.3. Chinese BMI calculator

the effect of presenting ILSI as a close ally of the government was to veil its industry roots. Through such activities, ILSI's industry-friendly version of the scientific truth about obesity became the public's version of the scientific truth. The effect of rebranding ILSI as a government agency was to keep the corporate forces that shaped everything out of public view. No one would guess that behind the message "this is your government speaking" was another message: "this is industry speaking."

The broader effects of these outreach programs and of the lifestyle campaign more generally are worth lingering over. What this means is that in the early twenty-first century, some 1.3 billion Chinese citizen-consumers were being informed that their government had concluded that the main solution to the raging epidemic of diet-related chronic diseases was not restricting the consumption of junk food, but balancing calories consumed and expended, with a particular focus on increasing exercise. (Of course not all 1.3 billion citizens heard the message, but the number was surely large.)

Governmental health-promotion messages such as these are especially important in a one-party authoritarian state like China, where a policy or position that has been endorsed by the party-state is essentially set in stone, backed by the weight of the top governing authorities. Again, and at the risk of repetition, this is not to suggest that official policies and interventions promoting energy balance were bad. Encouraging citizens to eat better and exercise more was surely beneficial to their health. But these policies were tilted in one direction, with the tilting hidden from view. And that posed real risks to the health of China's people.

A Critical but Invisible Force in Obesity Policy Making

Through the early 2010s, then, ILSI-China was the main force behind the creation of China's policies to combat obesity and related diseases. Yet a review of this and the last two chapters suggests its impact was even broader than that. Big Food's push to influence obesity science penetrated and transformed China's health sector. ILSI's China branch left its mark on the *political institutions* of chronic disease policy making, the CDC and the MOH, which became home to an institutionalized agent of foreign industry. This agent of Big Food also altered the *process of policy making* by ensuring the participation of foreign companies in a number of overt and covert ways. And it changed the *logics behind health policy* by supplementing the original logic of relieving human distress with a new logic of corporate profit making. To be sure, the Ministry of Health was already pursuing a market orientation. But, with its access to a rare source of power, corporate money, ILSI pushed the process and logics of policy making even further in that direction. The anthropologist Susan Erikson's contention that global health is becoming "a thing en route to, amid, and alongside making money" is only too apt here.[32]

My expert-informants agreed with this judgment about ILSI's political significance. "ILSI's work is mostly related to national policy; there it plays a key role." More specifically: "ILSI's work defining obesity and creating the guidelines was very important; no one else was doing that." Another angle: "The great bulk of obesity research done in China . . . deals with small data sets or concrete interventions. ILSI's work represents the only large-scale [that is, national] research on obesity done in the country." ILSI's contribution to getting obesity on the agenda was especially important, my informants stressed, because the government was too busy to pay attention to it. How concerned was it, I wondered? The responses ranged all the way from sympathy to veiled disgust, with political realism and cynicism in between:

CDC RESEARCHER: The government prioritizes food safety—an urgent trade and political issue—over nutrition, a long-term problem that doesn't kill people (at least not right away). It has too much to manage so must grasp the most important issues.

CHILD HEALTH SPECIALIST: Obesity is getting more and more serious, but not to the point of threatening the Communist Party's leadership or the national interest. For that reason, it doesn't get a lot of attention.

HOSPITAL-BASED NUTRITIONIST: The government has a consciousness of the importance of nutrition, but no laws or policies. It neither understands nor supports nutrition.

ACADEMIC SCIENTIST: On obesity, it's all words and no action.

Given the highly positive evaluation of ILSI's contributions by these knowledgeable insiders, I have been puzzled by the quizzical looks I have received over the years when I talk about my research findings. What is ILSI, people ask, stating they have never heard of it. How could ILSI's China branch be so important and yet so little known? An influential academic expert who had worked in the field for decades explained the paradox this way: "ILSI has two very famous people; it uses those names to promote ILSI projects. [Over the years] ILSI has supported or organized a lot of work that is invisible. ILSI has a special feature: those working for it are borrowed. The authors formally listed on publications belong to another unit. For that reason, you cannot see ILSI's influence from those papers" (paraphrased from interview). In other words, ILSI's role in creating obesity science and policy was truly important. And that fact was hidden from public view.

Chen was caught in a politically risky if not perilous contradiction. Government policy was supposed to be made by the MOH, not a nongovernmental think tank, and certainly not one that was essentially a branch of a foreign organization funded and largely governed by non-Chinese firms. For a foreign organization to be involved in making Chinese policy is "rare and weird," according to a leading Chinese scholar of the nation's scientific policy-making practices.[33] That it did here, he suggested, can only mean the two Chens had an extraordinary stock of political capital. Chen Chunming's response to the peril was to constantly downplay her organization's and her own importance. At the same time that she bragged about ILSI's achievements, she obscured them in politically safe language. On the one hand, she was rightfully proud of these accomplishments, emphasizing that once the government issues a policy document the matter is settled once and for all. Yet even as she claimed credit for influencing a decade of important policy decisions, again and again she drew on ILSI-Global language to stress that

ILSI "provides only evidence and recommendations ('tools'); this is our sole role." In this everyone-wins fairy tale, China benefits, ILSI plays a minor technical role, and the foreign organization and funding corporations are whisked out of sight. The effect if not the intention was to publicly hide the truth of ILSI's—and its food company supporters'—influence on China's efforts to stanch its obesity epidemic at all levels, from national policy to mass education programs. Did Chen see the contradiction between the two stories she told about ILSI's accomplishments? We will see in the next chapter.

There was yet another way in which ILSI was important, and this was even more deeply hidden from view. Just as ILSI-Global used hard-edged measures like bribery at the FAO/WHO meeting in Rome in 1997, ILSI-China seems to have played dirty too. There is persuasive evidence that Chen Chunming effectively blocked a relatively early push by another group, the Chinese Nutrition Society (CNS), to tax sugary drinks. The CNS had co-organized a big conference on sugar-sweetened beverages, inviting top global specialists on sugar and sugary drinks to speak. ILSI, however, made sure its recommendation that China tax sugar-sweetened beverages was not carried out. Although that effort lay outside the scope of my research, it is clear that ILSI actively obstructed measures that might shrink the profits of the soda industry. If ILSI's role in getting obesity on the agenda at the turn of the century aligned China with the best of international thinking at that time, its role in excluding regulatory measures aimed at junk food and sugary drinks in the 2000s and 2010s left China far behind international best practice in obesity policy.

∵

Having promoted soda science as the best in the world, ILSI now had to make that corrupt body of knowledge seem ethical to Chinese sensibilities. In that effort, untangled in the next chapter, it would be boxed in by the state, whose inability to curtail industry influence on science would leave everyone, scientists and state alike, vulnerable to the self-serving schemes of Big Food.

CHAPTER EIGHT

Doing Ethics

The Silent Scream

Funded by industry and fitted to its needs, the soda science lodged at the core of China's policies on diet-related chronic disease was unquestionably corrupted. For that science to be viable and durable—for the food industry's project of defending soda with science to succeed in China—it had to be framed as ethical in terms that made sense in the convoluted political culture of twenty-first-century China. In this chapter we take up that fundamental question of ethics, exploring how the Chinese scientists negotiated the ethical terrain to make their science appear ethically unproblematic, and the dilemmas that effort entailed. In our discussions and in her writings, ILSI-China's longtime head displayed pride in the progress she had made in solving the obesity problem, depicting her actions as eminently virtuous. The conversations with Chen and other experts were suffused with tacit ethics talk—talk about right and wrong—and layered reflections on ethical aspects of science. In this chapter, I draw on these conversations to understand how Chen and those working under her did ethics, how they sought to achieve that sense of science well done in partnership with industry, whose leaders were responsible for growing profits, not ensuring the integrity of the science they funded.

Doing ethics in China bears little resemblance to doing ethics in the United States. In the United States, Hill, Blair, and others on Team Coke simply declared they were not influenced by their corporate funders and proclaimed the benefits of partnerships with industry, saying little about the risks. In China, where sensitive issues are involved, silence, not voice, is the name of the game. So too is public obeisance to the state. In this one-party, highly authoritarian political system, then, our ethical inquiry must start with the state. We will see that the state's position on corporate ethics left China's scientists in an impossible bind. The official stance not only was minimalist in the extreme; it was contradictory, forcing the scientists to rely on corporate money but without providing effective ethical guidelines to manage corporate attempts to bias the science. Scientists laboring in their labs and offices were left to make things work.

Chen Chunming's solution was to rigidly adhere to the state's formal ethics rule while relegating everything that did not fit—including delicate questions about corporate funding—to a zone of silence and secrecy. While that allowed work on obesity to move forward, with so many important issues rendered unspeakable, nothing could be resolved, and the few who saw problems brewing dared not open their mouths. The state's failure to find a workable solution to the ethics question left everyone—citizens, scientists, and the state itself—vulnerable to the manipulations of giant corporations pushing product-defense science in the name of public health. This final triumph for the food industry posed real dangers for China and its people.

This story of the global travels of corporate science raises the question of where, in the end, corruption lies. In a favorite US media story line, compared to American science, Chinese science is especially susceptible to corruption because of the power of money and personal ties in the political culture.[1] Many Chinese share that view. This case study of soda science gets us to the bottom of the corruption of science that troubles so many. But it also shows that, in the case of corporate science, we need to turn the spotlight around and shine it on America and its system of giant corporations that dominate the global economy, extracting profit everywhere they go.

The State Stance on Science Ethics: Minimalist and Contradictory

Academic malfeasance exists in every society, but it takes different forms in different places. From major scandals that draw international attention to more humble types of everyday fraud, distortion of science has been rampant in contemporary China. Yet one kind of ethical problem—conflicts of interest in the industry funding of science and the warping of knowledge that often results—has received remarkably little attention. None of my informants mentioned it, despite dogged questioning on my part. In a 2016 survey of over 725 science and technology researchers at China's top universities, one-third cited excessive government intervention as a research challenge, yet no one mentioned pressure from foreign funders to produce favorable results.[2] To understand why the Chinese researchers I talked to did not imagine that the companies funding ILSI's work might have quietly distorted the science coming from America, we start where everything starts in China: with the party-state. The state had two norms on science ethics, Ethics-1 and Ethics-2. We will see that these norms not only were very basic, they were conflicting, leaving experts on diet-related chronic disease in an impossible bind they had to somehow resolve on their own.

ACADEMIC FRAUD RUN RAMPANT

Before we get to those norms, we need to understand the wider context in which ILSI and allied scientists were operating. When, tiptoeing carefully, I queried my expert-informants about scientific misconduct involving the corporate sector, I got an earful of stories about infamous cases that had come to light in the Chinese media. Some brought up megascandals like the 2013 case of the British pharmaceutical giant GSK (GlaxoSmith-Kline) caught bribing Chinese doctors and government officials to promote its drugs. Or the terrible milk scandal of 2008 when twenty-two Chinese companies were discovered adulterating infant milk formula with the toxin melamine to boost its protein content, sickening some three hundred thousand babies.[3] In the wake of widely publicized instances of ignominious behavior such as this, the Chinese government has issued sharp warnings and new regulations. Yet the state has found it difficult to stay ahead of canny businessmen who profit by making false scientific claims.

I also heard a litany of cases of everyday forms of corporate-science fraud. Pharma company bribery of doctors, a "very, very serious problem," was so mundane it hardly merited mention. A small sampling of the cases the experts shared hints at the ethical chaos in the world of Chinese (health) science in the mid-2010s. The following examples were provided by two scientists at some distance from ILSI, and thus more open to discussing such matters. One was a biomedical researcher at a top Chinese university. The other was an employee of an American public relations (PR) firm, whose job it was to help client companies sell their products using any means available, including manipulation of scientific claims:

> Every famous person in China has cooperative relations with companies. There's a certain professor of nutrition who is constantly in the media advising citizens on what to eat. She works for the food companies, which give her money and gifts like iPads. She will say, "it's good to eat organic vegetables," then the vegetable company will quote her in its press release, using her comment as an endorsement for its products.
>
> The makers of food and health products [*baojian pin*] have especially good marketing skills. They format some of their print ads like news items. It's called soft news [*ruan xinwen, ruan wen*]. The company pays someone—an expert, a star, a singer—to endorse the product. People in

remote areas can't distinguish news reports from ads. The material is provided by a commercial firm, but the newspaper is not necessarily responsible for printing that, so people think the product is recommended by a journalist or an editor at the publication.

These cases, and most of the instances these researchers mentioned, focused on false advertising, in which the company pays an expert to make its products appear to be supported by the latest scientific research. In a small number of cases, the corporate distortion affected the making or presentation of scientific results. The aim here was also ultimately to advertise a product:

> In some small universities things are not so scientific. I call these professors "businessmen" or "scientists doing business research." They do whatever the company wants, whether it's producing a certain result or publishing a certain article. "What I conclude is what you need"—that's their motto. Some steal data, others manufacture evidence—whatever the company needs.
>
> It's very common for a pharmaceutical company to pay a doctor US$1,000 (usually in kind) to attend a conference, either to give a paper written by company staff or to simply participate. The company promotes its products by having a poster or stand at the entrance to the hotel. A company employee will give a speech to promote the product, then the doctor will present the paper, leading participants to believe he is independently evaluating the drug.

Clearly, ILSI operated in a wider climate in which payment for science and the corporate corruption of science were pervasive and, despite mighty efforts by the state, remained underregulated. (Things reportedly have tightened up under Xi Jinping.) My informants, the great majority practicing scientists or clinicians, were keenly aware of the dangers of such activities and of the need to loudly condemn anything smacking of corruption. And yet, when I asked directly about the central problem with ILSI's own activities—the potential for conflicts of interests in industry funding, and the consequent biasing of science created not to market products but *to make public policy*—the response was mostly a blank stare. From their expressions, I got the sense that for many the question did not even register as meaningful. In this morass of fraudulent scientific activity, the problem that actually plagued ILSI had not yet been identified and named. As one savvy professor put it to me, people were fully aware of the influence of company-paid ads *on consumers*, but unaware of possible corporate influence *on researchers like themselves.* Indeed, he and several others offered the astonishing observation that China did not even have a concept of conflicts of interest in research funding. How could that be?

ETHICS-1 AND ETHICS-2

Like all Chinese citizens, Chinese scientists are subject to state controls on their speech. The state establishes the political discourse in which things are to be talked about, setting limits on what can be discussed and how. To be safe from (rarely specified) political problems, the scientists I got to know had to stay within the speech space of the state, taking its framing and limits as their own.[4] Perhaps because the political dangers were very real, these rules of the political game seemed to be front and center in the minds of some expert-informants who, when offering a comment that may have seemed a bit risky, would point to the state as the source of their ideas.

Since the early 2000s, the bureaucracies of the state concerned with science have grappled with many problems of scientific misconduct, introducing oversight committees, regulations, and other measures to address them. Industry funding has escaped oversight. By the mid-2010s, when I was canvassing members of the obesity research community, the state had articulated but two rudimentary ethical positions on corporate support for science: industry funding of science is good; industry bias of science is bad. Let's call them Ethics-1 and Ethics-2. Importantly, each was associated with one or more of the state's broad policy orientations. Ethics-1 (industry funding of science is good) followed from the broad policy of marketization, which urged experts to seek research money from the market. Ethics-2 (industry bias of science is bad) followed from the policy of scientization, which pushed for the rapid absorption and development of the best international science to inform a newly mandated process of scientific policy making. How did these play out in the community of obesity researchers? Let's start with Ethics-1.

As readers now know, from the early years of the reform era, the state has consistently urged scientists in health fields to obtain research funding from the market. ILSI's China branch was approved and long operated under that policy orientation. Given this strong state stance, it is not surprising that the use of industry money posed few if any ethical problems for the experts I talked to. This nutritionist may have spoken for many of my informants when he shared these thoughts:

NUTRITIONIST: Food companies are actively trying to build relationships with the government and doctors. A lot of money for public health comes from corporations.

SG: Do you worry about that?
NUTRITIONIST: It is neither good nor bad.

For him, accepting food company money seemed to be risk-free. The same was true for almost everyone I talked to. With the state both actively encouraging scientists to obtain support from commercial sources and leaving them few alternatives, doing so was deemed the height of ordinariness. It was necessary, commonplace, smart, and a point of pride for those who were successful in the money game. My informants were unanimous on this point:

BIOMEDICAL RESEARCHER: Everyone in China wants to get some support from companies. No matter if they work in the CDC or a university, given the high cost of research, if a researcher has an idea he must get support from a company to develop it and, even more, to bring it to market. Otherwise he can just do basic [simple] research. (paraphrased from interview)
PHILOSOPHER OF ETHICS: Everyone has to get money from the market.

So desirable was it to have corporate funds for science and health work that those whose organizations were too poor to attract the interest of profit-hungry businessmen secretly envied colleagues who were swimming in industry cash.

If securing and spending corporate money was ethically unproblematic, however, corporate bias of science was deemed wrong (Ethics-2). The ethical rules we have witnessed in operation—ILSI-China's no-commerce rule on company participation, for example, or the Chinese state's rule on the neutrality of think tank advice—reflect the underlying position that corporate influence on science is unacceptable, even corrupting. Yet while state opposition to industry biasing of science was very clear, the discourse of the state failed to recognize that the objectives of the two overarching party policies might be mutually contradictory, that is, that achieving one (industry funding via marketization) might undermine success in the other (scientific integrity to improve the quality of official policy via scientization). Perhaps because China's reform leaders were so deeply invested in both policy goals, it was politically impossible to acknowledge any conflict between them. But if the funding effect I described in chapter 1 is real—and much research suggests industry funding produces industry-friendly results by wide margins—those two positions on science ethics were indeed contradictory. Evidently, the international research on the funding effect that was a major source of concern in the United States was not part of the public discussion of science ethics in China. Moreover, aside from the very basic no-commerce rule, my informants reported that there were no concrete state regulations to keep corporate interests from tainting science funded by industry. As a result, the two incompatible principles and policies on science coexisted in uneasy tension.

Like scientists everywhere, the researchers I talked to were trained to believe science should be objective and unbiased. The state's minimalist and contradictory positions left them in a difficult quandary. They could not challenge the state, yet the state's silence about the ethical implications of corporate funding placed them in the unenviable position of having to rely on corporate money that quite likely biased their work, forcing them to violate their professional ethics. Facing this dilemma and with few official norms to guide their actions, the scientists had no choice but to innovate Chinese-style solutions to the general problem of knowledge corruption. As the leader of the chronic disease field, Chen Chunming was the innovator in chief, and so we start with her response.

Innovating Science Ethics: The Formal and the Informal

From our long conversation and Chen's many talks and writings, I was able to tease out a small collection of working ethical norms and practices that she dipped into. It had two parts. In the first, described in this section, she closely followed the *formal norms* set out by China's state (and by ILSI-Global). By formal, I mean not set out in a state document, but rather widely understood as applying to everyone and carrying sanctions for noncompliance. Those norms were fairly minimal, though. The many issues that were not covered would be handled according to *informal norms* guided by her own ethical sensibilities, which of course bore the imprint of the political culture. Potentially sensitive issues—such as funding arrangements, the crux of the ethical problem with corporate science—were placed in a zone of silence. In that zone Chen, protected by her status as a state-backed and morally upright leader who should be trusted not questioned, would become the keeper of the financial secrets to ILSI's success in fitting China's chronic disease science to corporate needs.[5] Working together, these practices placed ILSI's activities on the right side of the ethical line.

FORMAL ETHICS: FOLLOW THE RULES

ILSI's identity as a scientific organization and its legitimacy as a credible think tank depended crucially on its ability to create (at least the appearance of) solid, disinterested science and science-based policy advice. Its most important strategy for achieving sound science, we've seen, was to rigorously enforce the government's no-commerce rule allowing think-tank-type organizations to accept corporate funding as long as the companies don't benefit financially. As we know, though, ethics rules tend not to produce ethical behavior—which is what my informants were suggesting—but to allow deniability so scientific work can go on. Such rules are also subject to creative interpretation.

These assurances aside, sometimes the ethics rule did not quite fit the reality, producing uncomfortable junctures in conversations as my informants stretched the limits of creative interpretation. In the following conversation, J. S. Chen proved himself a master of flexible interpretation. I had asked him why the Swiss giant Roche Pharmaceuticals, maker of orlistat, had been willing to single-handedly fund virtually all the early work creating China's BMI cutoffs. Thinking of the history of Big Pharma in the United States, which after World War II sought to expand the market for drugs by defining disease precursor states (prediabetes, high cholesterol, and so on) as conditions needing treatment, I suspected that Roche was keen to have obesity defined as a disease in China—with low BMI cutoffs. Low cutoffs would maximize the number of Chinese who would be diagnosed with obesity and thus be in the market for the company's drugs.[6] Chen quickly rejected my implied scenario, which would have left an ethical cloud over his organization: "All ILSI members support all ILSI projects," he insisted with a touch of irritation. "There is no relation between the support of a particular project and a company's product development or marketing. This is in the ILSI bylaws." (Here he was talking about the standard annual payments by the supporting companies rather than these unusual Roche payments.) Yet then, perhaps recollecting the granular details of those payments, he began interpreting the rule more flexibly: "Roche's support for the BMI research is an exceptional case." Perhaps uncomfortable with that statement, which seemed to sidestep the ILSI-Global bylaw just mentioned, he changed directions, pointing out that Roche sold many other products, including supplements. The implication was that the firm's interests went far beyond obesity, and so its funding for the BMI work was not necessarily tied to its hopes for the obesity drug. In this case—and it was important, for it was Roche's funding that enabled ILSI to take control of the obesity issue—it was difficult to reconstitute reality around the no-commerce rule. Although making circumstances fit the no-commerce rule sometimes required considerable ingenuity, the very existence of the rule allowed ILSI's experts plausibly to deny any influence of corporate funding on their science and thus retain their confidence that all was well.

In these and other conversations, the main message my informants conveyed was that ILSI's activities were just business as usual. "The government has never had any problems with . . . ILSI's corporate support," J. S. Chen assured me. Mapping out the subtle politics of corporate funding for me, he continued: "One cannot say that the government *encourages* corporate funding, nor does it [proactively] *support* it." The most essential point came next: "You know, Susan, there is no criticism of research support by companies in China." In the delicate balancing act that is Chinese politics, what matters is that the state has not (yet?) targeted ILSI for surveillance and control ("there is no criticism"). There he was speaking of Ethics-1, telling me that the state promoted industry funding—as long as the recipient organization does not overstep an invisible line. As for Ethics-2, industry must not bias science, J. S. Chen explained; enforcement is in the hands of the receiving organization's leaders. Acknowledging the importance of the two Chens' guanxi ties to the health ministry, he added, pointedly, that the government "trusts that ILSI leaders will use the support well." In other words, the two Chens, with their stellar reputations, would vouch for, or personally guarantee, the correctness of the ILSI model of introducing international science using industry funds. In this system of state science, as long as the scientific activities that corporations are involved in *are overseen by trusted agents of the state* (the two Chens), there is no need for concern. Here is the magic of guanxi: anything goes as long as the leaders of the scientific field or organization have tight ties to important figures in the party-state. As field or organization leaders with the backing of the state, they operate on trust alone. There is no one overseeing what they do.

Another senior scientist underscored the importance of closely following the discourse established by the state. "Company support [of public health research] is not an issue in China," he assured me. "The assumption is that they [government agencies] have the rules and the company will do the right thing." In his telling, being ethical was simple: we have a rule, it provides plausible deniability, so we've done ethics. But he was also notifying me that the state was the ultimate arbiter of right and wrong, correct and incorrect. As long as it did not ban the practice of corporate funding of research, scientists could take advantage of the ambiguity by faithfully adhering to the ethics rule the government had laid down and then accepting company funds but doing so quietly. In operating that way, they would be safe from criticism and secure in their knowledge that their practices met an officially acceptable ethical bar. And by refraining from debating the issue among themselves or formulating shared professional norms, the scientists created a zone of silence and ethical secrecy in which each one could make decisions on the basis of their own culturally informed (and politically bounded) understandings of professional ethics. What was crucial was to stay inside the discourse or speech space of the state, and to constantly perform that correct act by stating that an important rule of ethical conduct exists and that they follow it to the letter. Adherence to this formal rule worked to legitimize ILSI-China's work and put the scientists' consciences to rest.

INFORMAL ETHICS: FOLLOW THE LEADER AND ASK NOT ABOUT FUNDING

Many practices were not covered by the formal rules, and here more informal ways of doing ethics came into play. These informal habits of acting ethically reflected China's hierarchical, leader-centric political culture. Although I was unable to observe these practices on the ground, it is useful to see which issues fell most clearly into Chen Chunming's zone of informal ethics. One was what the supporting companies received in exchange for their financial contributions to ILSI-China. Another was whether they ever tried to influence the science. Let's listen to her responses to my questions:

SG: Have there ever been . . . You know . . . you are using money from these big companies. ILSI has a very delicate role to play. Because you are using money from these companies but you don't want to be influenced by them. Have there ever been times when the companies have tried to exert influence?

CHEN: No. These companies do not try to influence ILSI. They are used to that. They know that there won't be commercial benefit for them. For example, we invited some Chinese companies to come [join us] as supporting companies. . . . We wrote a letter to them saying, if you want to be a supporting company of ILSI, we will have so much relationship. So they called us to say: "What's the benefit of joining?" We say: "No, you have no benefit [laughs]. You can only get scientific evidence from ILSI Focal Point." So, the Chinese companies are not coming to us. . . . They still don't understand the importance of the science for their companies.

SG: So why would any company want to donate the annual fee [of approximately $8,000]?

CHEN: They would do that because they can have this so-called evidence from ILSI Focal Point. They can use that. So they can report to their boss and say that when we join ILSI, we can get all these kind of benefits.

Here Chen responds to both questions by simply restating the formal norms established by the parent organization, ILSI-Global ("companies do not benefit," "the science is untouched"). In trying to stuff the realities of the branch into these preexisting boxes, she ends up making unpersuasive arguments ("they pay because they get evidence") and neglecting the more important question of how the norms are interpreted in practice.[7] For example, her joke about Chinese companies foolishly expecting something concrete from ILSI is meant to imply that foreign companies, by contrast, *do* understand the importance of science, and that explains why they join ILSI. The possibility that foreign firms could be supporting the branch to protect their products by fostering industry-friendly science and discouraging soda taxes cannot be articulated in the language of ILSI-Global. In simply repeating the official answers, Chen effectively placed virtually all questions that did not support the official narrative about the separation of business from science into that zone of unspeakability, whose borders I could not penetrate.

If ILSI-China's supporting companies were not supposed to materially benefit from their contributions, the whole matter of corporate funding— why the companies contributed, how it was arranged, and so forth—raised a multitude of ethical questions. Chen seems to have handled them by relegating them all to that zone of silence and secrecy, where she would be protected from intrusive questions by her status as a state-supported leader. She would also be protected by the workings of *Confucian virtue ethics*, which governed relations with those subordinate to her in the social hierarchy. It is here, in these practices of ethical leadership and followership, that the hierarchical dynamics of science mentioned in chapter 5 were most visible. Named for the ancient sage, Confucian virtue ethics is a set of cultural understandings in which relationships are hierarchical, leaders are expected to be virtuous and govern ethically, and followers, recognizing the leader's moral rectitude, are supposed to reciprocate with loyalty and obedience.[8] In this ethical system, Chen, the morally upright leader, was owed deference and respect—not doubtful questions about how she did things, especially concerning sensitive issues like funding.

These understandings about how followers should interact with their leaders appear to have operated seamlessly to protect Chen from nosy questions. Few of the scientists who had worked with ILSI over the years had any understanding of how financial matters were handled. When I asked, most sent me to Chen to find out. ILSI staff who had helped organize conferences "only know how the money was spent, not how it was obtained." These informants stressed to me that, as a good paternalistic leader, Chen took care of them, paying staff well and covering all the taxi, food, and other expenses of experts invited to help with ILSI events. The big picture remained obscure, however. Some senior scientists who had worked for many years in the nutrition field knew that ILSI was funded by large multinational food corporations, including Coca-Cola, but they knew none of the details. (It is possible they knew more than they were willing to divulge.) A senior nutritionist who was based in one of ILSI-China's supporting companies was tight-lipped in the extreme. She had just explained that ILSI gets an annual fee from its supporting companies. When I then asked how ILSI keeps those companies from influencing the science, she shot back impatiently: "I don't know, ask ILSI!" Did she truly not know, or was she unwilling to reveal that she knew? Or was it improper for her to tell me something only Chen Chunming, the leader, was authorized to say? It's impossible to know, but it's hard to escape the conclusion that the two Chens worked hard to keep the details to themselves. ILSI-Global's annual reports and ILSI-China's newsletters maintained the silence, listing the names of member and supporting companies (respectively), but no information on funding levels or uses. If no one knew the details, perhaps the logic went, no one would be able to spill ILSI's secrets.

Chen gained the loyalty and deference of her followers not only from her status as a virtuous paternalistic leader; equally important was her impressive position in the political and scientific hierarchies of the country. By pulling apart the tangled politics of status and hierarchy in Chinese science, we can find another clue to how ILSI's secrets remained so well hidden. From conversations with ILSI experts, I discovered that Chen did not volunteer details about ILSI's funding and others did not feel the need to know them. They also felt they had no right to ask since Chen, having obtained the funding, had the right to manage it in the way she saw fit. But the workings of hierarchy and position were more complex than that, for in China's political system, what followers should and should not know is largely determined by the leader. One senior researcher in a top university called Chen his "teacher" (*laoshi*), noting that "I did not dare ask, it was awkward to ask" (*bugan wen, buhao wen*).[9] Beyond those in the research community, Chen was held in high esteem by unnamed others who reportedly sought to use her famous name to support their health-promotion work. The reasons were many and varied. People respected her not only because of her political status ("she was the former leader of the CDC"), but also because of her scientific prowess ("she was a real expert in the field of basic nutrition"). Not only did she have the best education in nutritional science of her generation; because of the Cultural Revolution, those in the following generation had little chance to acquire formal expertise, leaving Chen with no rivals for expertise-based leadership in the field. And there was more. Chen was also respected because of her organizational skills ("very capable"), her vision ("exceptionally broad"), and even her energy ("young people could not keep up"). As an esteemed scientist, a high-level government official with the backing of the state, and the head of an influential think tank, Chen was a person of undeniable status, and so people seem to have trusted her to do the right thing. Even if Chen's colleagues did not fully trust her, they most likely feared what might happen if they questioned her. In that hierarchical political culture, two younger generation informants explained, to challenge a superior—to call them on accepting a big bribe, for example—is to challenge the whole system and risk one's future career within it.[10] Their only real option was to remain silent. In the stratified world of Chinese health science, it was neither acceptable *nor safe* to ask how things were done. And so the financial secrets behind ILSI's role in corrupting China's science of chronic disease remained hidden from all but a tiny handful in the know.

Virtue Narratives

Just as ILSI-Global's American advisers had created a narrative of themselves as ethical scientists, so too did ILSI-China's leaders articulate a narrative about their status as virtuous scientists. If for the Americans the ethical scientist was one who pursued the truth without bias, for these Chinese the good scientist was one who used science to serve the state and society. In Chen's account, she was the good scientist who was global minded, selfless, and a patriot dedicated to the nation's health. Interlinked with this—indeed, coconstituted with it—was a second story about the good company that had similar features: it was global, benefited the nation's health, and gained little in return. Working together, these two conceptions served as overarching *virtue narratives* that shaped the larger discourse, setting out what was permissible to say about ILSI and its supporting firms.

THE GOOD SCIENTIST

In her many communications, oral and written, Chen presented a clear picture of the good scientist she sought to be—and sought to be seen as. Three qualities stood out. All reflected values widely shared in the political culture and priorities in the state's reform agenda. First, presenting herself as globally oriented, Chen strived to move China out of its backward slot in the global hierarchy of science and into the ranks of nations producing top-quality science. This aspiration, which echoed the reform state's globalizing agenda, was a recurrent theme in Chen's writings and presentations, and I discussed it in previous chapters.

The good scientist was also a selfless scientist who devoted her energies to bettering her country with no expectation of material benefit. Hard work, plain living, sacrifice for the common good—these were enduring themes in party thought, and Chen's public performance of them in her everyday routines would resonate widely in her country. Although she paid ILSI's office staff from supporting-company funds, neither she nor J. S. Chen accepted a salary. (As an academician [*yuanshi*] of the Chinese Academy of Engineering, J. S. Chen was entitled to receive generous financial support from the government for life.)

Stories of Chen Chunming's exceptional selflessness and public generosity circulated in the public health community and among some of the Western experts I talked to. One American described her "modesty in dress and lifestyle." A Chinese colleague told me Chen was able to get scientists to contribute their carefully guarded, "private" data to the collective task of defining BMI cutoffs because people trusted that she would not use the data for personal benefit. In another account, Chen used personal funds to create ILSI-China: "I heard that actually an American supported Chen with much money for her own research, several tens of thousands of dollars. . . . But she used that money to build ILSI Focal Point in China." Although I was unable to corroborate this account, which seems improbable on the face of it, what's important is not its truth value but its circulation as a public narrative, which gave it an aura of truth. Another longtime colleague, who himself was a highly respected public health scientist, confided earnestly: "It is not easy to devote one's whole life to the cause of science, as Chen has done. I respect and admire her very much."

Finally, Chen's good scientist was a patriot who served her country by dedicating her life to solving critical problems of the day. Through strategic exploitation of global corporate resources, the story went, she overcame the state's inertia on chronic disease, helped it create policies and programs to address those conditions, and then spread the message to the broad public. As readers know, this is a simplified and sanitized version of the truth. Among other things, it omits ILSI's marginalization and exclusion of other scientists who were addressing obesity well before she did. In Chen's appraisal of her life's work, we can hear clearly a notion of science with long historical resonance in China. The aim of science was not the pursuit of the truth, but the development of practical tools to benefit the country in ways stipulated by the state.

THE GOOD MULTINATIONAL

Entwined with this was a narrative about the good company that was global, contributed to the nation's health, and received little in return beyond scientific evidence. The two figures not only had similar attributes; they were coconstructed, for the good scientist could not perform good works without the help of the good company and vice versa. Reflecting the party's globalizing agenda—which celebrated opening to the global market as vital to China's prosperity and foreign firms as key contributors—as well as the feeling that many local Chinese companies were untrustworthy, the good company was almost always foreign. (These interviews were conducted at the end of the first full year of Xi Jinping's administration, before the development of more aggressively nationalistic, China-first, and anti-West political sentiments.)

My informants shared this notion of the good company and saw Coca-Cola as a sparkling example.[11] Coca-Cola as a "good company" that favors China's interests over its own? American readers are likely to balk at that. But such views make sense if we consider how China saw itself in relation to the United States and the West more generally. Echoing the pervasive narrative about the backwardness of the country's public health science relative to that of the advanced West, some of my informants expressed appreciation for Coca-Cola's efforts to tackle the obesity problem in China. One researcher pointed to Coke's catchy slogans and clever ideas for exercise programs. Accepting the promise of corporate social responsibility at face value, several experts I talked to presented China as a lucky beneficiary of such programs. Reflecting the national glorification of science, my informants expressed faith in the quality of the foreign firms' health-related products, because their design was "based on scientific evidence." From much-needed funding to new technology, knowledge, knowhow, and even an ability to get things done, the researchers saw many benefits flowing from corporate investment in public health work. No one mentioned possible costs. A doctor in a big central Beijing hospital, for example, expounded on the benefits of big companies promoting their products at the door to conferences: "We get experience and knowledge from these big companies," she said enthusiastically, not considering that the science being presented in the meeting room might be subtly tilted in favor of the company's products. In our conversations, I kept pushing back on the idea that Western firms were giving money with no strings attached. Some sought to even the cost-benefit score by reminding me that acts of "giving back to society" boosted corporate reputations. Despite repeated questioning, though, I detected little wariness of potential harm or risk that might come from corporate funding of science—to health science, to the policies based on it, or to the public's health. Reflecting the wider pro-Western culture of the pre-Xi era, my informants saw multinational firms as largely free of corruption, trustworthy, and, given the shortage of state funding, essential contributors to China's public health work.

Did Chen Chunming and J. S. Chen secretly suspect the food industry was tilting the science? We can only speculate on that core question. In one view, the abundance of incentives provided by ILSI worked to make them see ILSI's arrangements as nothing other than beneficial. Just as Hill and other advisers to ILSI-Global had been incentivized to see the corporate funding of science at ILSI as advantageous but never risky, the two Chens had enjoyed ample professional benefits from their association with ILSI; it was clearly in their interests to *not* perceive problems. At the same time, the long-standing narrative of China as backward and the US as advanced made it easy for them to misconstrue problems by giving them an attractive alternative framing of things. In that uplifting story, their scientifically backward country, with the help of the generous foreign firms and their scientific nonprofit, was on its way to becoming an international leader in the field of chronic disease management. Working in harmony, the incentives and the narrative led ILSI-China's leaders to misperceive the larger dynamics of which they were a part.

Yet there is evidence in the email archive that they saw things only too clearly. In mid-2015, Malaspina was beside himself with worry about the WHO director-general Margaret Chan, who was accusing Big Soda of contributing to rising obesity rates among children globally and was backing regulations to restrict consumption of full-sugar soft drinks. Sharing his concern with ten close associates—including Chen Junshi—he wrote: "I am suggesting that collectively we must find a way to start a dialogue with Dr. [Chan]. If not, she will continue to blast us with significant negative consequences on a global basis. This threat to our business is serious." Malaspina then asked J. S. Chen for his advice and help in getting the WHO leader to back off Big Soda. (By this time, Chen Chunming was in the hospital with the brain tumor that eventually claimed her life.) Although this is the only instance I found of direct arm twisting of ILSI-China's leaders, it leaves little doubt that the two Chens were fully aware that when you worked for ILSI, Big Food was in charge. In an ILSI still under the sway of Malaspina fourteen years after his retirement, industry worries about profits overrode the policy directions established by one of the world's foremost health authorities.

These narratives about the good scientist and good company structured the broader ethical discourse surrounding ILSI by conveying what should and should not be said by members of this community. Outcomes that accorded with the narratives were to be stressed and praised. Observations and interpretations that did not fit or that challenged the narratives were less likely to be heard, at least in public. The result was that some larger truths about China's obesity science and policy remained inaudible, placed together with questions of money in a zone of silence. One such pair of truths that remained publicly unspoken in this community was that foreign firms often *do* benefit financially from funding chronic disease work, and that funding by interested companies inescapably taints the science.

The Silent Scream: Companies Absolutely Corrupt Science!

Though criticisms of industry influence on science had been pushed underground, by asking probing questions I discovered that concerns about the corruption of science were present, churning just below the surface of polite conversation. Three senior scientists who had lived through an earlier era when money did not dictate everything shared their worries about the perversion of science by talking about problems they had seen, or tried to avert, in their own careers. After the many interviews with scientists close to ILSI who, tight-lipped, said only "no big deal" or "not an issue," I was unprepared for the anguish my questions would provoke in these eminent scientists, who knew in their hearts that China's chronic disease science was being corrupted but were powerless to stop it.

I usually started this part of the conversation with a provocative question that would likely have disturbed familiar ways of thinking: "Do you think China's public health science has been marketized [*shichang huale*]"? There was no need for me to explain; everyone knew I was asking about the entanglement of science with market forces. As these informants—a head of a professional association, an educator, and a university-based researcher—struggled to come to terms with that unorthodox, faintly reprehensible idea, each filled in a different piece of the puzzle of the potential problems associated with corporate funding of health science.

THE ASSOCIATION LEADER: "IT'S IMPOSSIBLE TO SAY NO"

The first of these eminent scientists welcomed me into the sitting area of a cozy, bookshelf-lined office for what would turn out to be one of my most memorable conversations in Beijing. For many years he had led a major association in a related scientific field that, like ILSI, had little choice but to rely on corporate funds. Unlike Chen, who maintained that following the government's simple no-commerce rule protected ILSI science from corporate influence, this scientist-leader believed that it was necessary to defend against inevitable company attempts to exert influence by proactively creating clear structures, guidelines, and norms clarifying the nature of allowable company involvement.

I began with an innocuous question: "Does the government encourage corporate involvement in public health?" Yes, he said with a knowing look, before launching into the history of his association and how he defended against untoward influence:

> [For many years], as the head of a professional association, I took the initiative to attract company support. We had very strict rules [on the use of these funds]. . . . Academic people decided how to use the money. Only in leading small groups [*lingdao xiaozu*] were there company representatives, but decisions were made by the majority, not by the companies. Also [importantly], I never served as adviser [*guwen*] for any company. I never took money or accepted gifts from any company. The government gave me a salary. My lifestyle has been very simple, but I feel very happy. People under me followed the leader; if I didn't take money, they wouldn't dare do it, either.
>
> There were no government regulations covering any of this—support of associations, how to use the money, where it should come from, what to tell the public—so we had to decide how to manage things by ourselves. [In general] people do what they feel most comfortable with. But if you don't do things properly, you will be criticized. The leader must be careful and consult widely.

"So, is public health research being marketized?" I asked. Looking chagrined and unhappy, he replied with feeling: "Whenever you take money from a company, it will definitely affect the scientific character of your work. These days a lot of people use their personal power to benefit themselves with gifts. I never did that. But that stance is impossible today. No one can say no to corporate support." At this point, he became agitated, abruptly stopped the interview, and, perhaps realizing how much he had revealed, begged that I not use his name in connection with those remarks.

THE HEALTH EDUCATOR: "DON'T ASK ME THESE QUESTIONS!"

I also raised the question of marketization with a senior health educator/researcher with no connection to ILSI. We were meeting in a big, empty classroom, and the atmosphere was a bit tense. He had strong opinions he wanted to share, including that China "must not follow the road to fatness taken by Europe." My question about a topic he had not prepared to talk about caught him by surprise. As he tried to sort things out in his mind, he jumped from one concern to another, in the process conveying a good idea of the many and diverse issues the question of corporate influence raised for the practicing scientist:

> As a teacher I must speak the truth. You should not cooperate with companies. I'm in health education; we must put education in a holy position; we cannot tell a lie or speak falsehoods. [We must be] very careful.
>
> If a professional association takes money, it can't criticize the [company] providing it, although that source may have some influence on it. There's an old saying: "If you eat another's food, you need to say good words about them." If scientists take money from capitalists, they aren't scientists!
>
> [Ultimately,] how much the company will influence you is related to your own ethical standards. But not everyone thinks that way. Commercialization is the trend. In the US, money is first; it's impossible to ignore money. In China, there is also this tendency. I don't have much experience, though. Taking project money from companies—and especially having too many connections [guanxi]—is very dangerous for the scientist.

Carried along by his engagement with the topic, I pushed further: "So, then, where do Chinese researchers draw the line between acceptable and unacceptable forms of corporate support?" At this, he blew up, shouting that he had prepared to talk about obesity, not corporate influence: "Don't ask me these questions!! I don't know how to respond. You are off track! I agree with your idea but I have no evidence. You should ask the politicians and managers. They will know."

THE UNIVERSITY SCIENTIST: "THE GOVERNMENT NEEDS TO STEP UP AND CREATE GUIDELINES!"

All the thorny problems that came bubbling to the surface in the interviews—the frustrating lack of information, the need to devise ad hoc solutions to impossible problems, and so on—were rooted in one thing: the government's lackadaisical attitude toward industry regulation. On this everyone agreed: there were virtually no regulations, so everything was chaotic. Indeed, aside from the simple rule that scientific think tanks and associations using corporate money must not advertise the donor's products, the government had issued virtually no guidelines on the use of these funds.

I arrived early for my interview with a prominent obesity scientist at a major university and was ushered into a large and busy academic office with research assistants working diligently in the corner. Their presence added to a general atmosphere that was lighthearted, friendly, and full of laughter at the creative mistranslations between Chinese and English. With decades of experience sitting on standard-setting panels, serving on working groups, and attending conferences—all funded at least in part by industry—my informant had developed strong views about these matters and, as an academic, was eager to share his thoughts.

After a long conversation about the development of the obesity field in China, I launched into the questions about corporate influence: "Could we say the field is marketized?" Interrupting me before I finished, he said: "This is the first time I've heard this term, but it's a fact. They [companies] influence experts and government officials. They don't talk about making money; they talk about science and technology." He sighed loudly. "Their influence is gradual and indirect, not immediate and direct."

Giving a nod to the benefits of corporate support, I then wondered: "Are there also risks?" The response was vehement: "Definitely! Companies have their own ways of thinking. Take the case of nutritional standards. The government solicits the companies' views, and they are not supposed to have voting power, but they have tried to influence the standards—on milk, for example. I believe the influence is very big. Corporate support influences the research results, the whole research design, the numerical analysis—all of these. Definitely."

"Well then," I continued, "is there any concern that companies might bias or distort the science?"

Sounding frustrated, he replied, in effect, "Hardly":

Not many people complain about company influence on scientific results. Some people are concerned [*guanxin*], yes, there are some, but their voice is very small. China is unlike the US, where this is very sensitive. Risks—definitely there will be. In this area of management, [things are not good], China has not caught up to the US. There are no written regulations. Nor is there concern. [In fact,] there is no conflict-of-interest concept in China! Chinese are relatively casual [*suiyi*] about this.

The [government's] requirement that organizations accepting corporate funding just not advertise products—that is too low a standard![12] Maybe you can suggest [*fabiao*, literally issue] some policies for the Chinese government, because we are too weak [*cha*] in the area of corporate influence. I hope our leaders can hear this criticism, and can then [develop] some management [practices] that are stricter. Slowly, with modernization, government officials are going abroad to interview [and see how things are done]. Under Xi Jinping, things are a little better now.

The university scientist had finally gotten to the root of all the nightmarish problems people had described in the interviews. Without firm government leadership and clear guidelines, the companies (and unprincipled, or just underpaid, scientists) can never be reined in or held to account. China's public health scientists—from Chen Chunming and the others at ILSI, to all those I talked to in universities and hospitals around Beijing—will continue to do their best to fill the gap with personal ethics, but until the government draws the line and creates clear and comprehensive rules, little is likely to change. In an authoritarian political system, only the authorities could resolve the ethics problem plaguing corporate science.

The Dangers of Business as Usual

This up-close look at the ethical struggles in chronic disease science shows us how members of one community of scientists tried—with very partial success—to cope with the dilemma created by the failure of an authoritarian state to create either ethical guidelines or practical regulations on the corporate funding of scientific research. Their dual approach to the ethics of public health science was full of hazards—for the scientists and for China's people. With so many things rendered unsayable and so many alternative truths marginalized, important questions became unaskable. In a world of huge global firms pursuing profits with little regard for human health, the approach to ethics ILSI's China branch developed over the years seems to have left these researchers vulnerable to the hidden manipulations of corporate science, heedless of the possibility that an esteemed American company might be working quietly and systematically behind the scenes to shape China's science and policy to its own advantage, to the detriment of the health of China's people. As a result, corporate interests prevailed, and the corporate imprint on China's chronic disease policies and practices remained unsuspected and undetected.

As my informants constantly reminded me, this corporation-dominated structure was just business as usual in public health—not only in China, but in the United States and around the world. They certainly had a point; even the WHO has been subject to food-industry influence.[13] Yet China was—and still is—different in a critically important way. Unlike in the United States, where a freewheeling democracy has encouraged the emergence of watchdog organizations who protest when corporate greed threatens human health—the reaction to the Coke/GEBN scandal is a prime example—in China, there seems to be no one protecting the public's health from corporate overreach. While I did not investigate the relations between Coca-Cola and the Chinese party-state, ample evidence suggests that Coke has maintained a good business profile in China in part by cultivating ties throughout the official health sector. Today (late 2023) COFCO Coca-Cola Beverages Ltd., a joint enterprise of the state-owned COFCO Foods and Coca-Cola (65 and 35 percent ownership, respectively), operates twenty bottlers in nineteen provincial markets covering 81 percent of the geographic area of the country.[14] With this sort of tie, the state is unlikely to aggressively protect the public from corporate overreach by Coca-Cola.

Nor are other institutions available to do that work. Since Xi's rise to power in late 2012, NGO and press freedoms have been quashed.[15] Power has been increasingly centralized in the party and its leader. Threats to the party's power have been routinely suppressed, and the role of party organizations in civil society groups, private companies, and foreign firms has expanded. In this context, the existence of independent social watchdogs critiquing a quasi-governmental organization like ILSI-China is scarcely imaginable. In China today, companies like Coke, those with mutually beneficial connections to the state, appear able to wield inordinate amounts of indirect, subtle, and deliberately hidden influence that, without in-depth sleuth work of the sort reported here, would never come to light.

These findings force us to ask who, after all, is corrupt? Following the media, many Americans see Chinese science as especially prone to being crooked, bent this way and that by the force of money and connections in the political culture. American media delight in sensationalizing stories about scientific corruption in China: the factory-scale fake papers, the fraudulent peer reviews, the use of CRISPR to produce the world's first gene-edited baby. Such stories play into broader Western binaries in which China always ends up at the bottom: corrupt not honest, poor not rich, unfree not free. The story told here turns that binary upside down. American scientists working for a US-based institution created a corrupt form of obesity science, transported it to China, and helped get it endorsed as the authoritative science of obesity. Raw exploitation played a role too. The American researchers took advantage of their privileged status in the global hierarchy of science by exploiting the yearnings of colleagues from the Global South to climb the ladder of scientific success by emulating the West / the best, a label the Americans routinely emphasized. In America, a relatively open, democratic system of science allowed critical voices to root out the corruption and bring down the science. In China, a hierarchical, state-centric system of science did not allow open criticism. As a result,

China, the unwitting (if cooperative) recipient of corrupt knowledge, bore the costs. In this deeply researched case of corporate science, the primary responsibility for this wrong belongs to the American companies and the scientists who helped make their dream of scientific empire come true.

∴

In the United States, the fall of soda-defense science in late 2015 brought a grand reckoning. In China, it was met with a stark silence. What then has happened to the soda science that was built into Chinese understandings and practical approaches to obesity? Was it somehow plucked out, idea by idea, and removed? Or is it still there, lurking beneath the surface of Xi Jinping's signature program Healthy China 2030?

Soda Science Lives On

A Policy Brief

The news of Coke's massive funding of the GEBN that created such an up-roar in the United States in 2015 was largely ignored in China. And so ILSI's prescriptions for obesity policy—*exercise-first (soda-tax-never)*—have lived on, hidden out of sight, in the next generation of Chinese policies. Since the early to mid-2010s, China's efforts to combat obesity have increasingly been folded into a larger national initiative on chronic disease prevention and control, in which obesity is addressed as a disease and as a risk factor for other serious diseases. Although my in-depth research ended as of 2015, I conducted a systematic review of policies China adopted during the next five years (2016–20), looking for evidence of a lingering impact of energy balance ideas. The results—presented in the first part of this coda—reveal how the soda science introduced between 2004 and 2015 became the foundation of a wide array of state policies that are likely to be in place for a very long time. These findings can be read as evidence of the genius of the Coke/ILSI strategies of spreading soda science around the world.

The findings can also be read as a warning of the dangers facing one of the soda industry's most important target countries. In 2018 Chen Chunming, who had led China's work on obesity, passed away, and her organization's influence on chronic disease policy has waned as other actors have entered the policy arena.[1] The second part of the coda introduces a newer group of Chinese obesity experts who today are calling out the food industry for making outsized contributions to the epidemic of diet-related chronic dis-eases, and calling on their government to take action to rein it in and protect the public's health. Their work provides one answer to the critical question of how the distortions introduced into the science and policy of chronic disease by the global food industry might be undone, if the state were on board and the politics were properly aligned.

In chapter 5 I suggested that this in-depth story of science and policy gone awry would open a window on some fundamental flaws in China's governing system. In the third part of this coda I spell out those flaws and their implications for China's rise to scientific superpower status.

The Long Afterlife of Soda Science in China

A series of developments in the mid-2010s set the stage for a broadened approach to chronic disease. In 2012, China issued its very first national guidelines for chronic disease control, the 2012–15 National Plan on Chronic Disease Prevention and Control.[2] A few years later, in 2016, at the most important national meeting on health in twenty years, President and party head Xi Jinping gave an important speech declaring the health of the nation a top-level political priority.[3] Recognizing that the rising tide of chronic diseases was affecting not just the nation's health, but also its economic development and social stability, Xi announced that health must now be integrated into all policies and introduced the national strategic program known as Healthy China 2030. Issued by the party's Central Committee and the governmental State Council in October 2016, the Healthy China 2030 Planning Outline was the first medium- to long-term strategic plan in the health sector developed at the national level since the founding of the PRC. Between 2020 and 2022, COVID-19 pushed virtually all other health concerns off the agenda. Since then, chronic diseases, by far the most important cause of disease and death in China, have been getting serious high-level attention again.

Over the next few years (2016–20), the Chinese government issued a series of important policies, plans, and measures that sought to address interconnected dimensions of the crisis of chronic disease (the major ones are listed in table coda 2.1). Rather than distinct policy initiatives, these documents referred to, built on, and supported each other, fitting together into a multifaceted but relatively integrated approach to addressing the rapid rise in obesity and related conditions. Because ILSI had shaped the first generation of policies on diet-related chronic disease, and health policy was cumulative, with the second generation building on the first, the organization left its fingerprints on every one of these measures. In this way, ILSI's signature contributions to China's obesity policy—which ultimately served the interests of the global food industry—came to be incorporated into China's chronic disease policies at the highest level. ILSI's imprint today can be seen in four dimensions of the policies adopted between 2016 and 2020.

TRADEMARK PROGRAMS: HLFA AND EIM

ILSI's most direct influence on current policy can be seen in the incorporation of two of its trademark programs into China's national policies: the

TABLE CODA 2.1. Main Policies on the Prevention and Control of Chronic (or Noncommunicable [NCD]) Diseases in China Enacted 2016–20 (Includes First Plan for the Prevention and Treatment of NCDs, 2012–15)

Date Issued	Name	Issuing Agency
May 2012	China National Plan for the Prevention and Treatment of Chronic Diseases (2012–15)[1]	Ministry of Health of the PRC
May 2016	Dietary Guidelines for Chinese Residents (2016)[2]	Chinese Nutrition Society
Oct. 2016	Outline of the Plan for Healthy China 2030[3]	Central Committee of the CCP and State Council
Jan. 2017	China Medium- and Long-Term Plan for the Prevention and Treatment of Chronic Diseases (2017–25)[4]	General Office of the State Council
June 2017	National Nutrition Plan (2017–30)[5]	General Office of the State Council
July 2019	Healthy China Action (2019–30)[6]	Healthy China Promotion Committee, National Health Commission

1. Ministry of Health and Other 15 Departments. "Notice from the Ministry of Health and Other 15 Departments on the Issuance of the 'China Chronic Disease Prevention and Treatment Work Plan (2012–15).'"

2. Chinese Nutrition Society, Dietary Guidelines for Chinese Residents (2016).

3. "Central Committee of the CCP and State Council Issue the 'Outline of the Healthy China 2030 Plan.'"

4. General Office of the State Council, "The General Office of the State Council's Notice on the Prevention and Treatment of Chronic Diseases in China, Notice of the Medium and Long-Term Planning (2017–25), State Council General Office Document (2017) No. 12."

5. National Nutrition Plan (2017–30), in General Office of the State Council, "General Office of the State Council Issued."

6. Healthy China Promotion Committee. "Healthy China Action (2019–30)."

Healthy Lifestyle for All Action, inspired by America on the Move (AOM) in the United States, and Exercise Is Medicine (EIM), generously supported and widely promoted by Coca-Cola. Both national plans to prevent and control noncommunicable diseases (2012–15 and 2017–25) take the promotion of "healthy lifestyles for all" as a core strategy. If the current plan is successfully fulfilled, by 2025, 95 percent of China's counties and districts will be carrying out annual healthy lifestyle campaigns.

Virtually all the policies to curb the rise of chronic disease also strongly encourage the promotion of physical exercise. Three of the major measures call for the EIM goal of "integrating physical activity and sports into medical care" (*tiyi ronghe*).[4] The Healthy China 2030 Plan, for example, urges the creation of an innovative mode of disease management by combining sports and exercise with medical care. With this, the EIM program, which Blair was instrumental in creating in the 2000s and that ILSI brought to China in 2012, has become a cornerstone of Chinese national policy. As noted many times, these activity-promotion programs are not "bad" in themselves; to the contrary, they advocate changes in lifestyle behaviors beneficial to the prevention of many chronic diseases. Yet they form part of a larger package of policies that place the onus on individuals and may tilt the country's approach toward exercise over other effective approaches.

OBESITY CONCEPTS: ENERGY BALANCE, EATING-MOVING EQUILIBRIUM

ILSI's (and Coke's) influence can also be seen in the conceptual framework that undergirds China's policies on chronic disease. With all the policies I reviewed calling for promoting "healthy lifestyles," and with "healthy lifestyles" routinely defined as "balancing eating and moving" (*chidong pingheng*), the energy balance framework at the heart of ILSI-Global's corporate science of obesity forms the conceptual foundation of China's policies as well. Keeping eating and moving in equilibrium is not only the essence of a healthy lifestyle; it is also the strategy for achieving a healthy weight. As the 2016 Dietary Guidelines instruct: "A balance between energy intake and energy expenditure is the key to maintaining a healthy weight. People of all ages should know how to select foods for healthy eating and should exercise every day to maintain energy balance and a healthy weight."[5] The approach long promoted by Hill, ILSI-Global, and the food industry is now the official Chinese strategy for maintaining a healthy weight.

While this formula for weight control may sound eminently reasonable, it rests on the assumption that all calories ingested through eating are equal, that is, that the macronutrient content of the foods we eat—protein, fat, carbohydrates (especially sugar and saturated fats)—does not matter. Nor does the amount of processing. Eating hot dogs is little different from eating broiled chicken (though one would have to eat two medium-sized servings of chicken breast to get the number of calories in one hot dog with bun, a bit over three hundred).[6] It also assumes that eating less and moving more are equally important to weight management. Or, as the food industry has long coached us, calories from unhealthy, highly processed foods can be quickly exercised away. Few experts support these positions. There is also the practical question of how citizens are supposed to measure calories consumed and expended. Without some kind of very convenient calorie-counting tool, it's hard to see how the approach can be successfully implemented. And then there is the question of whether we can adequately control our energy expenditure through exercising, an issue I return to in the conclusion.

The national-level plans and actions introduced since 2015 map out comprehensive strategies to reduce the burden of chronic disease through concerted efforts at the environmental, community, family, and individual levels. These policies spell out in great detail the measures needed to improve both diet and physical fitness. Given this abundance of recommendations on both sides of the equation, it would be meaningless to try to assess whether exercise is emphasized over diet, as was the case under the exercise-first solution promoted earlier when ILSI was a major force behind chronic disease policy. Nonetheless, a close study of the plan targets suggests some interesting observations. In the Medium and Long-Term Plan for the Prevention and Control of Chronic Diseases (2017–25),[7] for example, eleven of the sixteen main planning goals deal with specific chronic diseases (prevalence, medical management, or death rate). A handful deal with healthy lifestyle behaviors. While one of these is for overall activity levels (regular participation in physical exercise), the only indicator for diet deals with daily salt intake. Limits on the consumption of oil and sugar are *recommended* as part of China's health education and promotion efforts— especially under the new slogan *sanjian sanjian* (three reductions and three healthies: reduced salt, oil, and sugar; healthy oral cavity, weight, and bones)—but not *required* as an indicator of successful plan fulfillment.

MARKET ORIENTATION, INDIVIDUAL RESPONSIBILITY

ILSI-China's approach to chronic disease was heavily market oriented; so too is the Chinese government's. Official policies on chronic disease call for market-based solutions in which individuals are responsible for chronic disease prevention and control, industry is to be stimulated not regulated, and the state's role is to further develop the market and educate the public, not introduce structural changes that limit industry's freedom.[8] The plans envision a "decisive role for the market" in the allocation of nutrition resources and provision of services[9] and call for "rapidly developing" or "significantly expanding" the food, nutrition, and health industries.[10] Assuming that education alone will produce significant behavioral change, all the policies and plans give enormous weight to popularizing health-related education and include raising the level of health literacy as a core indicator of successful implementation. Some measures take the notion of personal responsibility further, "advocating the concept that 'everyone is the first responsible person for their own health'" (*changdao "meige ren shi ziji jiankang diyi zeren ren" de linian*).[11] Of course, a market-friendly, neoliberal-type orientation of this sort—in which the state limits its support for social welfare, the market is seen as the source of solutions, and individuals are made responsible for themselves—was not just ILSI's orientation; it was China's predominant approach to governing its society in the era of reform, reflected in the over-arching policies of marketization and globalization. As the leading force on the obesity issue for the first decade-plus of its policy life, however, ILSI-China had played a decisive role in championing and normalizing the market orientation of chronic disease policy.

STILL NO SODA TAX

The ultimate goal of soda-defense science was to protect sugary soda from government regulation, and on this front ILSI and Coke appear to have had lasting success in China. Reflecting the intensifying concern in the international community about the health effects of added sugars, the measures adopted after 2015 recommend that individuals limit the intake of added sugars (usually to no more than 10 percent of daily caloric intake) and urge food companies to reduce the sugar content in processed foods. But there are no teeth to these recommendations, no taxes, warning labels, marketing bans, or other policies shown to have impact elsewhere. Instead, these recommendations give individuals and other elements of society primary responsibility for limiting sugar in the Chinese diet; the government's role is to guide these behaviors by mobilizing society and educating the citizenry. Despite strong calls from the WHO for governments to tax soda and regulate the marketing of sugary drinks and energy-dense foods, hard-hitting governmental policies of this sort were still missing from China's policies in 2020, and in December 2023, when this book went into production.

Chinese chronic disease specialists I spoke with in the late 2010s expressed little surprise. One, choosing his words carefully, described the government's attitude toward global fast-food and beverage companies as "relaxed." Another researcher who had worked closely with ILSI and the CDC in China, when asked whether official thinking was moving toward taxing soda or regulating food advertising for children, shot back: "No way! The government does not even regulate tobacco, the number 1 killer; how could it regulate soda?" A Chinese nutrition researcher, writing in 2023, put the soda tax issue in larger context. On public health issues like smoking and unhealthy lifestyle practices, she wrote, "China['s] government would only use mild education or noncompulsory policy. . . . [It] would not issue any hard policy on industry, *for the sake of economic growth*" (email from Yixi Wang, May 8, 2023; emphasis added). All the hype around Healthy China 2030 notwithstanding, economic growth still matters more than human health.

Listening to the *Lancet* Ten: Avoiding the American Road to Being Rich and Fat

If there was one thing my informants were adamant about, it was that China *must not follow the American road to obesity*, the road that led to a rich society in which over 70 percent of adults carry unhealthy amounts of fat on their bodies. But by the time my Chinese confidants shared that concern with me, American science—at least one industry-friendly, US-made corporate science—had already been built into their country's policy. Without more effective measures, especially measures that weaken the influence of soda science, China may not be able to find the exit ramp.

By the early 2020s, a new group of critical obesity experts had begun to speak out strongly about the role of junk food and sugary drinks in the relentless rise in China's obesity rates. In mid-2021, the *Lancet* published an important three-part series on obesity in China.[12] Warning of a deepening public health catastrophe, the articles, by ten leading voices in China's obesity research community—hence my name for them, the *Lancet* Ten— issued an urgent plea to the Chinese government to finally get serious and tackle this slow-motion disaster with hard-edged measures that have proven successful elsewhere.

The articles make three main points. Since the beginning of the reform era, the prevalence of overweight and obesity has skyrocketed—from 5.5 percent of adults in 1982 to 50.7 percent in 2015–19 (by Chinese criteria)— giving China the dubious distinction of having the largest number of adults with obesity in the world.[13] Second, although China has been trying to tackle the obesity issue since the turn of the millennium, "existing policies primarily address people's diet and physical activity, with a focus on children"; "these efforts are clearly inadequate."[14] And third, one of the main obstacles to reducing death and disease from chronic disease is the power of the ultraprocessed-food industry and the Chinese government's failure to curb its power. In understated, apolitical language, they write: "Western-style fast-food outlets and the market for packaged and processed foods have expanded rapidly in China in the past few decades, [with] sales of ultra-processed foods and drinks [doubling or trebling from 2002 to 2016]."

The insufficient policy attention [to] . . . the wider determinants of obesity might have hampered progress in obesity control. . . . Some of the potentially effective policy mechanisms implemented in other countries . . .
such as restrictions on the marketing of unhealthy foods to children, food
nutrition labeling, and taxes for SSBs [sugar-sweetened beverages], are
understudied in China. . . . Whereas the vested interests and influences of
the food industry in food governance and policies are increasingly recognized worldwide, *there is little public awareness and an absence of established institutional mechanisms in China to prevent such adverse influences
and balance the commercial and public health interests in the policy-making
processes.*[15] (emphasis added)

Drawing on international research on best practices, the authors urge action on multiple fronts. Two recommendations stand out most sharply:
"As proposed by international organizations such as the World Bank and
UNICEF . . . promoting reductions of ultra-processed food and beverage
consumption through laws and regulations should be one of the most urgent actions to be taken in China. . . . Obesity prevention in China requires
high-level government support and leadership."[16]

Many of the policy gaps identified by the *Lancet* Ten can be traced to
the convoluted history of Chinese obesity science and policy unraveled in
this book. Indeed, this book can be read as a history of what went wrong in
China. Can it now be made right?

∴

Several years before the *Lancet* Ten published their results, Margaret Chan,
MD, director-general of the World Health Organization (2006–17) told a
remarkably similar story about the global obesity epidemic in an impas-
sioned address to the US National Academy of Medicine. The powerful
food industry, she said, with its globalized marketing of unhealthy foods
and drinks, bears primary responsibility, yet governments have failed to
take it on. "Ladies and gentlemen," she said, "I have a final comment. When
crafting preventive strategies, government officials must recognize that the
widespread occurrence of obesity and diabetes throughout a population
is not a failure of individual willpower to resist fats and sweets or exercise
more. It is a failure of political will to take on powerful economic operators,
like the food and soda industries."[17]

To get policy right, the WHO director declared, governments must
get the politics right. Can China get the politics of obesity right now? Can
China's powerful Communist Party–state take on the global food and soda

industry that is marketing unhealthy, indeed, sickening food to China; hold it accountable; impose the needed regulatory, legal, and other measures; and prioritize the public's health over economic and trade issues in its making of chronic disease policy? The *Lancet* scientists hailed President Xi's Healthy China initiative as a historic opportunity for China to create a comprehensive, coordinated, and supportive policy system to tackle the obesity question.[18] The time may now be ripe for China to rethink its approach and follow both its own obesity experts and the global health leaders on diet-related chronic disease.

The history relayed in these pages suggests that a policy adjustment of this sort would face major challenges. Redirecting policy to target sugary drinks is not a matter of simply adding a few provisions to existing policy. For China's policies to be coherent, it may require rethinking the all-calories-equal science of energy balance on which current policy is based. It may necessitate recruiting into the policy process experts with virtually no ties to the food industry. Curbing industry influence would also involve renegotiating the difficult terrain of science ethics, recognizing that industry funding of science usually brings industry influence on science, and rearranging the relations between industry and government at fairly high levels of decision making.

Flaws in China's System of Governing through Science

This intimate story of obesity policy sheds light on some larger problems with the way China's state governs its society. In the reform era, the legitimacy of the party-state has been based not on the consent of the governed, but on sustained economic performance and the quality of its governance.[19] Scientific policy making was supposed to produce higher-quality policy and governance. We have seen that the science of chronic disease, now the leading cause of death in the country, was tainted and the policy compromised, falling short of international best practice, which includes regulating the food industry. A number of factors conspired to produce this unfortunate outcome. Among them were the narrative of America's unquestionable scientific superiority, the state's close ties to industry, and its preference for industry-friendly policy. But the country's system of scientific policy making—the institutional setup and the characteristic dynamics—was also at fault.

To think of scientific policy making in China as simply the making of policy with the benefit of scientists and scientific facts would be mistaken. For science itself is a domain of politics. The most fundamental weakness in China's system of governing through science is that science, rather than being a relatively independent domain, is subordinate to the party-state, which can exert influence through its control over scientific organizations, funding, and publication outlets, among other means. As a constituent part of party-state politics, science is subject to the rules of the political game. In this hierarchical political system, once science gets subordinated to political forces, the competition between scientific viewpoints that helps the best ideas rise to the top elsewhere (at least ideally) gets suppressed, as political forces intervene in the scientific process. To see how scientific policy making worked on the ground, I viewed science not just as content (numbers, facts, arguments)—the conventional approach—but also as a human and eminently political process. A key phase in the making of policy scientifically involves identifying the science that will inform policy making and certifying its quality. In this case, the science the experts brought into the policy arena was distorted by its association with the transnational food industry. A system of verification—norms to discourage bias in the science, checks to eliminate bias and ensure quality—should have ensured that the science was sound. State norms were indeed in place specifying that industry must not bias science. But I uncovered no concrete means to enforce them.

Once the flawed science got inserted into China's Leninist system, the bureaucratic pathologies that political scientists have documented at the level of elite politics—a leader-centric politics, personalization of power, a lack of accountability, personal relationships counting more than rules and institutions—were reproduced on a miniature scale, with predictable effects on public policy.[20] Certain features of the Chinese-style think tank made it easy for researchers to bend, ignore, or cover up rule violations, thus hiding the bias in the science. In a common pattern, the think tank we tracked was located within a governmental ministry and was headed by a former ministry official, whose policy experience and personal ties to current ministry leaders earned her the trust she needed to informally lead the policy process. With the dynamics of policy making governed more by guanxi than by formal rules, exploiting ambiguities in the rules could be easily accomplished. In this case, a leader who had dual loyalties (to the state and to ILSI) and who was well incentivized to promote ILSI's industry-friendly science was able to quietly conceal the proindustry slant by keeping ILSI's funding arrangements confidential. Practices of performing ethics and concealing secrets worked to veil the contradictions between the rules and the political reality and keep the bias in the science hidden.

In China's hierarchical, leader-centric political culture, in which power is concentrated in the hands of a leader like Chen, one with personal authority and state backing, there were virtually no parties with the political space to perform the service of independent review and verification of her scientific claims and political actions. And once the policies she helped craft were accepted and formally issued by the state, they were essentially set in stone, unquestionable by anyone. In this way an industry-skewed science got embedded in a policy that gained the endorsement of the state. The label "science" had worked to give the policy making political legitimacy, without performing the substantive role of guaranteeing scientific (rational, data-based) quality. The policy was based on science, to be sure, but it was not the best science. Rather, it was the science that best fit the needs of the political actors pushing it: the world's biggest junk-food companies and their agents in the Chinese state.

China aspires to become a global science and technology power by 2049, and countless indexes—dollars invested, numbers of researchers, counts of journals and publications—suggest it is well on its way to success. But the distinctive features of the Chinese context require us to think about science in China differently. In-depth accounts like this should prompt us to look beyond the aggregate numbers and ask new questions about where the science is institutionally located, how it is made in cooperation with party-state officials, and whether public policy made "scientifically" is actually better policy. Beyond that, we should ask what such indexes mean in a context here, a growing body of evidence suggests, science making is influenced and sometimes actively obstructed by political forces.[21] No doubt the numbers will continue to rise, but without answers to these questions, we won't know what those numbers mean.

[CONCLUSION]

So What, and What Now?

In the last few chapters I've shown why Big Food and its science to defend soda matter to China and its people. What about the United States, home to the creators of this sprawling project to make the world safe for Coca-Cola by spreading a distorted science of obesity and chronic disease around the world? What effects has soda science had on ordinary Americans and their ideas about which lifestyles will make them healthy? And what can we do about this kind of corporate influence? How can we productively respond to the aggressive actions of the global food industry to twist our ideas and risk our health for corporate profit?

Soda science may seem a thing of the past, but it lives on, shaping Americans' everyday lives in ways that remain invisible. In the first part of this conclusion I bring those effects to light by placing the short history of soda science within the longer history of America's corporate-dominated fitness culture. The soda scientists we've come to know built on and promoted that commercially driven fitness culture, while extending the list of purported benefits to include solving the obesity problem. In doing so, Coke's obesity scientists became key progenitors of today's society of step-counting, fitness-tracking, exercise- and weight-obsessed citizens. That is a problem because some of the core beliefs they helped build into that fitness culture are deeply problematic.

While writing about the politics of corporate science, I've had a political agenda of my own. I've sought to illuminate the hidden effects of one influential product-defense science and then join others in calling for greater accountability on the part of the companies, nonprofits, and academic scientists involved in the scheme. The early publication of some of the results allows me to share the reactions of the two central actors, Coca-Cola and ILSI, to this research. Have they acknowledged the findings or taken responsibility in any way? The answer, we see in the second section, is a surprising yes.

Soda science may no longer exist as an active project of the soda industry, but the producers of ultraprocessed foods and drinks continue to mold the science of nutrition and chronic disease, inventing novel tactics and targeting ever more parts of the world. The global mapping of Big Food and its scientific subterfuges—charting the who, how, when, why, and so-what of corporate science—has a long way to go. Seeking to point the next wave of researchers in critical new directions, the final section tracks the political strategies the industry is using to great success in penetrating national governments and co-opting local officials into their projects of product defense and profit maximization.

Soda Science and the Making of America's Self-Help, Step-Counting Fitness Culture

Exercising for weight loss is cultural common sense in America today. Some 55 percent of Americans would like to lose weight, and nearly 65 percent of those on a weight-reduction program are using exercise.[1] So accustomed are we to the daily flood of articles, blogposts, and videos telling us which sports, what equipment, and which workout routines take off the fat that we hardly stop to ask whether the claims are true.

The pairing of exercise with weight loss that we take for granted is actually quite recent. In *Getting Physical: The Rise of Fitness Culture in America* (2013), Shelly McKenzie shows that since the 1950s, when exercise was invented as a national solution to the problem of softening bodies in an era of postwar affluence and suburban lifestyles, exercise and weight loss have functioned as separate concepts, occupying distinct domains of thought and practice.[2] In the 1960s large numbers of women seeking the perfect body began trying to lose weight, but their method was dieting. (Some blamed the lack of activity for the rise in obesity, but the solution was almost always dieting.) In that same decade, men began to exercise en masse, but not for weight loss but to ward off heart attacks and other degenerative diseases of modern life. The separation between the notions "exercise" and "weight loss" continued through the 1970s and 1980s and into the 1990s, when fitness grew into a mass culture. Although some individuals in every decade hoped working out would help take off pounds—"Slenderella" reducing salons and "trimnastics" programs spoke to that wish—that was secondary to other culture-wide goals Americans pursued through exercise: gaining energy and a happy demeanor (1950s), achieving a toned or muscled body (for women and men, 1960s on), expressing a personal identity based on self-health (1970s), demonstrating personal and professional competence (1980s), and attaining the lifestyle of the affluent and virtuous (1990s on).[3]

Before the late 1990s and early 2000s, there was little agreement on the value of exercise for weight loss. The science had not yet been done. The first time exercise and weight loss were officially linked in a major public statement was in 2001, when the surgeon general's Call to Action on obesity urged Americans to balance calories consumed and burned by eating more healthfully and exercising more regularly.[4] The vagueness of the message— how much exercise? of what kind?—opened the way for commercial entities to promote their workout equipment, exercise classes, gym chains, and so on as scientifically proven to contribute to weight reduction, tapping into the fears of a public newly anxious about the epidemic of obesity that was undermining their health.

The 2008 physical activity (PA) guidelines for Americans—the first ever—added specificity but muddied the message by distinguishing three aspects of weight management: prevention of weight gain, weight loss, and maintenance of weight following weight loss. The guidelines reported the existence of strong evidence that PA helped prevent weight gain, strong evidence that PA *with caloric restriction* fostered weight loss, and moderate evidence that it worked for weight maintenance following weight loss.[5] Out of these details, the average person who, we can reasonably assume, is primarily concerned about dropping pounds, is likely to hear a single message: "exercise is good for losing weight."

The complex and confusing directives, as well as measurement difficulties that continued to hamper the formulation of widely accepted guidelines, left loads of room for companies in the fitness industry to amplify the message about exercise and weight loss to their commercial advantage. As we've seen, companies in the food industry saw the same opportunity and jumped on it. When in 1999 ILSI's food-industry CEOs declared the promotion of exercise the only acceptable solution to the obesity crisis, it was a small step for Hill and his colleagues to quietly convert "exercise is good for energy balance and maintaining healthy weight" into "exercise-first for weight," using linguistic tricks like diet tokenism to make it appear they were giving equivalent attention to exercise and diet. The mantra of activity for weight loss exploded in 2003–4, when Coke and a bevy of other giant food companies and trade associations—McDonald's, PepsiCo, the Grocery Manufacturers Association—began widely promoting "active healthy lifestyles," equating "healthy" with "active," and creating highly visible programs to spread the message to weight-anxious consumers far and wide. In 2004 Hill and Peters's book taught us how to "count steps, not calories, to lose weight and keep it off forever." Since around that time, the message that exercise takes off pounds, invariably promoted as a truth of science, has been used to sell countless step counters, fitness trackers, gym memberships, stationary bikes, workout videos, and sports gear, helping create the $30–40 billion per year fitness industry we have today. Through their diet books (a second one, by Hill and Wyatt, appeared in 2013),[6] articles, public health campaigns, and media presence, Hill, Peters, Blair, and the other soda scientists helped turn us into the health-obsessed, weight-preoccupied, Fitbit-wearing, self-responsible citizen-consumers of today. That belief that exercise by itself takes off pounds and the bodily practices associated with it are among the most significant legacies of soda science today.

This legacy matters because many of our beliefs about exercise and weight are not well supported by the science. My reading of recent literature reviews suggests that, in the mid-2020s, most experts agree that regular exercise is an important part of a weight-loss program, but it is secondary to caloric restriction, which has more profound and consistent effects than exercise alone.[7] Moderate-intensity aerobic exercise alone (brisk walking, for example) is of little use to weight loss; at most one can expect to lose a few pounds. Despite the popularity of step counting, there is nothing magic about ten thousand steps, a common daily goal. Like so much of what we "know" about lifestyle and health, the number was invented by a company, this one a manufacturer of pedometers.[8] Walking interventions of this sort can produce 1 to 1.5 percent weight loss; brisk step counters may lose 2 to 3 percent of their body weight.[9] To lose the clinically significant benchmark of at least 5 percent of initial weight with exercise alone, we need to engage in 225–420 minutes (3.75 to 7 hours) a week of moderate-intensity exercise, and more for those with uncooperative metabolisms. Even then, any benefit is likely to be temporary. Whether because our metabolisms adapt or because the exercise programs are too intense for us to sustain, 80 percent of those who lose weight are unable to maintain the loss.[10] Though regular exercise has countless health benefits, for most people exercise alone will not help much with their weight.

Meantime, in the last decade, the pathbreaking research of evolutionary anthropologist Herman Pontzer has been challenging the most fundamental assumption of soda science: that by exercising more, we as individuals can micromanage our energy expenditure to achieve the energy balance needed for a healthy weight. In a meticulous study of caloric expenditure among the Hazda, which forms the centerpiece of Pontzer's 2021 book *Burn*, a group of hunter-gatherers in Tanzania, Pontzer and his colleagues found that high levels of activity did not translate into high caloric expenditure, as he and everyone else expected. Instead, the Hazda's bodies compensated by using less energy on other things (the functioning of their immune and endocrine systems, for example). Despite being extremely active, the Hazda and other hunter-gathering groups burned the same number of calories per day as sedentary office workers in the United States and Europe, roughly twenty-five hundred calories. Other studies, brought together in Pontzer's 2021 book *Burn*, showed that daily energy expenditure in the United States is virtually the same now as it was before the rise of the obesity epidemic, undercutting the common view that the epidemic is heavily rooted in modern sedentary lifestyles.[11] Pontzer argues that obesity is indeed due to an imbalance in energy (or calories) in and out, but we've gotten the energy-out side of the equation wrong. Energy out is not something we can consciously control through exercise. Despite the rise of our modern fitness-obsessed culture, the culprit clearly lies on the energy-*in* side of the equation: our unhealthy, calorie-packed diets. These remarkable findings not only help us understand why we can't lose much weight simply by getting more exercise; they challenge the entire conceptual apparatus of soda science, from the mantra of exercise-first to the figure of the self-calculating, weight-managing citizen-consumer.

Throughout this book I have emphasized the dangers of the corporate control of the obesity issue. In part 2, I spelled out the risks of China's market-centered, corporate-heavy approach, but America's approach is not that different. In their books on American fitness culture, both Shelly McKenzie and Natalia Mehlman Petrzela, author of *Fit Nation: The Gains and Pains of America's Exercise Obsession* (2022), emphasize the central role of American business. Since it began to flourish in the 1950s and 1960s, the national fitness culture that the soda scientists helped to expand and redirect toward obesity reduction has been dominated by private-sector entrepreneurs, whose main motivation is making money, not advancing human health. The federal government largely abdicated its responsibilities for investing in the infrastructure needed to make exercise available to all Americans, and for tracking down and penalizing advertisers who push deceptive claims. The result is a corporate-dominated fitness and weight-loss culture that reflects our underlying belief that, as Petrzela puts it, "issues of health [are] appropriately solved by personal willpower and savvy product development."[12] Health designed by the market is hardly optimal health.

As individuals made responsible for our own health, we need to constantly remind ourselves that our ideas about exercise and weight are partly if not largely corporate constructions. Whether Coke is twisting the science of obesity or Peloton is marketing its stationary bikes and cardio classes as fat busting, it's all the same, just business as usual in the American fitness market. Dieter, beware!

Holding Industry to Account: Coke and ILSI Change Directions

In doggedly pursuing the roots of soda science over so many years, my aim was not just to learn the truth. I also wanted to hold the actors responsible for manipulating science and public policy for corporate benefit. The early publication of some key results allows us to ask whether in-depth social research such as this can make a difference in a world dominated by rich, powerful companies who can buy science, transport it around the world, and get it embedded in the policies of a major foreign country.

Hoping to make my key findings quickly available to those with the greatest immediate interest—the public health community—between 2019 and 2021, I published three articles in journals in the health sciences. The first two, a related pair, documented the impact of Coca-Cola, working through ILSI, in skewing China's obesity policy toward physical activity solutions. Published simultaneously in early 2019 in the prominent journal *BMJ* (formerly *British Medical Journal*) (as investigative journalism) and the US-based public health publication *Journal of Public Health Policy* (as a more conventional scientific article), the work, especially the *BMJ* piece, attracted wide attention globally (China was an exception).[13] The third article unearthed the institutional mechanisms by which Coke and ILSI came to shape every step in the making of China's obesity policies. Pointedly titled "Inside ILSI: How Coca-Cola, Working through Its Scientific Nonprofit, Created a Global Science of Exercise for Obesity and Got It Embedded in Chinese Policy (1995–2015)," this article was published online in late 2020 and in print in early 2021 by the *Journal of Health Politics, Policy and Law*.[14]

Before the publication of those first two articles, I was deeply fearful of the legal or other actions that ILSI and, even more so, Coca-Cola, might take against me. Family and friends worried about my safety. There was reason for concern. In 2016, Mexican advocates of a soda tax began receiving menacing messages on their cell phones.[15] Their phones were laced with spyware evidently intended to track their advocacy work and perhaps derail their careers. Though there is no concrete evidence Coke was involved, Mexico was Coke's biggest consumer market by per capita consumption. The company had actively lobbied against the tax and poured billions of dollars' worth of investment into the country. Were these surveillance tools being used to advance the soda industry's commercial interests in Mexico?

Would I be next? Inquiries with my university revealed that it would not offer legal protection, since "faculty choose their own research topics." As a measure of protection, I asked the publishers of the *BMJ* and the *Journal of Health Politics, Policy and Law* to conduct a legal review in advance of publication. Neither review uncovered any potential legal exposure. Even so, I was intensely anxious to learn how Coke and ILSI would respond.

DECISIVE ACTION BY COCA-COLA

We saw in the first coda, "Reckoning," that after the 2015 exposé in the *New York Times*, Coca-Cola moved quickly to undo the damage from the public revelations about its "mercenary science" of obesity. Three and a half years later, in response to the *BMJ* article on China, the company issued a confident press release. It said the company agreed too much sugar was bad, its earlier handling of the scandal surrounding energy balance science was correct, and it was staying the course.[16]

Despite the company's earnest efforts to do the right thing and to change the subject, in 2019 a new wave of unfavorable press emerged in the form of articles by critical public health scholars, many based in the UK and Australia, examining the secrets of the past—confidential conversations between Coke, ILSI, the US CDC, and so forth—in embarrassing detail. Most were based on the same set of revealing emails that I used in chapters 3 and 4. With the publication of each article (including "Inside ILSI"), the media flooded the internet with a fresh round of stories reporting the unsavory goings-on.

Perhaps weary of the incessant barrage of criticisms sullying its good-company image, and perhaps seeing the connection to ILSI as having served its purpose, on January 1, 2021, Coca-Cola took the drastic step of severing all ties to ILSI at the global, regional, and national levels. According to Bloomberg.com, "the decision was made after a routine review";[17] no further details were released. The decision apparently was a response to the growing number of critiques by watchdog groups and public health scholars depicting ILSI as a scientific front group for the food-and-beverage industry. The company's decision seems to have been made in October 2020, just weeks after the online publication of "Inside ILSI," whose subtitle strongly implicated Coke in the distortion of Chinese policy. Coca-Cola was the founding company and one of the dominant forces in ILSI for nearly thirty years. Its decision to end the association—apparently signaling a desire to put the exercise-for-obesity episode behind it once and for all—was a very big deal. For ILSI, Coke's withdrawal was a major political and symbolic blow, especially because it followed the departures of Mars in 2018, which cited the growing criticism, and Nestlé, which cut ties to focus on other initiatives at the beginning of 2020. ILSI was bleeding corporate members.

DENUNCIATION BY ILSI

If the China findings posed reputational risks for Coke, they presented an even greater threat to ILSI. ILSI's public identity (and legal status) is that of a 501(c)(3) nonprofit, an organization that creates science for the public good and refrains from lobbying and policy advocacy. Without directly countering its claims to this legal identity, those three articles challenged that foundational myth by providing extensive evidence that, on the obesity issue, ILSI science served private corporate interests and, in one big country, engaged in covert policy making. In February 2019, one month after the first two articles appeared, ILSI posted a lengthy rebuttal on its website. Titled "Setting the Record Straight: ILSI Focal Point in China Response to *BMJ* and *Journal of Public Health Policy*," the statement describes how my conclusion "significantly deviates from fact, and it has serious misleading effects."[18]

The original version of the rebuttal contained two serious distortions of my work. First was the claim that I argued that China focused almost entirely on physical activity (it even included a manufactured quote of me saying that). In fact, my work argued that China offered both dietary and activity solutions, but the balance shifted over time. In the second distortion, ILSI-China implied that I stated that its programs promoted consumption of soft drinks. Of course I did not make that preposterous claim; my point was that China's policies fail to lower soda consumption through taxation. I asked ILSI to correct these distortions, and it did so promptly. The revised version of the rebuttal, which remains on the ILSI website at the time of writing (late 2023), continues to make a number of claims about ILSI's work in China that my research challenges.

Published online a year and a half later, "Inside ILSI," which delved into the hidden workings of the organization, was potentially more damaging. Within a few weeks of the article's online publication, the director of operations for ILSI-Global sent a letter to the editor of the journal denouncing the article in scathing terms. Instead of taking up the arguments or questioning the data, the director simply inveighed against my conclusions, denied that I substantiated my findings, and repeated the defense of ILSI's procedures published on its website. The letter merits quoting at length:

> Greenhalgh [has] . . . advanced new libelous unsubstantiated misinformation about ILSI. . . . [The paper] posits a conspiracy that does not

exist, applies innuendo and supposition to innocent facts and totally misrepresents the truth on ILSI's efforts to support public health in China and globally. . . . ILSI vehemently refutes the misinformation contained throughout this report and stands firmly behind its process, approach, scientific integrity, and credibility. The credibility of science is judged on the merits of study design, methodology, and validity of the conclusions—not funding. . . . In contrast, the paper published by Greenhalgh is a series of unsubstantiated statements, suggestions and conclusions that are untraceable and not verifiable. . . . [ILSI conducts science] in a neutral and transparent manner and with a very high level of scientific rigor.[19]

The operations director strongly questioned the journal's peer-review process and ended by hinting vaguely at legal or other action if the article was not withdrawn. The journal's editor did not answer the letter, and the article appeared in print a few months later.

ILSI REORGANIZES, AND ILSI-CHINA DISSOLVES

The critical public health articles that targeted Coke also pointed the finger at ILSI, demonstrating with email evidence that its claims to be a neutral nonprofit were not borne out by the communications of its leaders.[20] Evidently in response to such criticisms, the organization began quietly changing its profile. In the last few years it has revamped its website again and again. In the early 2020s, the organization added a list headed "Common Misconceptions" to its web page on frequently asked questions. One misconception concerned the organization's founder. "Dr. Alex Malaspina," the page states, "ceased to be an ILSI Trustee, Officer, or representative of any kind in 2001. He has no position with ILSI and any comments he has made subsequently are as a long-retired private citizen with no authority to direct or influence ILSI's actions."[21] All the critiques must have hit close to home for ILSI to distance itself from its founder in this very public way. The organization also began sloughing off affiliated entities. In 2020 the ILSI Research Foundation renamed itself the Agriculture and Food Systems Institute. In 2021 ILSI North America, a major player in the story of soda science, became the Institute for the Advancement of Food and Nutrition Sciences.

With the loss of three major member companies and all their subsidiaries, and facing a rapidly changing media environment that made secrets ever harder to keep, in May 2022 ILSI announced a major reorganization and rebranding.[22] ILSI-Global dropped the full name and began using only the acronym ILSI. Under a new management team, the organization created a new US and Canada Research Program with a novel model designed to eliminate corporate membership fees and encourage alternative grant support.

In 2022 I began noticing that the China and South Andean flags were no longer flying on the organization's map of the world dotted with ILSI flags. I emailed the ILSI-Global communications director asking what information she could share, especially about ILSI Focal Point in China. In a gracious and informative reply, she told me that several ILSI branches (including the South Andean one) have closed shop. Several others—including ILSI-China, ILSI North America, ILSI Argentina, and the Research Foundation—have become independent bodies no longer connected to the ILSI Federation. After a quarter century, ILSI-China was no longer part of ILSI! These entities, she wrote, "chose to re-establish themselves as separate, independent organizations" in the context of the global organization's renewal of its commitment to research conducted "within a framework of the highest principles of scientific integrity" (email from Katherine Broendel, December 30, 2022).

Seeking clarification from China, I learned from one of the directors that ILSI-China had stopped functioning in 2019 (email from Wenhua Zhao, January 3, 2023). Some evidence suggests the branch was still operating to some degree in 2020 and 2021, but, despite several inquiries, I was unable to learn more about when and under what circumstances it closed shop. Perhaps the sharp criticisms of its work with Coke, coming on top of the mid-2018 loss of Chen Chunming, who had always been the heart of the organization, convinced the current heads it was time to dissolve the organization. This is an extraordinary development that, among other consequences, deprives the giant food companies operating in China of their trusted friend in the Ministry of Health (now named the National Commission on Health).

On paper these changes appear to be sweeping. Coke dropped its ties to ILSI. ILSI-China ceased functioning, and ILSI-Global underwent major reorganization, severing ties to entities that have been part of the organization for decades. Even in our corporate-dominated world, it seems that in-depth social research can make a real difference. By exposing exactly how companies have bent the science of health and meticulously documenting the process, social research can produce concrete evidence of corporate malfeasance that is hard to challenge. This evidence creates the potential for real reputational damage, compelling organizational change.

What Next for Product-Defense Science? New
Targets and Tricks of the Food Industry

The encouraging changes at Coke and ILSI notwithstanding, our work on corporate science is far from done. Though China's unusual political system made it an exceptional case, ILSI-Global actively promoted soda science throughout its worldwide network, suggesting that the impact on obesity science and policy might have been notable in other ILSI branch countries as well. And indeed, a review of the branch achievement posters displayed at the annual ILSI meetings reveals the existence of Take 10! and other soda-science programs and working groups in the South Andean, Mesoamerican, Argentinian, Brazilian, and Japan branches.[23] There can be no doubt that exercise-first projects were—and may still be—promoted in many ILSI branch countries and regions.

Today the food industry is focusing its energies on defending ultraprocessed food and drinks by *opposing regulation*, the second part of its strategy in China (*soda-tax-never*). As the most rapidly growing sector of diet worldwide and a major source of revenue for global food companies, ultraprocessed foods and beverages, designed to be addictive and overconsumed, have been linked to obesity, diabetes, hypertension, cardiovascular diseases, cancers, and mortality from all causes, representing a major threat to human health. Since 2010, a growing body of research has concluded that ultraprocessed foods are "very probably the primary cause of . . . the rapid global rise in obesity."[24] To protect their citizens against these unhealthy manufactured foods, governments in countries like Mexico, Chile, and South Africa have been working to create mandatory anti-junk-food policies that have been shown to work: taxes on sugary drinks and energy-dense packaged foods, front-of-package warning labels, marketing restrictions, and school food policies.[25]

But Big Food and its scientific agents have been moving aggressively to weaken, delay, and block these policies. New research reveals that the kinds of political dynamics we have witnessed in China are working to undermine health in virtually all the major emerging economies. In *Junk Food Politics* (2023), the political scientist Eduardo J. Gómez shows how, in Brazil, China, India, Indonesia, Mexico, and South Africa—which together make up 44 percent of the world's population—global beverage and fast-food companies are energetically fighting such measures, while promoting themselves as part of the solution to the exploding levels of chronic disease.[26] Although several of these countries have impressive chronic-disease-prevention programs, the food industry has been deploying a wide range of political strategies to chip away at their effectiveness. By proactively cultivating top leaders, forming partnerships with health officials, strategically infiltrating bureaucratic institutions, and contributing generously to corporate social responsibility programs, the companies not only have motivated leaders to protect them from anti-junk-food policies, but they have established industry-friendly policy views as legitimate and effective, and instilled the notion that Big Food is a genuine partner and contributor to national health and welfare. By now, Gómez contends, the interests of the junk-food industry and national governments are deeply intertwined and mutually reinforcing, leaving the industry more powerful today than ever.[27] Meantime, levels of obesity and diabetes continue to rise.

While pursuing these larger political strategies for commercial success, the food companies have also been innovating powerful new discursive techniques—arguments, narratives, images—to protect their markets. Such tactics appear especially elaborate in Latin America. There the food industry has been presenting ultraprocessed foods as essential products; making health claims based on macronutrients added to its junk products; stressing its own vital role in local economies; depicting regulations as unnecessary, harmful to trade, or in violation of trade agreements; developing novel legal claims; and making corporate responsibility pledges that are more show than substance.[28] As in China, in Latin America and, indeed, much of the developing world, the growing power of corporations to manipulate health science and policy is rooted in a familiar neoliberal order that everyone from policy makers to researchers to members of the public simply accepts.[29]

These findings raise critical questions. How many other countries have been subjected to political pressures to partner with and champion the interests of Big Food? Perhaps more to the point: is there anywhere in the world that is *free* from the hazardous products and seductive strategies of the junk-food industry? In a world where industry "routinely attempts to stop mandatory regulation," Barry Popkin and his colleagues recently declared, "most countries crucially need a consortium to systematically study, expose, and neutralize industry interference in health policy making."[30] Here is where social research, especially of a collaborative and comparative sort, can play a vital role. I hope this story of the hidden power of soda science will inspire readers—whether student, professional researcher, investigative journalist, or armchair sleuth—to join the effort. In these pages I've developed a set of conceptual tools that I hope others might find useful in investigating corporate science. The core concepts can be found in appendix 1.

Acknowledgments

Over the years that this story has been coming together, I have accumulated countless debts.

This book was made thinkable by Sergio Sismondo's brilliant writing on epistemic corruption, David Michaels's powerful analysis of product-defense science, Barry Popkin's trenchant research on public health nutrition, and Marion Nestle's fierce critiques of the corporate distortion of nutrition science.

Without my expert-informants, I would know nothing about soda science and its travels, and so my biggest debts are to the roughly two dozen Chinese and one dozen Euro-American experts who spent precious hours sharing their work and views with me. As a way of acknowledging their vital input, I want to include their names here. I list my Chinese colleagues in Chinese style (surname first), along with their affiliation at the time of the interview. In Beijing, in addition to Chen Chunming and Chen Junshi of ILSI, I met with Dong Zhong (Beijing CDC), Ge Keyou (China Nutrition Society), Gu Zhongyi (Beijing Friendship Hospital), Hu Zhiping (Palmer) (Peking University People's Hospital), Ji Chengye (School of Public Health, Peking University), Jia Mei (Nestlé), Jiang Jingxiong (National Center for Women's and Children's Health, China CDC), Jing Jun (Research Center for Public Health, Tsinghua University), Li Keji (School of Public Health, Peking University), Liu Qingfei (School of Medicine, Tsinghua University), Ma Guansheng (National Institute for Nutrition and Health, China CDC), Chadwick Wang (Institute of Science, Technology, and Society, Tsinghua University), Wang Ke'an (ThinkTank Center for Health Development), Wang Mei (China Institute of Sport Science, General Administration of Sport of China), Wen Lingling (Wendy) (American PR Company), Xia Meng (An Zhen Hospital of Capital Medical University Hospital), Xiao Wei (Philosophy Department, Tsinghua University), Zhai Fengying (China Nutrition Society), Zhang Lin (Peking Union Medical College Hospital), and Zhao Lin (People's Liberation Army General Hospital). While in Beijing, I

268 ‹ ACKNOWLEDGMENTS

also spoke briefly with two Americans, Rhona Applebaum (Coca-Cola) and Michael Engelgau (US CDC).

In Europe I was hosted by Chizuru Nishida of the WHO in Geneva; Philip (W. P. T.) James and Jean James, formerly of the International Obesity Task Force (IOTF), in London; and Neville Rigby, also formerly of IOTF, in Inverness, Scotland. In the United States, I had the pleasure of lengthy discussions with William (Bill) Dietz (George Washington University, formerly US CDC), Katherine Flegal (National Center for Health Data, US CDC), James (Jim) Hill (University of Colorado), William (Bill) Hsiao (Harvard), Frank Hu (Harvard), Shiriki Kumanyika (University of Pennsylvania), Yanping Li (Harvard), and Barry Popkin (University of North Carolina, Chapel Hill). I am especially grateful to Suzanne Harris, then executive director at ILSI-Global, for helping me understand ILSI's mission and operations. Though these three dozen specialists filled my notebooks with fascinating stories and observations, none of them is responsible for what I have written about them and the worlds they brought to life for me. All interpretations in this book are mine alone.

Several institutions provided vital support to my research. I am indebted to Jing Jun for hosting me as a visiting scholar at the Tsinghua Research Center for Public Health during my field research. His skilled staff helped me navigate the bureaucracies of the university in more ways than I can describe. At the Harvard Business School Library, Alex Caracuzzo of the Baker Research and Data Services led me to GuideStar, which opened my eyes to the data-packed tax forms used by all of ILSI's US-based entities. Very special appreciation goes to Gary Ruskin, cofounder and executive director of US Right to Know, a nonprofit investigative public health research and journalism group that works to expose corporate wrongdoing and government failures. I am grateful to Gary and his team for filing the Freedom of Information Act requests that resulted in a huge archive of emails between Coke and its friends in academia, and to the University of California San Francisco Library for making these available to researchers such as myself. Cynthia Ogden, the NHANES Branch Chief at the CDC, kindly prepared for me the most recent estimate of overweight status among American adults. Without the generous help of these colleagues and their organizations, many of Coke's secrets and their real-world effects would have remain buried.

Over the years I was digging into Coke's secret science projects, I was joined in the effort by a number of talented research assistants, who brought not only superb research skills but also many fresh insights and much good humor to the project. Wu Xiaoqi served as my able field guide and companion in Beijing. In the United States, Angela Lee helped track down corporate funders of research, Yun Zhu unearthed the history of Roche Pharmaceuticals in China, and Sukie (Suqi) Zhang served as my COVID-era assistant, uncovering China-based scientific and policy resources from her spot in Shanghai when foreigners could not travel to China. Isabelle Yiyi Jiang and Han Zhang served as translators and transcribers, rendering the interviews with Chinese experts into excellent English and transforming my *BMJ* article into sparkling Chinese.

For substantive engagement with the book's arguments about China, nothing could beat the two days of discussion with three special colleague-friends who work on the politics of science in contemporary China. Joining me for a book workshop in the summer of 2022, Cong Cao, Abigail (Abby) Coplin, and Yanzhong Huang helped me unpack the unusual science/policy divide in China, shared highly relevant comparative cases, and raised critical questions about China's place in global science.

I have treasured my interactions with members of the Lise Meitner Research Group on China in the Global System of Science (within the Max Planck Institute for the History of Science in Berlin). Anna Lisa Ahlers, the group's founder and leader, along with scholars Andrea Braun Strelcová, Jelena Gross-Bley, Yishu Mao, Cheryl Mei-ting Schmitz, and others have pushed my thinking in novel directions, and I look forward to continuing the conversations.

This project has been nurtured over many years by ongoing conversations with some very insightful colleagues. Wei Hong unraveled the social dynamics of China's distinctive science formation; her answers to my endless questions fill the pages of this book. Special thanks to Jim (James) Flowers, Suzanne Gottschang, Barbara Katzenberg, Kate Mason, Janet Steins, Marty Alexander Zeve, and Li Zhang, among many others. I also had memorable one-time meetings with Tsong O. Cheng at George Washington University, Jiuheng He of Cornell University, Yixi Wang of Monash University (via email), and a host of colleagues at the Harvard T. H. Chan School of Public Health, all of whom took valuable time to talk with me about corporate distortion of science and how it might play out in China. These include Steve Gortmaker, Bill Hsiao, Frank Hu, Yanping Li, Walter Willett, and Winnie Yip.

I have had the good fortune to present this work to various audiences. These include invited lectures or keynote talks at the Conference on Epistemic Corruption at Queen's University in Kingston, Ontario; the Unit for Biocultural Variation and Obesity at the University of Oxford; the Lise Meitner Research Group in Berlin; the Conference on Lifestyle and Kinds of Living: Opportunities, Conditions, and Biology at the University of Copenhagen; the Conference on Chinese Techno Futures at the China Center,

University of Oxford; the Workshop on China in the Urban Age: Health, Food, Waste in the Chinese City at the University of Sydney; and at both the Research Center on Public Health and the Institute for Science, Technology, and Society of Tsinghua University, in Beijing. Thanks to all for the lively Q&A sessions at the end of the talks.

Closer to home, I am grateful for opportunities to share my work with faculty and students in anthropology during the Annual Kassen Lecture at Case Western Research University in Cleveland, in the Sustainability Showcase Series at Pennsylvania State University, and in the Center for Chinese Studies at the University of Michigan. At Harvard I have presented my research to audiences at the STS Circle, the Department of Anthropology, the T. H. Chan School of Public Health, and the Fairbank Center for Chinese Studies. I have also presented this work at the annual meetings of the American Anthropological Association, the American Ethnological Society, the 4S (Society for the Social Study of Science), and the Association for Asian Studies. The responses of copanelists and audience members are gratefully acknowledged.

Students often ask the best—the bravest and most unfiltered—questions, and I am grateful for the opportunities I've had to engage with them about this project. Among the most memorable are the lectures given in Marty Alexander Zeve and Salmaan Keshavjee's course Social Medicine Methods at the Harvard Medical School, Elanah Uretsky's course Global Health at Brandeis University, and Jennifer Heung's course Global China at St. Mary's College of California.

Hoping to quickly reach those with the greatest interest in my findings, I decided early on to publish the first results on Coca-Cola's influence in China in a medical journal. That I eventually succeeded is in good part due to the generous assistance of Donald Halstead, director of writing programs (and lecturer on epidemiology) at the Harvard School of Public Health, who provided valuable editorial suggestions, publishing tips, and leads; and to the *BMJ*'s editor in chief, Fiona Godlee, and senior editor Peter Doshi for inviting me to publish my work as investigative journalism, and then working with me to convert the prose into the style of a feature article. Thanks too to Anthony Robbins and Phyllis Freeman for working nonstop to arrange the simultaneous publication of the companion, more scientific, piece in the journal they edited, the *Journal of Public Health Policy*. And, before that, to Marion Nestle for bringing my article to the attention of Robbins and Freeman.

At the completed book manuscript stage, my longtime friend John Manter read the entire book for me, red pencil in hand, marking everything he found fascinating, strange ("invisibilize"), or otherwise noteworthy. The East Asian historian of science Jim Flowers read the full manuscript online, sharing comparative reactions from the perspective of Korean, Japanese, and Chinese history, as well as his experiences as a traditional Chinese medicine provider in China. Marion Nestle read the first (US) part of the completed manuscript, offering her insights as an insider and longtime critic, and keeping me from making a few embarrassing mistakes. Barry Popkin read the second (China) part, sharing insights on how the nutrition transition and shift to chronic diseases played out in China, where he worked in the 1980s and 1990s, and on how China's nutrition policies today look in global perspective. Shelly McKenzie kindly read the sections on the culture of exercise in the US. Two anonymous reviewers for the University of Chicago Press read the full manuscript and offered extremely useful suggestions. Although it may be hard for those who read the manuscript for me to decipher their impact, I can assure them their imprint has been very real. One example: the comments of the press's reviewers led me to expand the project by placing the story of obesity science in the context of the larger story of the rise of fitness in the United States. That was a great idea, for which I am truly grateful.

I am indebted to my dear friend the artist Gary Sohl who once again responded with grace and good humor to my plea for help with the artwork. He created the infographic on Coke's aftereffects on the body and used his technical wizardry to redraw and customize several other images to fit press requirements.

At the press, a huge thanks to Karen Merikangas Darling not only for her sharp editorial skills but also for her enthusiasm for and belief in this project. Thanks also to Fabiola Enríquez Flores, the press's able editorial assistant who can magically turn a rough manuscript into a polished Chicago book. My incredible team—including Elizabeth Ellingboe, in charge of manuscript editing, Anne Strother, promotions manager, and Bill Wheaton, my indexer—was instrumental in more ways than I can spell out here.

The research for this book was supported by the US National Science Foundation (NSF), research grant SES-0217508 (Science, Technology, and Society Program); grants from Harvard's Asia Center, Fairbank Center for Chinese Studies, and China Fund; and a fellowship from the John Simon Guggenheim Memorial Foundation. The NSF grant covered the 2013 fieldwork in China. The Asia Center grant supported internet research and the collection and preliminary analysis of the ILSI-China newsletter data. The Fairbank Center grant allowed me to host the book workshop with colleagues in China studies. A grant from the Harvard China Fund helped defray the expenses of a research assistant during the COVID lockdown. Finally, the Guggenheim Fellowship supported in-depth analysis of the newsletters and other data, as well as preparation and revision of the early articles. The generosity of these organizations made this book possible.

Core Concepts

In this book I have developed a set of conceptual tools that, while created to guide the understanding of soda science, should be useful in investigating other cases as well. I offer these not as fixed understandings, but as working concepts or general hypotheses to be modified as needed.

Product-Defense Science

Product-defense science is a kind of corporate science (science funded by industry and made for its benefit) that is often created by influential scientists, looks like ordinary science, and appears in mainstream journals, yet is unusual in that its mission is to defend the profits and reputations of companies making products that are harmful to human health or the environment or both. It does that by manufacturing doubt about mainstream science or by creating an alternative scientific story, sometimes both. I uncovered six unusual features of the science and its making that are likely to be common to many if not all product-defense sciences.

1. Industry-friendly character. All core elements (facts, frameworks, concepts, arguments, rhetorics) support industry interests.
2. Corrupted knowledge requiring careful management. For the science to be credible and durable, the corruption must be hidden via twin practices of doing ethics and burying secrets, especially concerning funding. All parties engage in variants of such practices.
3. Reversed (conclusion first) science-making process. Often if not always the science is made in inverse order, with conclusions first, and translation of science into concrete programs before the programs' efficacy is established.
4. Strong science rhetoric. Product-defense science tends to be characterized by a strong rhetoric of science (emphasizing precision, accuracy,

and other worthy attributes) that is strategically deployed to under-
score the supposedly high scientific quality of the knowledge.

5. Challenging terrains requiring shrewd treatment. In product-defense
   science there are terrains of delicacy in which industry's interests
   in protecting its harmful products (e.g., junk food) collide with
   mainstream science and popular views (in this case, that junk food
   is unhealthy). The product-defense scientists must produce canny
   solutions that protect industry interests while appearing to conform to
   mainstream views (e.g., diet tokenism). Most if not all product-defense
   sciences are likely to have comparably challenging terrains requiring
   clever scientific responses.
6. Short life. In democratic contexts where science is relatively independent and open debate is possible, product-defense science has a
   relatively short life span. In authoritarian contexts where open debate is
   suppressed, its life span is potentially quite long.

Product-Defense Nonprofit

Product-defense science is created by a diverse set of parties that include
companies, public-private partnerships, nonprofit organizations, and academic and government scientists. Each party plays a different role that is
shaped by the culture and political economy in which the party exists. The
central agent in the making and promoting of a corporate science to defend
soda, ILSI, is a complex, sophisticated, and highly secretive entity skillfully
designed to accomplish its product defense and other goals. Its effectiveness is rooted in six features of structure and operation that are likely to be
found in many if not most product-defense nonprofits:

1. Dual structure and mode of operation (visible/invisible, formal/
   informal);
2. Hidden pathways that allow industry to quietly influence the science
   and nonprofit leaders to discreetly influence public policy;
3. Nominal ethics rule purporting to ensure scientific integrity;
4. Recruitment and incentivization systems by which key personnel (academic advisers, heads of nonprofit branches) are turned into reliable
   allies of the organization;
5. Invitation-only rules of participation that create a quasi-closed scientific world of mostly if not solely industry-friendly voices;
6. For organizations that are global in scope, a hierarchical global structure centering power in the global core, combined with mechanisms to

facilitate the top-down flow of ideas and practices to branch countries around the world.

Sociology of Corporate Recruitment and Cultivation of Academic Advisers

Corporations outsource much of their product-defense science making to (nominally) nonprofit organizations, but they also have science officers who work to turn academic scientists into quasi-corporate or virtually corporate scientists willing to do industry's bidding. The dynamics of that process amount to a sociology of corporate recruitment and cultivation of external (noncorporate) scientists that works in at least three distinctive ways:

1. Incentivization of academic and governmental scientists through provision of research grants and other nonmonetary professional benefits;
2. Cultivation of loyal allies through the development of multidimensional relationships via the exchange of favors, humor, personal charm, and us-versus-them narratives of scientific warfare;
3. Collaboration between the company and various scientists through organization of activities promoting product-defense science around the world (e.g., conference sessions).

Methods

This appendix provides a narrative description of my research, tracing the broad outlines of the process by which I came to know what I know. It then describes the use of specific methods mentioned in the text, lists my expert-informants by institution and generic job title (for Chinese informants) or name (American and European informants), and outlines the limitations of the research. The final section presents excerpts from ILSI North America's bylaws concerning technical committees.

Looking for Big Food in China

In this book I've told a tidy, if complex, story of how Coca-Cola and ILSI created, spread, and translated soda science into public policy. When I began these inquiries in 2013, the field I hoped to study was mostly a blank. I started with a clear plan for filling in the blanks but soon discovered that many things I wanted to know were wrapped in layer upon layer of secrecy. Unearthing those secrets turned out to involve a long, convoluted process that took me to places I never imagined going. This sort of serpentine research process is unexceptional in anthropology, as many if not most projects start by asking one question and, after multiple detours, end by answering a different one. Yet for those wishing to try their hand at investigating corporate science, or those merely curious about how one might uncover the secrets of corporate science, I want to back up to the beginning and relay, briefly, how I discovered all the pieces that make up this story.

BEIJING FIELDWORK

Given the lack of basic information about almost everything, it was necessary to start with open-ended questions and work inductively, slowly building up an account of how science and policy were made, actor by actor,

episode by episode. My plan was to construct a timeline of the key moments in China's obesity science making and policy making from around
1990, and then trace important developments in that history back to key
forces, events, and actors in the larger political and economic environment,
first at the Chinese level, and then at the global level. At the outset, the
plan seemed straightforward, if a bit ambitious. Before leaving for fieldwork
in China, I carefully examined the published research on Chinese obesity,
concentrating on the applied or public health branch of the field. (There
was virtually no basic research on the condition.) With the information
gathered on key scientists, institutions, and events, I began putting together
a working timeline of the development of China's science and policy on
obesity.

In late fall 2013, I spent ten weeks in Beijing conducting wide-ranging
ethnographic fieldwork on the post-1990 history of obesity science and
policy. I had received human subjects approval from Harvard, my home
institution, as well as from Tsinghua University, whose Research Center
for Public Health, led by Professor Jing Jun, kindly hosted me as a visiting
scholar.[1] The ethnography involved informal research in various conferences, fast-food restaurants, drugstores, and bookstores. The heart of the
project, though, was a set of in-depth, semistructured interviews, one to
four hours in length, with twenty-five experts on obesity or related topics,
including most of the top researchers in my target field of applied obesity
research. Most of the interviews were conducted in the informant's office,
with a Chinese research assistant present. The interviews were conducted
in Chinese, English, or both, depending on my informant's language preference. With two exceptions, all the core interviews were taped (the exceptions were due to equipment failure). I met with nutritionists, sport
scientists, CDC researchers, corporate scientists, and even a public health
ethicist, each providing a different insider perspective. As the interviews
proceeded, however, I discovered that what I most wanted to know (details
about corporate funding of science) no one wanted to—or was able to—tell
me. Instead, they sent me to an organization that I had never heard of but,
according to my informants, was immensely powerful.

Within two weeks of starting fieldwork, I learned of the existence of the
industry-funded scientific nonprofit International Life Sciences Institute
(which had a very low profile at the time), based in Washington, DC, and
discovered that its China branch was the dominant organization working
on obesity in the country. ILSI-China's founding director described it as a
company-funded organization devoted to making disinterested science for
use by China's policy makers. This mission statement left me profoundly
puzzled. I simply could not fathom how an organization funded by the

food industry could produce objective science. Might that statement mean something different in the Chinese context? I also discovered that the science circulating in China was made not in China, but in the United States, and then exported to China and other countries through ILSI's global channels. That whole process would need to be mapped out. ILSI would turn out to be the predominant vehicle through which Coca-Cola influenced the understandings and policies on obesity around the world. Figuring out how it worked was the key to the success of my research, and it claimed an enormous amount of my time.

Serendipity played a huge role in my research. A chance conversation led to some of the most important discoveries of the whole project. While chatting with one of ILSI's heads one day, I learned of the newsletters the China branch issued twice a year in Chinese and English documenting all its activities. Because ILSI was the primary sponsor of obesity research and policy activities in China, those newsletters contained a written record of virtually all the most important events in the history of obesity management in the country. They would become treasured sources of information; indeed, it was these publications, strategically analyzed, that led to the discovery of the remarkable impact of soda science on China's approach to obesity. ILSI-China had inadvertently let me in on its secrets. Near the end of the fieldwork period, I overheard discussion of a big conference on obesity prevention and control in China, and I asked if I could attend. ILSI generously agreed. Over two days in a hotel ballroom in Beijing, I had my first encounter with the American principals of the GEBN and was able to observe firsthand how obesity science and policy were produced in practice.

US AND EUROPEAN INTERVIEWS, INTERNET RESEARCH ON KEY ORGANIZATIONS

The fieldwork generated much of the local history and context I had sought to map out, but it also raised a multitude of new questions. Back in the United States, the project began to snowball. In late 2015, when the *New York Times* published its exposé of Coke's promotion of exercise for obesity, I realized Coca-Cola's project was global, and I was investigating one of the biggest national targets of that effort. I resolved to use the in-depth methods at my disposal to get to the bottom of what Coke and its partners were doing, no matter what it took. As I tried to understand the larger forces at work, it became clear that the management of China's obesity epidemic was embedded in, and to some extent shaped by, four large, internally complex institutional systems, all international in scale. Understanding how they worked, what they did, and how they were entangled with each other was a critical task. In this way the project grew in scope and complexity to include not just ILSI-Global, mentioned above, but also the Coca-Cola Company, the Exercise Is Medicine (EIM) program, and the World Health Organization (WHO), all of which were active in China. For each of these institutional systems, there were countless resources requiring study. None of this was in the research plan.

Corporations are notoriously secretive, but so too, I quickly learned, are scientific nonprofits like ILSI that are funded by rich and powerful companies to quietly do industry's shady work. To get beneath the surface of ILSI's public narrative about itself as a charitable organization making science for public benefit, I dug into the many materials archived on the organization's website, including legal documents, annual reports of key ILSI entities, and conference programs and papers from ILSI-sponsored meetings. I discovered the IRS 990 tax forms available for all US-based ILSI entities and mined them for information on board membership, activities, and funding, all valuable clues to how ILSI worked. In 2016, I was granted an interview with the executive director of ILSI-Global in Washington, DC. A key insider at ILSI since the 1980s, her descriptions of the work of the organization and her cameos of key personalities proved indispensable.

I was developing a clearer understanding of the management of obesity in China, but what was the larger context in which China's experience was unfolding? How was China's approach viewed by the community of global obesity experts? For insight into those questions, in 2015–16, I visited twelve internationally prominent obesity specialists in the United States and Europe. Lengthy conversations with these leaders of the field associated with the WHO, the International Obesity Task Force (IOTF), the US CDC, and leading American universities helped me see that food-industry influence was endemic and that the narrative of scientific harmony under ILSI-China's leadership was achieved in part by the exclusion of dissidents. About half the Euro-American experts had worked in China *before* the soda industry came on the scientific scene (around 2004). Their perceptions of how things changed once the industry became involved added a disturbing new episode to my story.

ONLINE RESEARCH ON SCIENCE, POLICY, AND EMAIL COMMUNICATIONS

To understand the distinctive features of soda science, I turned to the internet, one of the most crucial sources available for the social study of scientific communities and their practices. With everything from the professional biographies of the leading scientists, to conference talks, to publications fully archived online, the internet was an essential source of information on soda science. Using online ethnography, I was able to trace soda science's American history, step by step, from its birth to its death over some twenty years, tracking who made it, how, and so forth. After August 2015, when the science-buying scandal silenced the principals at Coke and ILSI, it became virtually the only available source of data. I had intended to return to China to reinterview my expert-informants, but after the exposé of Coke's funding of obesity science, it became clear such interviews would not be welcomed.

Because a key objective of soda science was to influence policy, especially by discouraging regulation or legislation that might threaten profits, it was important to examine China's policies on obesity and chronic disease more generally. To see whether soda science and its industry-friendly ideas left their fingerprint on China's official policies, during these years I regularly monitored the websites of China's Ministry of Health (variously named), the China CDC, and the WHO for news and policy developments. Then, in 2021, I conducted a systematic examination of all of China's policies, programs, and guidelines on chronic disease prevention and control from the late 1980s to 2020 to assess the nature and extent of food-industry influence.

For an intimate look at the complex ties between Coca-Cola and its US-based academic collaborators, there is no better source than the emails they exchanged over many years, not realizing their messages might be subject to document requests through the Freedom of Information Act. Copies of their emails were indeed requested by the nonprofit US Right to Know (and other organizations), then archived on the website of the University of California San Francisco Library. In 2022 I dug into these emails in search of the nitty-gritties of the relationship between Coke and its academic advisers. That research, spanning the years 2009–15, put faces on the people making soda science in the United States and revealed the fears and bravado that drove all parties to take risks that ultimately brought the entire project down.

By the time I had to bring my inquiries to an end, I had conducted two phases of (nonconsecutive) research stretching over nearly ten years: a first, more intensive phase of in-person data gathering (2013–16), followed by a second phase of mostly online research and data analysis (2017–22). By the end, I had gathered a large body of empirical material, which I assembled into a set of data archives and histories. Nonetheless, the data are necessarily incomplete. I provide an assessment of their limitations just below.

Specific Methods and Expert-Informants

BEIJING INTERVIEWS

While in Beijing I was very keen to talk to as many people as possible involved in the making of China's obesity science and policy. I started with a list of key informants I had created in advance and interviewed everyone I was able to locate with the help of my capable research assistant, a recent graduate in communications from Tsinghua. I talked to the two main leaders of ILSI-China and thirteen government- and university-based researchers who had been involved in ILSI research projects or conferences (or both) on obesity. These fifteen individuals involved with ILSI-China formed the core group of interviewees. To gain a broader perspective, we located ten other experts through snowball sampling; these are listed below. I also had brief discussions with the two American academics involved with the GEBN; these are not included on the list of interviewees.

In this list of those I talked to, aside from the two ILSI-China directors, I use generic job titles ("officer," "senior researcher") to protect my informants' identities. Affiliation and position are those at the time of the interview. The numbers here do not correspond to the numbers in the Interview File (IF) that are cited in the notes to the text.

Core Interviewees

1. Senior Adviser, ILSI-China (2004–18), Founding Director, ILSI-China (1993–2004); formerly Founding President, Chinese Academy of Preventive Medicine; interviewed in Beijing (1:35)
2. Director, ILSI-China (2004–ca. 2019), Deputy Director, ILSI-China (1993–2004); Senior Research Professor, China National Center for Food Safety Risk Assessment, and Institute for Nutrition and Food Safety, China CDC; interviewed in Beijing (1:50)
3. Officer, ThinkTank Center for Health Development (1:45)
4. Professor, Institute of Child and Adolescent Health, School of Public Health, Peking University Health Science Center (3:30)
5. Professor, Department of Nutrition and Food Hygiene, School of Public Health, Peking University Health Science Center (2:00)
6. Senior Researcher, Institute for Nutrition and Food Safety, Chinese Center for Disease Control and Prevention (China CDC) (1:45)
7. Senior Researcher, China Institute of Sport Science, General Administration of Sport of China (1:55)

8. Senior Researcher, National Center for Women's and Children's Health, China CDC (1:25)
9. Officer, Chinese Nutrition Society (CNS) (3:00)
10. Officer, Chinese Nutrition Society; formerly Senior Researcher, National Institute of Nutrition and Food Safety, China CDC; before that, Chinese Academy of Preventive Medicine (1:50)
11. Senior Researcher and Director of Nutrition and Health, Greater China branch of a major Euro-American food corporation (1:05)
12. Physician and Clinical Nutritionist, Department of Clinical Nutrition, Beijing An Zhen Hospital of Capital Medical University (1:40)
13. Officer, Institute for Noncommunicable Disease Control and Prevention, Beijing CDC (1:05)
14. Officer, US CDC China Country Office (several short conversations)
15. Officer and Chief Science and Health Officer, the Coca-Cola Company (short conversation)

Other Specialists on Nutrition, Obesity, and Public Health

16. Senior Researcher and Educator, Department of Nutrition, Chinese People's Liberation Army (PLA) General Hospital (1:40)
17. Associate Professor, Department of Pharmacology and Pharmacological Sciences, School of Medicine, Tsinghua University (1:35)
18. Physician, Peking University People's Hospital (1:00)
19. Physician, Peking Union Medical College Hospital (3:00)
20. Clinical Nutritionist, Beijing Friendship Hospital (4:00)

Other Informants Knowledgeable about Science and Industry-Science Relations in China

21. Director, Research Center for Public Health and Professor of Sociology, Tsinghua University (multiple conversations of varying lengths; same for informants 22 to 25)
22. Associate Professor, Institute of Science, Technology, and Society, Tsinghua University
23. Assistant Professor, Institute of Science, Technology, and Society, Tsinghua University
24. Professor, Department of Philosophy, Tsinghua University
25. Employee of American public relations company, working for food and other companies

INTERVIEWS IN THE UNITED STATES
AND WESTERN EUROPE

WHO and International Obesity Task Force (IOTF)

1. Chizuru Nishida, Coordinator of Nutrition Policy and Scientific Advice Unit, Department of Nutrition for Health and Development, World Health Organization (WHO); interviewed in Geneva, Switzerland (3:05)
2. Philip (W. P. T.) James, Physician, Founding Director of the International Obesity Task Force (IOTF); interviewed along with Jean James in London, England (4:00+)
3. Neville Rigby, formerly Director of Policy and Public Affairs and Media Expert, International Obesity Task Force (IOTF); interviewed in Inverness, Scotland (3:30+)

US CDC

4. William Dietz, Professor, Director, and Sumner M. Redstone Center Chair, Sumner M. Redstone Global Center for Prevention and Wellness, Milken Institute School of Public Health, George Washington University, and formerly Director of the Division of Nutrition, Physical Activity, and Obesity at the Center for Chronic Disease Prevention and Health Promotion of the CDC; interviewed in Washington, DC (2:50)
5. Katherine M. Flegal, Epidemiologist, Senior Scientist at National Center for Health Data, US CDC; interviewed in Washington, DC, and by telephone and email (10:00+)

US Universities

6. James O. Hill, Professor of Pediatrics and Medicine, Executive Director of Anschutz Health and Wellness Center, University of Colorado; interviewed in Boulder, CO (1:10)
7. Shiriki K. Kumanyika, Professor of Epidemiology, Emeritus, Department of Biostatistics and Epidemiology, University of Pennsylvania Perelman School of Medicine; interviewed in Washington, DC (2:15)
8. Barry M. Popkin, Distinguished Professor of Nutrition, Gillings School of Global Public Health, University of North Carolina, Chapel Hill; interviewed by telephone (1:10)
9. William C. Hsiao, K. T. Li Research Professor of Economics, Emeritus,

Departments of Health Policy and Management, and Global Health and Population, Harvard T. H. Chan School of Public Health, Harvard University; interviewed in Cambridge, MA (2:00)

10. Frank B. Hu, Professor of Nutrition and Epidemiology, Department of Nutrition, Harvard T. H. Chan School of Public Health, Harvard University; interviewed in Boston, MA (1:00)

11. Li Yanping, Senior Research Scientist, Department of Nutrition, Harvard T. H. Chan School of Public Health, Harvard University; interviewed in Boston, MA (1:45)

ILSI-Global

12. Suzanne Harris, Executive Director, ILSI-Global; interviewed in Washington, DC (1:30)

ANALYSIS OF ILSI FOCAL POINT IN CHINA NEWSLETTERS

To understand trends in obesity-related activities managed by ILSI-China, I analyzed the content of its semiannual newsletters between 1999 and 2015. I first identified all the articles on activities concerning obesity or obesity and other chronic diseases. In the thirty-four issues published between 1999 and 2015, seventy-two news items (representing 25 percent of the total) fit that criterion. I then selected for further analysis all items reporting concrete activities (not including talks) held in China. Next I coded the seventy-two articles for emphasis: physical activity, dietary strategies, neither, or both. This was based in the first instance on the name of the event, which was usually also the title of the news item. In 92 percent of the cases, the emphasis was unambiguous. For the other 8 percent of activities, I read the articles closely, paying particular attention to the description of the event's aim. I categorized them as neither if they focused on measurement issues (rather than intervention strategies) or both if they emphasized exercise and dietary strategies to some extent. All classifications were checked three times. Examining changes in emphasis over time (1999–2003, 2004–9, 2010–15) revealed a remarkable increase in obesity-related events emphasizing exercise and a corresponding decline in events emphasizing nutrition. These results are discussed in chapter 6. I then created a subset of articles on activities emphasizing physical activity and mined those for information on featured speakers, company speakers, key scientific concepts and conclusions of these speakers, and, where available, funding. This information was readily located in the text of the news items. It is used in chapters 6 and 7.

RESEARCH LIMITATIONS

The sensitivity of the issues examined made it difficult to interview some key Chinese informants. Officials at the Ministry of Health (in 2013 reorganized and renamed the National Health and Family Planning Commission; in 2018 renamed the National Health Commission) did not respond to interview requests. It also proved difficult to talk to scientists associated with the multinational food companies in China. In the end, I managed to interview one company scientist, who requested that the conversation not be recorded and their name not be used.

After August 2015, when the *New York Times* story broke about the Coca-Cola Company paying leading exercise scientists to promote physical activity, it became nearly impossible to interview on this subject. The three Americans at the heart of that scandal (Applebaum, Blair, Hill) had all been involved in work on China's obesity science for many years. Understandably, they were reluctant to talk to me. Hill agreed to an interview in late 2015 but kept the conversation very general; the other two did not respond to requests for interviews in late 2016. My attempt to arrange an interview with Alex Malaspina did not succeed. It would have been helpful to learn my Chinese informants' reactions to the findings of this research, but it became clear such an effort would not be welcome.

ILSI-China's semiannual newsletters were a superb source of information on virtually all the activities China organized to combat obesity. They had limitations, however. The articles touched only on the highlights of the events described. Unsurprisingly, they also presented a uniformly positive image of the organization, no doubt representing what Chen Chunming wanted the outside world to think. For information on controversies and conflicts I had to rely on other sources, primarily interviews. Apparently mandated (or merely expected) by ILSI-Global, the newsletters may also have been biased toward coverage of events that conformed to the policies and preferences of ILSI headquarters at the time. There is no way to check that. Nonetheless, since all activities had to have funding, and virtually all the activities on obesity were supported (at least in part) by ILSI or its member and supporting companies (other than what the MOH contributed), one might expect the activities reported in the newsletters to align with ILSI-Global preferences. I attended the big 2013 obesity conference, and a comparison of what I witnessed with the account of the conference in the newsletter shows a close correspondence. That encourages a favorable view of the newsletters as sources of information. Chen Chunming took this responsibility very seriously, and my experience with using the publications over many years leads me to believe the information presented is accurate.

Excerpts from ILSI North America Bylaws Regarding Technical Committees (2015)

ARTICLE XI: TECHNICAL COMMITTEES

SECTION 1. Within the Programs, the Board of Trustees shall establish Technical Committees which will investigate matters of public health or safety, sponsor research, obtain financing and assume such other duties as may be assigned by the Assembly of Members or the Board of Trustees. Membership in a Technical Committee will ordinarily be open to all active Members of ILSI N.A. in good standing who have a good faith interest in the mission and objectives of the Technical Committee, and who are willing to pay their equitable share of the cost of the Technical Committee's activities. Each Technical Committee shall establish its own nondiscriminatory membership format and its own program.

SECTION 2. Each Technical Committee shall be responsible for all expenses it may incur, and for obtaining the funds necessary to meet those expenses. Funds shall normally be obtained by means of an assessment of the members of such Technical Committee, devised according to a formula adopted by that Committee and approved by the Board of Trustees. Such assessment shall then be imposed upon Committee members.

SECTION 3. The Chairman of a Technical Committee shall be elected by the members of the Committee and shall be responsible to the President and the Board of Trustees for the management of the Committee. He shall (i) preside at all meetings of the Committee; and (ii) see to the carrying out of all orders and resolutions of the Assembly of Members and the Board of Trustees pertaining to the Committee. In consultation with the other members of the Technical Committee, the Chairman may (i) appoint observers to the Committee, as appropriate; (ii) determine the membership of the Committee, as may be appropriate and consistent with the purposes of the Committee; and (iii) manage the dissemination of information produced by the Committee as is consistent with policies of ILSI N.A. regarding the dissemination of information. In the Chairman's absence, the Vice Chairman elected by the members of the Committee or a designated member shall have the responsibilities of the Chairman and shall assume all of the duties of the Chairman during the latter's absence, disability, refusal to act or resignation.

SECTION 4. Scientists from companies, in conjunction with qualified non-industry scientific advisors appointed by the Technical Committee, shall, in the normal course of business, guide the activities of each Technical Committee.

[Section 5 omitted.]

SECTION 6. Each Technical Committee shall submit a report of the Committee's operational activities at the annual meeting of the Assembly of Members. Each Technical Committee shall also submit a report of its operational and financial activities at the annual meeting of the Board of Trustees at which time the Board of Trustees shall approve the Technical Committee's budget.

SECTION 7. Technical Committees shall operate according to the general policies, procedures and guidelines of ILSI N.A.[2]

Notes

Introduction

1. I wrote this book very much for my fellow Americans, though, in fact, I hope that it will find a much wider readership. Please be assured that the familiar "we" I use to address other Americans is not meant to exclude anyone from any part of the world from the discussions in which this book takes part.

2. Surgeon General of the United States, *Surgeon General's Call to Action to Prevent and Decrease Overweight and Obesity*, v.

3. This story is told in the first four chapters of Critser, *Fat Land*, 7–108.

4. A compelling history of how fatness came to be so deeply stigmatized is A. Farrell, *Fat Shame*.

5. The percent of adults (those aged twenty and older) classified as overweight or with obesity rose from 56 (1988–94) to 64.5 (1999–2000) to 73.1 (2017–18), an increase that is huge by any measure. These and other statistics can be found in Fryar, Carroll, and Afful, "Prevalence of Overweight."

6. The arc of this fifteen-year panic over obesity can be seen in the answers to a Gallup poll question about Americans' health concerns. The proportion naming obesity as America's most urgent health problem rose steadily between 2000 and 2007 (from 3 to 10 percent) and peaked between 2010 and 2015 (at 16 percent), before declining between 2016 and 2021 (to 4 to 8 percent). During those peak years, obesity was deemed a more urgent problem than cancer almost every year. See Gallup News Service, "Healthcare System."

7. The figures on weight-loss practices and perceptions in this paragraph are drawn primarily from CDC and Gallup Poll data. See Martin et al., "Attempts to Lose Weight"; Brenan, "What Percentage of Americans Consider Themselves Overweight?" A useful compendium of related statistics is *Insight*, "81 Weight Loss Statistics and Health Benefits in 2023."

8. Hall, "Energy Compensation and Metabolic Adaptation." See also, Hall, "Diet versus Exercise in 'The Biggest Loser.'"

9. New drugs introduced since 2021 (Wegovy, Ozempic) have been remarkably successful in lowering weight, but their high price, limited availability, and common gastrointestinal side effects make them a poor choice for many. Another deterrent is that stopping the drug generally leads to regaining most of the pretreatment weight.

10. Freudenberg, *Lethal but Legal*.

11. On the making and unmaking of the US-centered neoliberal order, see Gerstle, *Rise and Fall of the Neoliberal Order*.

12. Freudenberg, *Lethal but Legal*. Freudenberg develops the arguments further in *At What Cost*.

13. Swinburn et al., "Global Syndemic of Obesity, Undernutrition, and Climate Change." The World Health Organization (WHO) reports that in 2016, 52 percent of adults worldwide (age eighteen and older) were overweight or had obesity; between 1975 and 2016 the prevalence of obesity globally nearly tripled. America is far from alone in its obesity problem. See WHO, "Obesity and Overweight Fact Sheet." See also Hall, "Did the Food Environment Cause the Obesity Epidemic?" Hall argues that while disentangling the contributions of various environmental factors is difficult, it seems clear that the food environment is likely the primary driver.

14. Moss, *Salt, Sugar, Fat*.

15. Moss, *Hooked*.

16. Hall, Ayuketah, et al., "Ultra-processed Diets Cause Excess Calorie Intake and Weight Gain."

17. Monteiro et al., "Ultra-processed Foods." Van Tulleken's important book *Ultra-processed People* marshals extensive evidence showing that what's harmful about these foods is not that they contain too much of the bad things (salt, sugar, fats) but that they are hyperprocessed.

18. WHO, "Obesity and Overweight Fact Sheet."

19. WHO, "Global Health Observatory, Noncommunicable Diseases."

20. Thacker, "Coca-Cola's Secret Influence."

21. For a more scientific perspective on the health effects of drinking sugary soda, see Harvard T. H. Chan School of Public Health, "Sugary Drinks."

22. O'Connor, "Coca-Cola Funds Scientists."

23. Michaels, 247–49.

24. Classic histories of science include Brandt, *Cigarette Century*; Oreskes and Conway, *Merchants of Doubt*; Proctor, *Golden Holocaust*.

25. Nestle, *Soda Politics*, 104.

26. M. Simon, *Appetite for Profit*, 13. Also relevant is her pamphlet *Best Public Relations Money Can Buy*.

27. Since 2018, the work of these scholars has appeared with increasing frequency in *Globalization and Health*, *Public Health Nutrition*, and other public health journals. The studies, which rely primarily on internal emails and documents, tend to support the findings of my more in-depth historical research, and I cite a number of them in the text and notes.

28. If, as science studies suggests, scientific knowledge (and technological artifacts) are humanly constructed, then it's important to view scientists and engineers in action, manipulating things, making claims, gaining allies, and so forth. As a human product, science comes to bear the stamp of its human makers, as well as of the institutional and broader political-economic and cultural contexts in which it is made. These ideas are

fleshed out by Knorr-Cetina, *Manufacture of Knowledge*; and Hacking, *Social Construction of What?*

29. J. Greene, *Prescribing by Numbers*; Petryna, *When Experiments Travel*; Sismondo, *Ghost-Managed Medicine*; Dumit, *Drugs for Life*.

30. Adams, "Metrics of the Global Sovereign," 25.

31. Erikson, "Metrics and Market Logics of Global Health," 158.

32. Nestle, *Unsavory Truth*, 29–32. Additionally, studies of diet and health are largely observational, rendering them especially vulnerable to biases in design and interpretation.

33. Michaels, *Triumph of Doubt*, 15–26.

34. I call this a *fact* rather than merely a *claim* because it was presented to the food industry's scientific nonprofit as an unchallengeable basis for a new industry-friendly science of obesity that had to be made.

35. Translational research is that aimed at translating the results of basic research into results that directly benefit humans. The term is usually used in medicine but works well for public health too.

36. My use of this term was inspired by Anderson, *Imagined Communities*. Elaboration must be left for elsewhere.

37. Sismondo, "Epistemic Corruption."

38. Michaels, *Triumph of Doubt*, 11–12.

39. Sismondo, *Ghost-Managed Medicine*, 178, 179.

40. The analysis here was inspired by Sismondo's discussion of the work performed by rules of ethical conduct among publication planners working to benefit the pharmaceutical industry. In the case I studied, such ethics rules operated in virtually every phase of the making, circulating, and translating of science, leading me to generalize the ideas about ethics rules to the full scientific process. See Sismondo, *Ghost-Managed Medicine*, 107–9.

Chapter One

1. Kuczmarski et al., "Increasing Prevalence of Overweight," 207. At the time this article was published, "overweight" was defined as BMI equal to or greater than 27.8 and 27.3 (for men and women, respectively). In 1998 overweight was redefined as BMI 25–29.9, and differentiated from obesity (30+).

2. Callahan, "Q&A—Katherine Flegal."

3. Critser, *Fat Land*, 4.

4. Pi-Sunyer, "Fattening of America." 238.

5. Brownell, "Get Slim with Higher Taxes."

6. Malaspina's background is from Toxicology Forum, "History of the Toxicology Forum."

7. On ILSI's successful protection of caffeine from regulation, see James, "'Third-Party' Threats"; and James, "Caffeine, Health, and Commercial Interests."

8. Profile in this paragraph based on Hill, Curriculum Vitae.

9. Critser, *Fat Land*, 3.

10. The pervasiveness of industry connections in the field of nutrition science is amply documented in Nestle, *Unsavory Truth*, esp. 1–12.

11. Litman et al., "Source of Bias in Sugar-Sweetened Beverage Research."

12. Marion Nestle documents the struggles within the nutrition field over these issues, concluding that "nutrition professionals still have a long way to go to create a culture that views conflicts of interest as a problem worth serious attention." Her work also suggests that the kinds of rhetorical practices Hill used are common in the field. See her *Unsavory Truth*, 164–65, 216.

13. Sheldon Krimsky, one of the first to systematically investigate the funding effect—the correlation between a research outcome and the funding source— maintains that it provides not definitive but prima facie evidence that bias may exist. He also suggests that ethnographic research can be useful in determining when scientists manipulate data, methods, findings, and so forth in ways favorable to funder interests. See Krimsky, *Conflicts of Interest in Science*, esp. the chapter "Do Financial Conflicts of Interest Bias Research? An Inquiry into the 'Funding Effect' Hypothesis," 232–53.

14. CEO in the subheading and elsewhere stands for chief executive officer. This and other abbreviations in more general use are not included in the list of abbreviations in the frontmatter; only those particular to this work are included in the abbreviations list. Koplan and Dietz, "Caloric Imbalance and Public Health Policy."

15. A well-known set of four maps (1991–98) was published in A. H. Mokdad et al., "Spread of the Obesity Epidemic." Their impact is stressed in Dietz, "Response," a history of the official public health response to obesity. On the power of visual representations in science generally, it's hard to beat Lynch and Woolgar, *Representation in Scientific Practice*.

16. Jacobson, *Liquid Candy*.

17. Nestle, *Food Politics*, 198–99.

18. In 1999, Coca-Cola held 44.1 percent of the US soft-drink market. That was down half a percentage point from the prior year and the first drop in market share in six years. See Bloomberg News, "Coca-Cola's Market Share."

19. Jacobson and Brownell, "Small Taxes on Soft Drinks and Snack Foods."

20. For the CDC's effort, see Dietz, "Response." The US Surgeon General did indeed issue a major report: Surgeon General of the United States, *Surgeon General's Call to Action to Prevent and Decrease Overweight and Obesity*.

21. The Pillsbury executive was James R. Behnke. An online biography indicates that he served on the ILSI board for many years and was president of the ILSI North America board from January 2000 to January 2003. My source is Behnke, "Profile of James Behnke."

22. The passages below are drawn from the typed version of the talk: Mudd and Hill, "Remarks for ILSI CEO Dinner."

23. Experts distinguish between *physical activity*, any bodily movement produced by skeletal muscles that requires the expenditure of energy, and *exercise*, structured activity aimed at maintaining or improving physical fitness. Physical activity is what really

matters. Since exercise is the more colloquial term and the difference is not important to our story, I use the terms interchangeably.

24. Hill and Trowbridge, "Childhood Obesity."

25. Moss, *Salt, Sugar, Fat*, esp. xxi–xxii.

26. Hill and Peters, "Environmental Contributions"; Hill and Trowbridge, "Childhood Obesity"; and Klem et al., "Descriptive Study."

27. Of course, the sequencing of scientific activities is often idiosyncratic; in the case of product-defense science, the irregularity bears study because it appears to be common to a class of sciences.

28. Moss, *Salt, Sugar, Fat*, xi.

29. ILSI, "Early Commitment," states: "In 1999, ILSI approached the CEOs of food companies—a major source of ILSI funding—with evidence showing the public health threat and outlining comprehensive ways to address both food intake and physical inactivity of the energy imbalance equation."

30. Information from ILSI North America, *ILSI North America 2015 Annual Report*.

31. The key documents are ILSI, "Code of Ethics and Organizational Standards of Conduct" (dated 2009), and ILSI "Bylaws, Conformed as of January 17, 2015."

32. ILSI-Global, "Mission and Operating Principles." On the international focus discussed just below, see ILSI, "Global Network."

33. ILSI, "Bylaws, Conformed as of January 17, 2015."

34. The numbers in this paragraph are from ILSI North America, *ILSI North America 2015 Annual Report*, 30; ILSI, *Global Partnerships for a Healthier World*; data are for 2015.

35. ILSI, "Media FAQs."

36. Boseley, "WHO 'Infiltrated by Food Industry.'"

37. ILSI, "Mission and Operating Principles."

38. Internal Revenue Service, District Director, "1985 IRS Tax Code Determination." On the legal definition of business leagues, see Internal Revenue Service, "Business Leagues."

39. Coca-Cola Journey Staff, "Coca-Cola Honors 10 Young Scientists from around the World"; see also Dews, "Foreword," v.

40. Internal Revenue Service, "Exemption Requirements—501(c)(3) Organizations."

41. The idea that norms serve as ideological resources to be flexibly interpreted and manipulated by actors is a staple of constructivist thought in science studies. A compelling case for this approach can be found in Gieryn, *Cultural Boundaries*. On interpretive flexibility generally, see Pinch and Bijker, "Social Construction of Facts and Artifacts."

42. Sismondo, *Ghost-Managed Medicine*, 108–9.

43. ILSI, "Code of Ethics and Organizational Standards of Conduct."

44. In November 2015, the ILSI Board Executive Committee suspended the Mexico branch for engaging in activities that appeared to support the repeal of the country's tax on sugar-sweetened beverages. Those activities appeared to violate ILSI, "Code of

Ethics and Organizational Standards of Conduct," which proscribed lobbying and advocacy. The branch was reinstated and later merged with ILSI-Mesoamerica. See ILSI, "ILSI Response to Globalization and Health."

45. The CHP also promoted Malaspina's second pet project, the micronutrient fortification of foods to support child health.

46. Malaspina, "Celebrating 20 Years."

47. ILSI-CHP, IRS Form 990, 1999, 16.

48. Malaspina, "Global Partnerships."

49. This story is told by the journalist Gary Taubes in *The Case against Sugar*, 107–43. See also Kearns, Schmidt, and Glantz, "Sugar Industry and Coronary Heart Disease."

50. Hill, "Child Obesity," 3–4.

51. Quotations in this paragraph are from Hill, "Child Obesity," 5 and 6.

52. McKenzie, *Getting Physical*.

53. Pate et al., "Physical Activity and Public Health."

54. US Department of Agriculture and US Department of Health and Human Services, *Nutrition and Your Health: Dietary Guidelines*, 15.

55. Surgeon General of the United States, *Physical Activity and Health*.

56. Surgeon General of the United States, 135.

57. Surgeon General of the United States, 134.

58. Hill and Peters, "Environmental Contributions."

59. Klem et al., "Descriptive Study."

60. Wing and Phelan, "Long-Term Weight Loss Maintenance."

61. Kibbe et al., "Ten Years." This broad-based review of the literature shows that integrating classroom-based physical activity with academics is feasible, and it helps students concentrate on learning, raise academic scores, and achieve moderate energy expenditure. Some studies suggest that students undergoing fifteen minutes of Take 10! daily may realize BMI maintenance over time.

62. Peregrin, "Take 10!"

63. Peregrin.

64. Kibbe et al., "Ten Years."

UNMARKED CITATIONS

Interview

CORPORATE MEMBERSHIP AND FUNDING

Para. 4, IF-33, September 27, 2016, Washington, DC.

Chapter Two

1. Hill and Peters, with Jortberg, *Step Diet Book*, xiv.

2. M. Simon, *Appetite for Profit*, 169.

3. Simon, 167.

4. Hill, Wyatt, Reed, and Peters, "Obesity and the Environment." The concepts and approach are clarified in Hill, Peters, and Wyatt, "Using the Energy Gap."

5. More support for these views is presented in Peters, Lindstrom, and Hill, "Stepping Up across America."

6. A fascinating history and evaluation of step counting is Bassett et al., "Step Counting."

7. That large-scale campaign is discouraged in Hill, Wyatt, Reed, and Peters, "Obesity and the Environment," 854; and in Hill, Peters, and Wyatt, "Using the Energy Gap."

8. This paragraph draws heavily on McKenzie, *Getting Physical*; and Petrzela, *Fit Nation*.

9. The ideas here draw primarily from Robert Crawford's important essay "Healthism and the Medicalization of Everyday Life."

10. Crawford, "Health as a Meaningful Social Practice." The ideology and political economy underlying the corporate manufacture of chronic disease are brilliantly mapped out in Freudenberg, *Lethal but Legal*, and Freudenberg, *At What Cost*.

11. The desperation of many young Americans to achieve the thin, fit body comes through in the stories they tell about their lives. An account of these struggles and their roots in the rise of the obesity epidemic and its treatment of their bodies as a national disaster is my earlier book *Fat-Talk Nation*.

12. Quotations in this paragraph from Hill and Peters, with Jortberg, *Step Diet Book*, 19, 88.

13. Hill and Peters, with Jortberg, 123.

14. Quotations in this paragraph from Hill and Peters, with Jortberg, v.

15. A wide-ranging discussion of this by a longtime insider is Nestle, *Unsavory Truth*.

16. Peeke, "Foreword," ix, xi.

17. Schüll, "Data-Based Self."

18. This paragraph draws from Hill and Peters, with Jortberg, *Step Diet Book*, quotations on 30.

19. Ikeda et al., "National Weight Control Registry." More recently, others have argued that the value of the registry data is frequently overstated, given its methodological limitations. They point to the self-reporting of activity, nutrition, and weight/height information; the self-selected nature of the study population; and the associational rather than causal nature of the findings. See Chin, Kahathuduwa, and Binks, "Physical Activity and Obesity."

20. Hill, Wyatt, Phelan, and Wing, "National Weight Control Registry."

21. The new nonprofit was formed in the wake of a December 1997 meeting, the Tufts University Dialogue Conference on the Role of Fat-Modified Foods in Dietary Change, funded by the Proctor and Gamble Company. The details can be found in Rosenberg and Taylor, "Foreword."

22. Hill, Goldberg, Pate, and Peters, "Introduction."

23. The board membership can be found in Partnership to Promote Healthy Eating

and Active Living, IRS Form 990, 2001, 4. The summit funders are listed in Hill, Gold-berg, Pate, and Peters, "Acknowledgement."

24. For example, at the end of the diet book the authors provided a list of "useful re-sources for weight management." Of the dozens they might have included under general resources, they listed only six, three of which were programs the authors were person-ally associated with. A fourth was the International Food Information Council (IFIC), the industry-funded public outreach companion of ILSI, which the authors promoted as having a great website. See Hill and Peters with Jortberg, *Step Diet Book*, 271–72.

25. See Hill, Goldberg, Pate, and Peters, "Introduction."

26. Inferred from the absence of discussion of soda taxes in any of the materials I reviewed on soda science written between 2000 and 2015.

27. Hill, Goldberg, Pate, and Peters, "Introduction," S5.

28. Hill, Goldberg, Pate, and Peters, "Discussion," S59.

29. Mayo, Guidant Corporation, and Cargill also provided some funding. See Part-nership to Promote Healthy Eating, "Partnership to Promote Healthy Eating."

30. America on the Move Foundation, IRS Form 990, 2008, Attachment, 24; fund-ing details in the rest of this paragraph from same form, for the years 2004–13.

31. Partnership to Promote Healthy Eating and Active Living, "Partnership to Pro-mote Healthy Eating." The approach is evaluated by the Hill team in Rodearmel et al., "Small Changes in Dietary Sugar and Physical Activity."

32. Partnership to Promote Healthy Eating and Active Living, "Partnership to Pro-mote Healthy Eating."

33. Hill, Curriculum Vitae. The YMCA-AOM campaigns were held every September for at least five years (2006–10). See America on the Move Foundation, "America on the Move and More Than 1,500 YMCAs Kick Off Steptember 2007."

Chapter Three

1. Surgeon General of the United States, *Surgeon General's Call to Action to Prevent and Decrease Overweight and Obesity*, 15–25. Michele Simon describes the report as a huge wake-up call for the media, policy makers, and the general public. See M. Simon, *Appetite for Profit*, xiii.

2. Dimassa and Hayasaki, "L.A. Schools Set to Can Soda Sales."

3. Ludwig, Peterson, and Gortmaker, "Relationship between Consumption of Sugar-Sweetened Drinks and Childhood Obesity."

4. Coca-Cola, "Frequently Asked Questions."

5. Pan and Hu, "Effects of Carbohydrates on Satiety."

6. Nestle, *Soda Politics*, 104.

7. M. Simon, *Appetite for Profit*, 21–44.

8. M. Simon, 32.

9. Pendergrast, *For God, Country and Coca-Cola*, 430.

10. Coca-Cola Company, Form 10-K, Annual Report to the US Security and Ex-change Commission, 27.

11. Pendergrast, *For God, Country and Coca-Cola*; Foster, *Coca-Globalization*.

12. Pendergrast, *For God, Country and Coca-Cola*, 430. For the details and the wider context of such efforts in the company, see Short, "When Science Met the Consumer."

13. Coca-Cola Company, *2006 Corporate Responsibility Review*, 22.

14. Applebaum, Curriculum Vitae.

15. An analysis of email exchanges among Coke executives also suggests a deliberate and coordinated approach to influencing scientific evidence and opinion. See Sacks et al., "How Food Companies Influence Evidence and Opinion."

16. Applebaum, "Balancing the Debate."

17. Coca-Cola Company, *2010/11 Sustainability Report*, 55.

18. Short, "When Science Met the Consumer."

19. Buyckx, LinkedIn profile.

20. M. Simon, *Appetite for Profit*, 189.

21. The solutions also include offering a greater variety of beverages, including no- and low-sugar products.

22. Sallis, "Exercise Is Medicine," 3, 4.

23. Quoted in R. Greene, "Coca-Cola/ACSM 'Exercise Is Medicine' Scheme."

24. R. Greene.

25. Calculated from Coca-Cola Company, "Our Commitment to Transparency." EIM critics claim that Coca-Cola's influence is even more pervasive, since the company pays universities, individual scientists, and the founder and chair of EIM, as well as the ACSM itself. See R. Greene, "Coca-Cola/ACSM 'Exercise Is Medicine' Scheme."

26. For the ISCOLE funding, see Stuckler et al., "Complexity and Conflicts of Interest Statements." The estimated Coke funding for the University of South Carolina is calculated from the budgets in the research agreements (and continuing research agreements) included in the UCSFL email archive. Details are available from the author. In his reporting for the *New York Times* before the emails became available, Anahad O'Connor wrote that Blair had received over $3.5 million between 2008 and 2015, while Hand received $806,500 for the energy flux study, for a total of $4,306,500. See O'Connor, "Coca-Cola Funds Scientists."

27. This is only a subset of the articles reporting Coke funding. Analysis of the full network of researchers relying at least partly on Coke funding involved 1,496 researchers and 461 publications. Coke's transparency list, while useful, was far from complete. See Serodio, McKee, and Stuckler, "Coca-Cola—a Model of Transparency?"

28. Rowe et al., "Funding Food Science," 1286.

29. This paragraph is based on Rowe et al.

30. Rowe et al.

31. Nestle, *Unsavory Truth*, esp. 173–87.

32. Depending on the issue, members of this outer circle included Robert E. Sallis (EIM leader), Angel Gil (University of Granada), Kenneth R. Fox (University of Bristol), and Edward Archer (University of South Carolina), among others.

33. Blair, Curriculum Vitae.

34. Blair, Kohl, et al., "Physical Fitness and All-Cause Mortality."

35. Barlow et al., "Physical Fitness, Mortality and Obesity."

36. Critser, *Fat Land*, 102–8. The terms in quotation marks are on 103.

37. Critser, 103.

38. CNN, "US Scientist."

39. W. Zhu, "If You are Physically Fit."

40. Pomeranz and Brownell, "Portion Sizes and Beyond."

UNMARKED CITATIONS

Food Industry Documents (UCSFL)

ETHICS IN ACTION: DEVELOPING A PROTOCOL

Para. 2, Email with responses, Rhona Applebaum to Steven Blair, cc Karen Cunningham, Susan Roberts, "Budget Planning—Energy Balance Study," June 1, 2010.

Para. 3, Email with responses, Rhona Applebaum to Susan A. Roberts, Karen Cunningham, cc Steven Blair, "Research Planning.docx," July 19, 21, 22, 25, 28, 29, 30, August 2, 3, 2010. Participants added later: Meghan Baruth, Trudy T. Shawn, Christmus Gaye. Quotations from July 28, 29, 30, 2010.

FUNDS FOR RESEARCH

Para. 1, The Coca-Cola Company Research Agreement, between the Coca-Cola Company and the South Carolina Research Foundation, filed on Food Industry Documents site (UCSFL), October 28, 2010 (Agreement for original Energy Balance Study, signed October 28, 2010, and November 4, 2010).

Para. 2, The Coca-Cola Company Research Agreement, between the Coca-Cola Company and the South Carolina Research Foundation, filed on Food Industry Documents site (UCSFL) December 13, 2011 (Agreement for Energy Flux Study, signed December 13, 2011, and December 16, 2011).

Para. 2, The Coca-Cola Company Research Agreement, between the Coca-Cola Company and the South Carolina Research Foundation, filed on Food Industry Documents site (UCSFL) April 11, 2010 (misfiled) (Energy Balance, Continuation through Year 3, signed June 18, 2014, by USC participants; continuation through Year 2 signed November 19, 2013, by Coca-Cola officer).

Para. 3, Email with responses, Rhona Applebaum to Steven Blair and Carl (Chip) Lavie, "33rd International Symposium of Diabetes and Nutrition, Toronto, ON, June 10–12, 2015," May 15, 16, October 21, November 23, 2014. Quotation from November 23, 2014.

FACTS ON DEMAND

Para. 3, Email with responses, Rhona Applebaum to Steven Blair, "Heartburn on the Rise—and Scientists Aren't Sure Why—Today Health," November 15, 21, 2010. Participant added later: Mei Sui. Quotations from November 15, 21, 2010.

Para. 1, Email with responses, Rhona Applebaum to Robert E. Sallis, Steve Blair, cc Adrian Hutber, Jim Whitehead, Mindy Millard Stafford, Michael Pratt, "Re: IUNS Meeting in BKK [Bangkok]," September 16, 2009; Email with responses, Chip Lavie to Rhona Applebaum, Steve Blair, "Fwd: WCC 2012, invitation to speak—Medical Students course, Dr. Lavie," December 2, 2011.

Para. 2, Email with responses, Steven Blair to Rhona Applebaum, "RE: CSEP and Exercise Is Medicine," December 4, 2011; Email with responses, Rhona Applebaum to Steven Blair, K. R. Fox, "RE: ICSEMIS 2012 Sub-Themes (FINAL).doc," September 25, 2011.

Para. 3, Email with responses, Rhona Applebaum to Steven Blair, K. R. Fox, cc Karen Cunningham, "RE: ICSEMIS 2012 Sub-Themes (FINAL).doc," September 25, 2011; Email, Steven Blair to L. Celeste Bottorff, "RE: Olympic Torch Relay," August 17, 2011.

Para. 5, Email, Rhona Applebaum to Steven Blair, Greg Hand, K. R. Fox, Peter Katzmarzyk, Michelle Carfrae, Jim Moshovelis, Rosalyn Kennedy, Johanna Rangel, "TCCC [the Coca-Cola Company] and ICPAPH QA.docx," October 30, 2012.

Para. 6, Email, Rhona Applebaum to James Hill, Maxime Buyckx, cc Trudy T. Strawn, "FW: IUNS Meeting in BKK," September 16, 2009; Email, Rhona Applebaum to Steven Blair, K. R. Fox, cc Karen Cunningham, "RE: ECSEMIS 2012 Sub-Themes (FINAL).doc," September 25, 2011; Email, Steven Blair to Rhona Applebaum, "RE: Coke and Moscow—Media event," August 24, 2011.

Para. 7, Email, Rhona Applebaum to Steven Blair, Greg Hand, K. R. Fox, Peter Katzmarzyk, Michelle Carfrae, Jim Moshovelis, Rosalyn Kennedy, Johanna Rangel, "TCCC [the Coca-Cola Company] AND ICPAPH QA.docs," October 30, 2012.

Para. 2, Email with responses, Steven Blair to Rhona Applebaum, Russell Pate, Timothy Church, David Allison, Greg Hand, James Hebert, James Hill, "Emailing: Portion Sizes and beyond—Government's Legal Authority to Regulate Food-Industry Practices—NEJM.htm," October 11, 12, 13, 2012. Quotations from all three days.

Para. 3, Email, Rhona Applebaum to Steven Blair, Greg Hand, James Hill, David Allison, Timothy Church, Peter Katzmarzyk, Peter Boyle, "Science communication/Science/guardian.co.uk," November 7, 2012.

Para. 3, Email, Rhona Applebaum to Steven Blair, cc Tim Church, Marc Hamilton, "Fw: NYT: Is Sitting a Lethal Activity?," April 20, 2011.

Chapter Four

1. In 2005 the North America branch had forty-nine members, in 2015, thirty-seven. ILSI North America, *ILSI North America 2005 Annual Report*; ILSI North America, *ILSI North America 2015 Annual Report*. Information on board membership just below is drawn from the annual reports like these.

2. The branch is also home to program committees, working groups, and task forces.

3. ILSI North America, "Bylaws, Conformed as of June 17, 2015."

4. ILSI North America, *ILSI North America 2015 Annual Report*, 10.

5. Information on membership is from Tancredi and Milner, "Energy Balance"; ILSI North America, "Energy Balance 2015 Fact Sheet."

6. Based on study of committee meeting agendas retrieved from the Food Industry Documents collection maintained by the University of California San Francisco Library (UCSFL). A typical meeting was the conference call on October 15, 2014. Of the seventeen participants, fifteen (88 percent) were corporate scientists, while two (12 percent) were university or government-based advisers. The agenda called for welcoming a new government liaison, providing updates on existing projects, and discussing new activities, all falling within the scope of their task to "set the membership format and program." These agendas suggest that industry had a major role in the work of the committee.

7. Jakicic, Curriculum Vitae. Jakicic served as a key adviser to the Energy Balance Study carried out under Steven Blair at the University of South Carolina, suggesting he was part of the network of industry-friendly obesity researchers.

8. Allison's ties to ILSI ran deeper than this. According to his curriculum vitae, he served on the food-safety and nutrition committee of the ILSI Research Foundation (1997–2000), and on the expert committee of the International Food Information Council (IFIC), an ILSI sibling organization (1995–2010). See Allison, Curriculum Vitae.

9. Pratt's activities are detailed in Gillam, "What Is Going On at the CDC?" A San Diego reporter notes that Pratt has authored at least seven papers acknowledging Coke funding. See Cook, "UCSD Hires Coke-Funded Health Researcher."

10. They were Carson Chow of the National Institutes of Health (service to the committee 2012), Krista Varady of the University of Illinois at Chicago (2015), and Stella L. Volpe of Drexel University (2015). See ILSI North America, *ILSI North America 2015 Annual Report*, 27.

11. Hall, Heymsfield, et al., "Energy Balance and Its Components." The discussion of measurement difficulties is on 993: "Our ability to measure precisely individual components of energy expenditure or energy intake is relatively poor. . . . The accuracy and precision of energy intake measurements by self-report in free-living individuals are much worse." Urging colleagues "not to have unreasonable expectations about the impact of [small lifestyle changes such as those promoted by Hill] on body weight," they write: "This potential error prevents evaluation of the benefits of interventions that have a small benefit on weight change over time."

12. The consensus statement is Hall, Heymsfield, et al. The new paradigm label comes from Tancredi and Milner, "Energy Balance." The EBAL-sponsored conference, called "Energy Balance: A New Paradigm," was held as part of the Experimental Biology 2012 annual meeting in San Diego, CA, April 21, 2012. For more, see ILSI North America website, http://ilsina.org/event/energy-balance-a-new-paradigm/, accessed December 2, 2017.

13. Hall, Heymsfield, et al., "Energy Balance and Its Components," 989.

14. Blair, Hand, and Hill, "Energy Balance: A Crucial Issue." See also Hand and Blair, "Energy Flux."

15. The arguments in this paragraph are elaborated in Hand and Blair, "Energy Flux"; Hill, Wyatt, and Peters, "Importance of Energy Balance"; and Hill, Wyatt, and Peters, "Energy Balance and Obesity."

16. Smith Edge, "Understanding the Consumer's Knowledge."

17. The arguments in this paragraph are presented in Hill, Wyatt, and Peters, "Energy Balance and Obesity"; Blair, Hand, and Hill, "Energy Balance: A Crucial Issue."

18. In a 2015 report, the WHO called for the adoption of fiscal policies that led to at least a 20 percent increase in the retail price of sugary drinks to limit caloric consumption and reduce the number of people suffering from overweight, obesity, diabetes, and tooth decay. See WHO, *Fiscal Policies for Diet and Prevention of Noncommunicable Diseases*.

19. Nestle, *Unsavory Truth*, esp. 157–216. In an earlier work she wrote that food-industry sponsorship of education and research was so pervasive in the field that it was impossible for nutrition academics not to be involved. That and other insights can be found in Nestle, *Food Politics*, esp. 111–36.

20. On one occasion, core ILSI-NA researchers devoted an entire article to defending their sponsors against critics who portrayed the sponsors not as partners but as adversaries. See Hill, Peters, and Blair, "Reducing Obesity."

21. Hand et al., "Energy Balance Study."

22. Hand, "Energy Balance: Year 1 of a Longitudinal Study." These PowerPoint slides are no longer available, but a very similar PowerPoint presentation with the same acknowledgment of funding is available. See PowerPoint presentation, Gregory Hand, "Energy Balance Framework for Weight Management," December 28, 2015, Food Industry Documents Collection (UCSFL).

23. Arnold School of Public Health, "Blair Connects Energy Balance Experts." See also Blair, Hand, and Hill, "Energy Balance: A Crucial Issue."

24. Blair and Powell, "Physical Activity Movement Comes of Age," 9.

25. Blair, Hand, and Hill, "Energy Balance: A Crucial Issue."

26. Descriptions of the GEBN in this and the following paragraph are drawn from Anschutz Health and Wellness Center, "Energy Balance Experts from Six Continents"; Arnold School of Public Health, "Blair Connects Energy Balance Experts"; Global Energy Balance Network website, GEBN.org/about, accessed June 30, 2023.

27. Hill and Peters, with Jortberg, *Step Diet Book*, 55.

28. Blair, "Physical Inactivity: A Major Public Health Problem," 116.

29. In one article, he cited Coca-Cola's support for physical activity research as a positive development for the field of exercise science, expressing no qualms about a possible gap between the agendas of a giant food company and of science. See Blair and Powell, "Physical Activity Movement," 11.

30. Email with responses, James Hill to Steven Blair, John C. Peters, Gregory Hand, "NY Times Reporter," May 9, 2015, Food Industry Documents collection (UCSFL), excerpted from longer email chain.

31. Blair, "Dr. Steven Blair of Coca-Cola."

32. See Steven N. Blair, "Statement from Dr. Steven N. Blair."

33. For more informal modes of involvement, see Choi, "Emails Reveal Coke's Role in Anti-obesity Group."

34. The Q&A document referred to just above was created in response to this message.

UNMARKED CITATIONS
Author Fieldnotes
ESCALATING THREATS TO INDUSTRY . . .

Para. 5, Chat with SB at the 2013 Conference on Obesity Control and Prevention in China, held at the Beijing Guangxi Hotel, December 12, 2013. Hereafter, references to this conference in the endnotes will be abbreviated as Conference on Obesity Control.

Interview
DIETARY DISCORD

Para. 2, IF-35, October 28, 2016, Washington, DC.

Food Industry Documents (UCSFL)
DISCLOSING, DENYING, DISPLAYING . . .

Para. 3, Email, L. Celeste Bottorff to Rhona Applebaum, Steven Blair, Greg Hand, James Hill, John C. Peters, Lillian Smith, Amelia Quint, "Statement of Intent," August 5, 2014.

Para. 3, PowerPoint Slides, Gregory Hand, "Energy Balance Framework for Weight Management: A Longitudinal Study of Weight Change in Young Adults" (no date).

ESCALATING THREATS TO INDUSTRY . . .

Para. 1, Email, Rhona Applebaum to Steven Blair and James Hill, "Heat is On," September 14, 2012.

Para. 2, Email with responses, Rhona Applebaum to James Hill, Steven Blair, David B. Allison, Greg Hand, John C. Peters, Edward Archer, Carl (Chip) Lavie, Peter Katzmarzyk, "Urgent!!," January 19–20, 2014.

Para. 3, Email, James Hill to Rhona Applebaum, "RE: 5000 have signed up for your Webinar," September 14, 2012; Email, Rhona Applebaum to James Hill, "Ready for a stimulus pkg?," October 16, 2012; Email, Rhona Applebaum to James Hill, John C. Peters, Steven Blair, Gregory Hand, cc Celeste Bottorff, Debbie Wells, "EBIs," January 24, 2014.

Para. 4, Email, Rhona Applebaum, to Steve Blair, Greg Hand, John C. Peters, James Hill, Lillian Smith, Amelia Quint, L. Celeste Bottorff, "Proposal for Establishment of the Global Energy Balance Network.docx," July 9, 2014. All quotations in this paragraph are drawn from this document.

Para. 6, Email, Steven Blair to Gregory Hand, James Hill, John C. Peters, Clemens Drenowatz, Robin Shook, Rhona Applebaum, Celeste Bottorff, Susan A. Roberts, Beate Lloyd, Russell Pate, Michael Pratt, Amanda Paluch, Vivek Prasad, John M. Jakicic, Harold W. Kohl III, Timothy Church, Chip Lavie, "FW: Vol. 385 / Number 9985, June 13, 2015," June 15, 2015.

DIETARY DISCORD

Para. 1, Email, James Hill to Alex Malaspina, cc John C. Peters, "RE: Lancet commentaries," June 13, 2014. Later that year, he would argue strongly that some slots on the executive committee should be reserved for people with specific expertise in nutrition. See email, James Hill to Steven Blair, Rhona Applebaum, Bill Layden, Amelia Quint, Greg Hand, Kristina Jackson, Kathleen Jaynes, L. Celeste Bottorff, Lillian Smith, John C. Peters, cc Patrycja Mulewska, Joyce Cheatham, Debbie Wells, trnass@hsc.wvu .edu, "RE: ACTION: GEBN Director Recruiting," October 22, 2014.

PERILS OF ONE-COMPANY SPONSORSHIP

Para. 1, Email, L. Celeste Bottorff to Rhona Applebaum, Steven Blair, Greg Hand, James Hill, John C. Peters, Lillian Smith, Amelia Quint, "GEBN Statement of Intent Survey Results.docx; Copy of Survey summary.xlsx; FABdraft080514.docx," August 5, 2015. The responses in this paragraph are all listed in the survey results included with this email.

COKE INVOLVEMENT: PRIVATE AND PUBLIC TRUTHS

Para. 2, Brochure, Bridge Strategy Group LLC, *Laying a Foundation for the Global Energy Balance Network: Recap of Discussions and Decisions at Initial Working Session*, draft February 28, 2014. Recipients n.a., probably the four scientists and four Coke employees or consultants who attended the session. This document established the division of responsibilities among the members of the organizing group.

Para. 3, Email, John C. Peters to Arne Astrup, Greg Hand, James Hill, Marianella Herrera, Nahla Houalla, John C. Peters, Steven N. Blair, Wendy Brown, Wenhua Zhao, Willem Van Mechelen, cc Amelia Quint, "Heads up," includes as attachment confidential draft of GEBN Q&As, February 6, 2015.

Para. 4, Email, James Hill to Ross Hammond, cc Stacia Lupberger, John C. Peters, "reconnecting," May 20, 2015.

THE LIFE-OR-DEATH STRUGGLE TO FIND FUNDERS

Para. 1, Email with reply, ten recipients plus two cc'd, Steven Blair to Brian Hainline (of NCAA), cc Harris Pastides, "Global Energy Balance Network," March 29, 31, 2015.

Para. 2, Email with responses, James Hill to Bill Layden, "RE: Tweet from Yoni Freedhoff, MD (@YoniFreedhoff)," February 6, 2015. Participant added later: John C. Peters.

Para. 3, Email, Alex Malaspina to John C. Peters, "Re: RESEND: Thank You," February 3, 2015; Email, Alex Malaspina to James Behnke, cc James Hill, John C. Peters, Ed Hays, Clyde Tuggle, "Re: Happy Holidays," December 24, 2014 (for the quote on helping Coke).

Para. 4, Email, Alex Malaspina to James Behnke, cc James Hill, John C. Peters, Ed Hays, Clyde Tuggle, "Re: Happy Holidays," December 24, 2014; Email, Alex Malaspina to John C. Peters, "Re: RESEND: Thank You," February 3, 2015.

Para. 5, Email, Alex Malaspina to Ed Hays, Clyde Tuggle, cc James Hill, John C. Peters, Wamwari Waichungo, H. Zhang, "Fwd: GEBN materials," April 11, 2015; Email, Alex Malaspina to John C. Peters, "Subject [blank]," April 9, 2015; Email, Alex Malaspina to D Banati, cc John C. Peters, James Hill, MEK59100, Somogi Arpad, Clyde Tuggle, Ed Hays, Wamwari Waichungo, Suzanne Harris, "GEBN Materials," April 10, 2015.

Coda One

1. This is one of the important insights of David Michaels's *The Triumph of Doubt: Dark Money and the Science of Deception*, 5.

2. O'Connor, "Coca-Cola Funds Scientists."

3. Jacobson and Willett, "Letter to the Editor."

4. Among the most important is Choi, "Emails Reveal Coke's Role in Anti-obesity Group."

5. Kent, "Coca-Cola: We'll Do Better."

6. Douglas, "Letter from Sandy Douglas"; O'Connor, "Coke's Chief Scientist"; Coca-Cola Company, "Scientific Research Guiding Principles."

7. The Hill, Blair, and Hand quotations in this paragraph are from O'Connor, "Coca-Cola Funds Scientists."

8. Peters and Hill, "Conspiracy or Good Education?" The article to which this is a response is Thacker, "Coca-Cola's Secret Influence on Medical and Science Journalists."

9. Nestle, *Unsavory Truth*, 166.

10. Nestle.

11. The executives were Ed Hays, senior vice president and chief technical officer; Clyde Tuggle, senior vice president and chief public affairs and communications officer; Beate Lloyd, senior director, Nutrition Center of Expertise, Scientific and Regulatory Affairs; and Maxime Buyckx, director, Health and Wellness Programs; as well as Alex Malaspina, who was formally retired. All positions are for the time the email exchanges took place.

12. Olinger, "CU Expert Who Took Coca-Cola Money Steps Down"; Editor, "Gregory Hand Forced Out."

13. American Society for Nutrition, "ASN Fellows" and "American Society for Nutrition and ASN Foundation Awards Recipients."

14. Lobosco, "Med School Returns $1 Million."

15. Information on these changes comes from ILSI North America, *ILSI North America 2015 Annual Report*, 27; and ILSI North America, *ILSI North America 2014–18 Annual Reports*.

16. E. Farrell et al. (including Hand and Blair), "Associations between the Dietary Inflammatory Index"; Ross et al. (including Hill), "Small Change Approach to Prevent Long-Term Weight Gain"; Archer, "Guidelines on 'Added Sugar.'"

UNMARKED CITATIONS

Food Industry Documents (UCSFL)

THE ACADEMIC SCIENTISTS RESPOND . . .

Para. 2, Email, James Hill to Steven Blair, "Re: GEBN Queries from Reader's Digest," March 1, 2016.

Para. 7, Email with reply, James Hill to Ed Hays, "Re: energy balance network," June 4, 2014; Email with replies, James Hill to Maxime Buyckx, "RE: research idea," August 30 to September 2, 2014.

Para. 8, Email with reply, James Hill to Clyde Tuggle, "Re: Thank you," November 9–12, 2014.

Chapter Five

1. Coca-Cola Company, Coca-Cola Form 10-K to the Security and Exchange Commission.

2. This discussion of the state's scientific ambitions draws from Vogel, *Deng Xiaoping and the Transformation of China*, 321–23; Lampton, *Relationship Restored*; and Suttmeier, *Science, Technology, and China's Drive for Modernization*.

3. This statement is true for fields I have studied over long periods of time (population, public health), and some evidence suggests it may be for others as well. Little research has been done on the social structure of scientific communities in China; this is an important topic needing more attention.

4. Paragraph based on Kraus, "More Than Just a Soft Drink"; Cendrowski, "Opening Happiness."

5. This aspiration for the China market belongs to the CEO of Coca-Cola China Industries Ltd., Bottling Investments Group, region director for China-Singapore-Malaysia. His story is told in Coca-Cola Journey Australia Staff, "Getting a Grip on Calories." For lusting over China and Deng's critical role, see Kraus, "More Than Just a Soft Drink," 115, 118.

6. Jacobson, *Liquid Candy*, 2nd ed., 1–3.

7. Taylor and Jacobson, *Carbonating the World*, 9–13, 21. By 2015, Coca-Cola controlled 58 percent of the Chinese carbonated-soft-drinks market, while PepsiCo controlled 27 percent.

8. Y. Yan, "McDonald's in Beijing."

9. Quotation in Taylor and Jacobson, *Carbonating the World*, 21.

10. Malaspina, "Celebrating Twenty Years."

11. For Chinese names, in the bibliography, I have used commas following surnames of first author, whether or not the individual uses surname first or given name first; thus Chen Chunming is listed in the bibliography as "Chen, Chunming." The notes follow this same usage. In the text, I follow everyday usage for individuals living in China: Chen Chunming is referred to as Chen Chunming. Many Chinese living abroad anglicize their names, putting given name first and surname last. In the text, I refer to them the way they refer to themselves.

12. This introduction to Chen draws on ILSI, "Chen Chunming, the First President of the Chinese Academy of Preventive Medicine"; Li, "Asia Pacific Clinical Nutrition Society"; Zhu Ling, "Nutritionist Chen Chunming's Health Concepts."

13. Greenhalgh, "Introduction"; Shen and Williams, "Critique of China's Utilitarian View."

14. J. S. Chen's background is from the Center for Health Protection, "Prof. Chen Junshi."

15. Zhu Xufeng describes a virtual freeze in think tank formation after Tiananmen, followed by a mushrooming following Deng's Southern Tour in January 1992. ILSI-China was part of that rapid growth. See X. Zhu, *Rise of Think Tanks in China*, 27–29.

16. These major shifts in the health sector are described in Duckett, *Chinese State's Retreat*, 59–95 (on the abandonment of the Mao-era experiments) and 42–44 (on the rejuvenating role of the SARS crisis); and Huang, *Governing Health in Contemporary China*.

17. For this story, see Mason, *Infectious Change*.

18. This critical transition to "scientific policy making" was introduced in the early 1980s and more officially launched at the National Soft Science Research Symposium in 1986. There Vice Premier Wan Li announced that "democratic and scientific decision making is an important task in political system reform." For an in-depth treatment, see Halpern, "Scientific Decision Making." International exchange of ideas, through study tours for Chinese experts of Western centers of knowledge and lecture tours by foreign experts in China, was common across multiple policy sectors in the 1980s. Duckett describes health-sector tours in *Chinese State's Retreat*, 94. Cao, *GMO China*, tracks the change in strategy from following and catching up with the West in biotechnology in the 1980s, to moving up the global technology ladder in the 1990s.

19. The perception of China's backwardness in the field of public health is one of the most robust findings of researchers on Chinese health. An especially thoughtful analysis is Mason, *Infectious Change*. See also Song, *Biomedical Odysseys*.

20. Information from ILSI-China newsletters, various years.

21. ILSI-China, "Preface."

22. The CAPM (1983–2001), the forerunner of the CDC (2002 to present), was located at 29 Nanwei Road in Beijing. ILSI-China was housed there (the building in the photo). Later the CDC moved its headquarters to a larger complex at 155 Changbai Road. ILSI-China remained in that CDC building on Nanwei Road.

23. The hierarchical nature of ILSI-China emerged in the interviews with experts involved with it. In deciding on the members of important working groups, for example, ILSI experts would throw out names—oh, he's not bad, she's okay—but in the end the two Chens would make the decision in private. IF-17, December 10, 2016, Beijing.

24. "The One ILSI philosophy encourages ILSI entities to work together to identify and resolve outstanding scientific questions in four thematic areas." From ILSI-Global, "One ILSI."

25. The account in this and the next paragraph is based on Powys, "Trouble with Sugar."

26. This information on mechanisms of influence is drawn from fieldwork and documentary research described in text. The mechanisms of influence include both mechanisms of company influence on science and ILSI leader influence on policy.

27. I learned about the expert recommendations in the interviews, the financial transfers in ILSI's tax filings, and the leader visits from ILSI-China's newsletters.

28. This description of the annual meetings is based on newsletter items and annual meeting programs. See for example, ILSI, "Value of Annual Meeting"; ILSI Global, "2013 ILSI Annual Meeting"; and ILSI, "2019 ILSI Annual Meeting."

29. Xiong-fei et al., "Epidemiology and Determinants of Obesity in China," esp. 383–84. The speed of China's nutrition transition is stressed by Zhai et al., "What Is China Doing in Policy-Making?"

30. On the popularity of these traditional practices, see N. Chen, *Breathing Spaces*; Farquhar and Qicheng Zhang, *Ten Thousand Things*.

31. Popkin, "Nutrition Transition in Low-Income Countries"; Popkin and Gordon-Larsen, "Nutrition Transition: Worldwide Obesity."

32. See Carolina Population Center, "China Health and Nutrition Survey." The number of articles published comes from information on the site under the Publications tab.

33. Zhai et al., "What Is China Doing in Policy-Making?"

34. ILSI-CHP, IRS Form 990, 1999–2004.

35. Malaspina, "Welcome Address"; ILSI-China, "Third Asian Conference."

36. Zeng et al. note that in China obesity is not officially recognized as a medical condition unless there are major comorbidities; for that reason, treatment is not covered by health insurance. Pharmacotherapy for obesity is very rare in China. See Zeng et al., "Clinical Management and Treatment," esp. 396, 400.

37. On the moral values attaching to body weight in the United States, see A. Farrell, *Fat Shame*.

38. This is a favorite topic of journalists. See, for example, BBC News, "China's Latest Online Skinny Fad."

39. Roche was an ILSI-China supporting company from 1999 to 2003, the peak years of its promotion of orlistat. The company had successfully launched the drug in 1999 and in 2000 gained approval to sell it in China. Following a global trend, in China sales of Xenical peaked in 2002 before beginning to fall. In 2004 the company dropped its support for ILSI-China. Roche's actions strongly suggest that, the no-commerce rule notwithstanding, the company was using its ILSI-China membership to promote its commercial interests; once those interests disappeared it dropped the membership.

40. On BMI cut-point standards, see Flegal, "How Body Size Became a Disease."

41. WHO, *Obesity: Preventing and Managing the Global*. But, as interviews with the key actors revealed, the global movement too had significant corporate funding (from Big Pharma), arranged through the International Obesity Task Force (IOTF), a subject too complex to delve into here.

42. WHO, *Global Strategy on Diet, Physical Activity and Health*. In the Global Strategy, obesity and overweight are treated both as intermediate conditions that are worsened by poor diet and inactivity, and as key risk factors for the major chronic diseases.

43. This paragraph draws on ILSI-China, "Round-Table Meeting"; ILSI-China, "Industry Round Table Meeting." The ministry's larger strategy for addressing the challenge of diet-related chronic disease was "working from three angles": policy formulation, market orientation, and public education.

44. ILSI-China, "Food Industry Round Table."

45. Key mechanisms included an ILSI-China Healthy Lifestyle Fund, to sponsor relevant projects, and a Scientific Advisory Committee, to guide key scientific decisions. Coca-Cola was a committee member. My source is ILSI-China, "Round-Table Meeting"; ILSI-China, "Scientific Advisory Committee."

UNMARKED CITATIONS

Interviews

ILSI'S CHINA DREAM: CREATING A BRANCH IN CHINA

Para. 1, IF-33, September 27, 2016, Washington, DC.

ESTABLISHING ILSI-CHINA: "LEARNING FROM THE WEST"

Para. 2, Chen's decision explained in IF-3, November 4, 2013, Beijing.

Para. 4, IF-8, November 20, 2013, Beijing; IF-5, November 11, 2013, Beijing.

Para. 7, IF-3, November 4, 2013, Beijing.

AN AGENT OF CORPORATE SCIENCE . . .

Para. 1, This early history of ILSI-China from IF-3, November 4, 2013, Beijing.

Para. 2, IF-3, November 4, 2013, Beijing.

Para. 4, IF-3, November 4, 2013, Beijing; IF-31, September 23, 2016, via telephone.

ILSI-GLOBAL: MEANS OF TOP-DOWN CONTROL . . .

Para. 5, IF-33, September 27, 2016, Washington, DC.

ILSI-CHINA: FUNDING MECHANISMS . . .

Para. 1, IF-3, November 4, 2013, Beijing; IF-16, December 6, 2013, Beijing; IF-13, December 3, 2013, Beijing.

Para. 3, IF-3, November 4, 2013, Beijing; IF-8, November 20, 2013.

ILSI TAKES OWNERSHIP OF THE OBESITY ISSUE . . .

Para. 3, IF-4, November 8, 2013, Beijing; IF-18, December 11, 2013, Beijing; IF-2, November 2, 2013, Beijing.

THE TRANSFORMATION OF A PUBLIC OFFICIAL . . .

Para. 1, This history was shared in IF-4, November 8, 2013, Beijing; IF-2, November 2, 2013, Beijing.

Para. 3, Insights in this paragraph drawn from IF-31, September 23, 2016, via telephone; IF-4, November 8, 2013, Beijing; IF-2, November 2, 2013, Beijing; IF-16, December 6, 2013.

Para. 4, IF-30, September 6, 2016, Geneva; IF-31, September 23, 2016, via telephone.

PLACING OBESITY ON THE AGENDA . . .

Para. 1, IF-33, September 27, 2016, Washington, DC.

Para. 4, IF-33, September 27, 2016, Washington, DC.

Para. 5, The inside story told in this paragraph emerged from IF-31, September 23, 2016, via telephone; IF-3, November 4, 2013, Beijing; IF-27, May 21, 2016, London.

Para. 6, These insights into the culture of fatness shared in IF-11, November 29, 2013, Beijing; IF-1, October 29, 2013, Beijing. On Big Pharma's mistakes, IF-1, October 29, 2013, Beijing.

THE FIGHT OVER BMI CUTOFFS . . .

Para. 1, These different perspectives on the BMI battle from IF-3, November 4, 2013, Beijing; IF-27, May 21, 2016, London; IF-31, September 23, 2016, via telephone; IF-12, December 2, 2013, Beijing; IF-2, November 2, 2013, Beijing. Some close to the process believed there were more and better data available than the data that were used. I was not able to confirm or disconfirm that statement.

Para. 2, IF-16, December 6, 2013, Beijing.

Food Industry Documents (UCSFL)

PLACING OBESITY ON THE AGENDA . . .

Para. 1, Email, Alex Malaspina to Barbara Bowman, cc Suzanne Harris, "Re: Daily European News Flash," June 26, 2015.

Chapter Six

1. Based on analysis of news items in ILSI-China newsletters.

2. These attitudes were widespread in China. A powerful case is Mason, "Divergent Trust and Dissonant Truths."

3. This shift can be traced back to the global financial crisis of 2008–9, when China's statist approach to managing the economy proved more successful than the neoliberal approach of the advanced Western economies. For cogent analyses of the shifts under Xi Jinping, see Dickson, *Party and the People*; Fewsmith, *Rethinking Chinese Politics*, 177–82; Saich, *From Rebel to Ruler*.

4. ILSI-China, "[2011] Conference on Obesity Control and Prevention in China."

5. See, for example, Bray, Nielsen, and Popkin, "Consumption of High-Fructose Corn Syrup." This article was published around the time ILSI-China was beginning to develop public health strategies to combat the country's obesity epidemic.

6. The Global Strategy's focus on activity as well as diet may well have encouraged China to give it more attention and to create a new Working Group on Physical Activity

in 2005. Still, the WHO did not prioritize exercise, as Hill would do, and its focus was not just obesity, but the full range of debilitating chronic diseases, for which activity is indeed crucial.

7. Xu and Gao, "Physical Activity Guidelines."

8. ILSI-China, "Joint Meeting."

9. ILSI-China, "People to People International Delegation."

10. ILSI-China, "'[2006] Conference on Obesity and Related Diseases,'" 5.

11. The newsletters have some limitations as a source of information on the shifting discourse on obesity. They cover the highlights of major conferences, but not small-scale gatherings or person-to-person conversations. It is possible that discussion of these approaches was more widespread than I am able to show.

12. My focus here is on the organizing themes of the conferences, those advanced by the invited foreign experts. The presentations of local Chinese scientists covered a wider range of topics.

13. ILSI-China, "[2011] Conference on Obesity Control and Prevention."

14. ILSI-China, "2014 Conference on Physical Activity and Health."

15. ILSI-China, "2015 Conference on Obesity."

16. The other commitments were offering low- or no-calorie beverage options in every market, providing transparent nutrition information, and marketing responsibly. Coca-Cola Company, "Coca-Cola Announces Global Commitments to Help Fight Obesity."

17. ILSI-China, *2013 Conference on Obesity*.

18. ILSI-China. Organizations are listed in the order given in the program. The list uses official names of the organizations. In March 2013, a few months before this event, the Ministry of Health was restructured and renamed the National Health and Family Planning Commission. As mentioned earlier, the COFCO (China Oil and Foodstuffs Corporation) Group is China's largest food processor, manufacturer, and trader. COFCO is state owned and has worked with Coca-Cola since the company reentered the Chinese market in 1978.

19. Heber, "David Heber."

20. Paragraph based on Hill, "Reducing Obesity."

21. Paragraph based on Blair, "Studies of Energy Balance."

22. Paragraph based on Applebaum, "Promoting Active Healthy Lifestyles."

23. Heber, "Integrative Approach to Nutrition and Obesity."

24. The latest thinking on dietary strategies, including efforts to address the global food system and rein in the food industry, was not presented.

25. Chen also expressed support for consuming food with many colors, Heber's proposal.

26. ILSI-China, "'Happy Ten Minutes' Well Recognized."

27. ILSI-China, "'Happy 10' Project."

28. Coca-Cola Company, *2007/2008 Sustainability Review*, 55. The report describes Happy Playtime as "a program we support in China [that] was launched in 2004."

29. ILSI-China, "ILSI FP-China-Coca-Cola Food Safety." A second scholarship focused on food safety and risk assessment.

30. Blair's university received $32,700 in 2012 and $34,825 in 2013 for training Chinese professionals in the Coke-ILSI-China Scholarship Program. See ILSI, IRS Form 990, 2012 (p. 28), 2013 (p. 27).

31. Zhao, "EIM Globe and in China."

32. Applebaum, "Promoting Active Healthy Lifestyles," 75.

33. "News: Coca-Cola Researching Chinese Medicine." Thanks to James Flowers for bringing this to my attention.

34. ILSI-China sponsored many more nutritional events that dealt with food safety or fortification, other major priorities of ILSI-Global. I deal here only with meetings (nominally) concerned with food and obesity.

35. ILSI-China, "Hydration and the Health Symposium"; ILSI-China, "Symposium on the Importance of Water." 7.

36. ILSI-China, "Workshop on Managing Sweetness"; ILSI-China, "Top Experts Address Seminar on Sweetness."

37. ILSI-China, "1st Workshop on Restaurant Food and Balanced Diet"; ILSI-China, "Summary Report."

38. Among the most active in presenting themselves as sources of healthy eating options and nutrition information were Carrefour, Yum!, Nestle, Coke, and Mondelēz. In the early 2010s, these firms showcased their efforts to promote healthy lifestyles and eating-moving balance at a pavilion sponsored by ILSI at the ministry-run China Health Forums. See ILSI-China, "'Beijing Nutrition Week in 2006,'"; ILSI-China, "ILSI Focal Point in China Supports."

UNMARKED CITATIONS

Author Fieldnotes

WHICH EXPERTS WERE INVITED . . .

Para. 3, Chat with SB, Conference on Obesity Control, December 13, 2013.

"SMALL STEPS," "ENERGY GAP," AND "EXERCISE-FIRST" . . .

Para. 5, Conversations with Chinese experts, Conference on Obesity Control, December 12–13, 2013.

SPREADING SODA SCIENCE . . .

Para. 2, Notes on oral remarks by JH, Conference on Obesity Control, December 12, 2013.

Para. 3, Notes on oral remarks by SB, Conference on Obesity Control, December 12, 2013.

Para. 4, Notes on oral remarks by RA, Conference on Obesity Control, December 12, 2013.

BREAK-TIME CHATS

Para. 1, Chat with RA, Conference on Obesity Control, December 12, 2013.

Para. 2, Chat with DH, Conference on Obesity Control, December 12, 2013.

"PHYSICAL EXERCISE IS A MUST" . . .

Para. 1, Notes on oral remarks by JH and SB, Conference on Obesity Control, December 13, 2013.

Para. 2, Notes on presentation by Chen Chunming, Conference on Obesity Control, December 13, 2013.

Interviews

A HISTORIC SHIFT FROM NUTRITION TO PHYSICAL ACTIVITY

Para. 3, IF-8, November 20, 2013, Beijing; IF-15, December 4, 2013, Beijing.

A GLOBAL HIERARCHY OF SCIENCE

Para. 2, IF-29, June 16, 2016, Cambridge, MA; IF-4, November 8, 2013, Beijing.

WHICH EXPERTS WERE INVITED . . .

Para. 3, IF-26, November 21, 2015, Denver.

"SMALL STEPS," "ENERGY GAP," AND "EXERCISE-FIRST" . . .

Para. 1, IF-26, November 21, 2015, Denver.

COKE-FUNDED, ACTIVITY-THEMED PROGRAMS

Para. 2, IF-36, November 22, 2016, Boston.

Para. 4, IF-7, November 19, 2013, Beijing; IF-5, November 11, 2013, Beijing.

Para. 5, IF-1, October 29, 2013, Beijing.

MARKETING PRODUCTS AND CORPORATE CITIZENSHIP

Para. 3, IF-1, October 29, 2013, Beijing.

Chapter Seven

1. Thanks to Cong Cao and Yanzhong Huang for illuminating discussion of these points.

2. This discussion of Chinese think tanks draws heavily on Xue Lan, Xufeng Zhu, and Wanqu Han, "Embracing Scientific Decision Making"; X. Zhu, *Rise of Think Tanks in China*, esp. 32–45; X. Zhu, "Government Advisors or Public Advocates?"; Tanner, "Changing Windows on a Changing China."

3. Xue Lan, Xufeng Zhu, and Wanqu Han, "Embracing Scientific Decision Making," 54; the quotation on neutrality is from X. Zhu, *Rise of Think Tanks*, 19.

4. Yanjie Bian, "Prevalence and the Increasing Significance of Guanxi," 609. A classic account of guanxi is Yang, *Gifts, Favors, and Banquets*; another key source is Chang, "Path to Understanding."

5. X. Zhu, *Rise of Think Tanks*, 42–43, presents data on the important role of connections to government officials in a think tank's influence.

6. The gift is acknowledged in Malaspina, "Celebrating Twenty Years."

7. Chen also had a close working relationship with Fureng Dong, director of the Institute of Economics at the Chinese Academy of Social Sciences. Dong was a member of the National People's Congress (China's legislature) and the Chinese People's Political Consultative Conference. See Zhu, "Nutritionist Chen Chunming's Health Concepts."

8. Ministry of Health, *Zhongguo chengren chaozhong he feipangzheng yufang kongzhi zhinan* (Guidelines for prevention and control of overweight and obesity in Chinese adults).

9. Of course, the party-state is not monolithic, and the policy process involves extensive bargaining and politicking. See Dickson, *Party and the People*, 97–98 for an overview.

10. Dickson, *Party and the People*, 65–98. Illuminating case studies of the proliferation of nonstate actors in the early twenty-first century include Huang, *Toxic Politics*; Cao, *GMO China*; and Mertha, *China's Water Warriors*.

11. Cao, *GMO China*.

12. The process Chen described was independently verified by another high-level informant (IF-16, December 6, 2013, Beijing). Informants report that the Chinese Nutrition Society followed a similar process in creating the dietary guidelines (IF-4, November 8, 2013, Beijing). The creation and official acceptance of the obesity guidelines is tracked in ILSI-China, "China Ministry of Health Promulgated the First 'Guidelines.'"

13. ILSI-China, "'[2006] Conference on Obesity and Related Diseases,'" 5.

14. Introduced in the early days of the People's Republic, patriotic health campaigns were nationwide mass movements meant to spur the masses to improve sanitation, modernize hygiene, and eradicate major epidemic diseases such as cholera, hookworm, small pox, and snail fever. One of China's pioneering public health achievements, the patriotic health campaign is still in use today.

15. The launch is reported in Division of NCD Control and Community Health, "National Action on Healthy Lifestyle."

16. The process is also described in ILSI-China, "'Happy 10' Project"; ILSI-China, "'Happy Ten Minutes' Well Recognized."

17. See ILSI, "Code of Ethics and Organizational Standards of Conduct"; and, for its use in China, ILSI-China, "Preface." The notion that ILSI-China provided scientific evidence for policy makers to turn into government measures was part of its narrative about its unique working pattern, described as "symposium—program—action." This working style was introduced to supporting companies at meetings in early 2001, perhaps to encourage their involvement with the China branch. See ILSI-China, "ILSI Focal Point in China Conducts."

18. On the place of science in Chinese political discourse and reasoning, see Wang Yeu-farn, *China's Science and Technology Policy*; D. Simon and Goldman, *Science and Technology in Post-Mao China*. For the work performed by the word *science* in Chinese society more generally, see contributions in Greenhalgh and Li Zhang, *Can Science and Technology Save China?*

19. ILSI-China, "Media Forum on Obesity."

20. See for example ILSI-China, "Active Public Education."

21. ILSI-China, "Preface," unnumbered page 2.

22. C. Chen and Lu, "Guidelines for Prevention and Control," 5; see also 16.

23. As noted in chapter 2, the 1997 Dietary Guidelines had for the first time advocated balancing food and physical movement to maintain a healthy weight. In the Healthy Lifestyle for All Action introduced in 2007, the notion was reinforced, rephrased in more catchy language, and introduced as part of a package of ideas promoted by Hill and his associates and described in the text.

24. This target may reflect Chinese calculations of how to close the so-called energy gap (see chapter 2 for the details). More likely is that Chen and her colleagues simply used a number that had been circulating internationally for several years. As we saw in chapter 3, around 2004 Coke introduced the Step with It! program urging schoolchildren to take at least ten thousand steps a day.

25. See, for example, Liang, "Assessment on the Five-Year Implementation."

26. Some evidence suggests that Hill might even have helped design the healthy lifestyle action using AOM as the template. In our late 2015 conversation, he reported that in the early 2000s many ILSI branch heads simply "loved AOM." Happy to put his ideas into action, he worked with five or six of the branches to develop AOM-type health promotion campaigns for their countries (IF-26, November 21, 2015, Denver). In this way Hill served to spread his and ILSI's energy balance ideas around the world.

27. Some Western food companies sponsored nutrition education classes or activities, but the evidence suggests the messages were industry-friendly ones of the sort described in chapter 1.

28. National Action Office, "*Quanmin jiankang shenghuo fangshi xingdong jianjie*" (Introduction to the National Healthy Lifestyle for All initiative).

29. Limiting salt and oil consumption to lower risk of high blood pressure was a constant theme of the campaign slogans, educational materials, and conferences held annually from 2012 to 2015. This is a critically important public health message.

30. The WHO first recommended reducing free sugars to less than 10 percent of total daily energy intake in 1989. That recommendation was elaborated by the WHO/FAO Expert Consultation in 2002, and the Guidelines on Sugars Intake for Adults and Children in 2015. In 2016, the Commission on Ending Childhood Obesity went further to recommend taxing sugar-sweetened beverages and regulating marketing of high-energy foods and drinks to children, among other measures See WHO, *Report of the Commission on Ending Childhood Obesity*. The history of WHO recommendations on limiting the intake of free sugars is recapped in WHO, "WHO Calls on Countries."

31. ILSI-China's newsletters during 1999–2015 carried more than ten items on media salons to educate journalists about obesity and related diseases. The website for the ThinkTank Research Center for Health Development lists popularizing new knowledge and healthy lifestyles to the public as one of its tasks. See ThinkTank Research Center for Health Development, "Think Tank Research Center." See especially specific professional activities under its Center Introduction tab.

32. Erikson, "Metrics and Market Logics of Global Health," 158.

33. Cong Cao, personal communication, July 18, 2022.

UNMARKED CITATIONS

Interviews

Para. 1, IF-33, September 27, 2016, Washington, DC; IF-31, September 23, 2016, by telephone.

WHO WAS CHEN CHUNMING? . . .

Para. 2, IF-16, December 6, 2013, Beijing; IF-17, December 10, 2013, Beijing.

Para. 5, IF-31, September 23, 2016, via telephone; IF-27, May 21, 2016, London; IF-28, May 23, 2016, Inverness, Scotland; IF-35, October 28, 2016, Washington, DC; IF-2, November 2, 2013, Beijing.

Para. 6, IF-35, October 28, 2016, Washington, DC; IF-32, September 26, 2016, Boston; IF-29, June 16, 2016, Cambridge MA; IF-28, May 23, 2016, Inverness, Scotland (on Chen's professional attributes); IF-27, May 21, 2016, London (for the personal anecdote).

WHAT WAS ILSI-CHINA? . . .

Para. 1, IF-8, November 20, 2013, Beijing; IF-3, November 4, 2013, Beijing.

Para. 2, IF-3, November 4, 2013, Beijing; IF-4, November 8, 2013, Beijing.

TURNING SODA SCIENCE INTO POLICY . . .

Para. 1, IF-31, September 23, 2016, via telephone; IF-38, December 16, 2021, confidential.

CREATING CHINA'S OBESITY GUIDELINES

Para. 3, IF-3, November 4, 2013, Beijing; IF-4, November 8, 2013, Beijing.

INITIATING A NATIONAL CAMPAIGN . . .

Para. 2, IF-3, November 4, 2013, Beijing.

Para. 4, IF-27, May 21, 2016, London.

Para. 5, IF-5, November 11, 2013, Beijing.

Para. 2, IF-3, November 4, 2013, Beijing.

Para. 3, IF-8, November 20, 2013, Beijing; IF-3, November 4, 2013, Beijing.

Para. 4, IF-8, November 20, 2013, Beijing.

EMPHASIS ON ACTION, SILENCE ON SUGAR

Para. 3, IF-36, November 22, 2016, Boston; IF-5, November 11, 2013, Beijing; IF-15, December 4, 2013, Beijing.

Para. 4, IF-3, November 4, 2013, Beijing.

EDUCATING THE MEDIA AND THE PUBLIC . . .

Para. 2, IF-4, November 8, 2013, Beijing; IF-16, December 6, 2013, Beijing.

A CRITICAL BUT INVISIBLE FORCE IN OBESITY POLICY MAKING

Para. 2, IF-17, December 10, 2013, Beijing; IF-16, December 6, 2013, Beijing; IF-5, November 11, 2013, Beijing; IF-13, December 3, 2013, Beijing; IF-5, November 11, 2013, Beijing; IF-4, November 8, 2013, Beijing; IF-11, November 29, 2013, Beijing; IF-17, December 10, 2013, Beijing (last four are quotes in list).

Para. 3, IF-17, December 10, 2013, Beijing.

Para. 4, IF-3, November 4, 2013, Beijing.

Para. 5, IF-38, December 16, 2021, confidential.

Chapter Eight

1. See, for example, Qin, "Fraud Scandals"; Tatlow, "Scientific Ethical Divide." A similar storyline can be found in the British media. See, for example, Olcott, Cookson, and Smith, "China's Fake Science Industry"; Qiu, "Scientists Caught."

2. Han and Appelbaum, "China's Science."

3. On the GSK scandal, see Lipworth and Kerridge, "China's Pharma Scandal." For the milk scandal, see Huang, *Governing Health in Contemporary China*, 126–43.

4. On the importance of speech space and political restrictions on language use more generally, see J. Yan, "Nature of Chinese Authoritarianism."

5. It seems likely that Chen Chunming worked out how to handle these ethical issues in discussions with J. S. Chen. Because she continued to manage the obesity issue, and my conversations with ILSI-affiliated researchers centered on her actions, I restrict discussion to her ethical practices.

6. J. Greene, *Prescribing by Numbers*; see also Dumit, *Drugs for Life*.

7. I am indebted to Jiuheng He for illuminating discussion of the internal workings of China's political culture, and especially how it plays out in the China CDC. His ideas inform my analysis in this section.

8. On Confucian virtue ethics, Wong, "Chinese Ethics."

9. The term "teacher" carries various meanings, and it's not clear which one this

senior scientist was using. Conventionally, to call someone a teacher is to signal respect and trust, and from the context of the conversation I believe this is what he meant to convey.

10. Thanks once more to Jiuheng He for insightful discussion of these issues. His interpretation is supported by my interviews. A young doctor at a leading Beijing hospital confided in me that any junior colleague who openly criticized senior doctors who accepted bribes from drug companies for putting their names on company-written papers would be fired. The best strategy for nonleaders, he told me, is to stay in the middle on any issue and never stand out. IF-10, November 26, 2013, Beijing.

11. I did not ask Chen about Coke in particular, but the branch newsletters (which I believe she mostly wrote) described the company in terms that fit the cultural narrative about the good company. Stories about the Coke-funded Happy 10 Minutes invariably focused on how it improved the nation's health by preventing and controlling weight-related diseases among schoolchildren. News items about Coke's physical activity scholarships emphasized their contribution to upgrading and globalizing China's sport science.

12. The interviewee is referring here to the no-commerce rule.

13. Lauber, Rutter, and Gilmore, "Big Food and the World Health Organization."

14. COFCO, "COFCO Coca-Cola."

15. Under Xi, the media have increasingly come under state control, and investigative journalists have been censored and even imprisoned. Civil society groups that criticize party policies or leaders are suppressed. Foreign NGOs must find an official sponsor and register with the Public Security Bureau. These and related trends are summarized in Dickson, *Party and the People*; Fewsmith, *Rethinking Chinese Politics*, 131–56.

UNMARKED CITATIONS

Interviews

ACADEMIC FRAUD RUN RAMPANT

Para. 2, IF-14, December 4, 2013, Beijing; Paraphrased from IF-9, November 23, 2013, Beijing.

Para. 4, IF-17, December 10, 2013, Beijing.

ETHICS-1 AND ETHICS-2

Para. 3, IF-1, October 29, 2013, Beijing; IF-14, December 4, 2013, Beijing; IF-24, December 5, 2013, Beijing; IF-15, December 4, 2013, Beijing.

Para. 4, IF-17, December 10, 2013, Beijing; IF-1, October 29, 2013, Beijing.

FORMAL ETHICS: FOLLOW THE RULES

Para. 2, IF-8, November 20, 2013, Beijing.

Para. 3, IF-8, November 20, 2013, Beijing.

Para. 4, IF-12, December 2, 2013, Beijing.

INFORMAL ETHICS: FOLLOW THE LEADER . . .

Para. 1, IF-3, November 4, 2013, Beijing.

Para. 3, IF-36, November 22, 2016, Boston; IF-2, November 2, 2013, Beijing; IF-12, December 2, 2013, Beijing; IF-6, November 15, 2013, Beijing.

Para. 4, IF-4, November 8, 2013, Beijing; on the respect for Chen Chunming, IF-36, November 22, 2016, Boston; IF-17, December 10, 2013, Beijing; IF-2, November 2, 2013, Beijing.

THE GOOD SCIENTIST

Para. 3, IF-31, September 23, 2016, via telephone; IF-8, November 20, 2013, Beijing; IF-4, November 8, 2013, Beijing; IF-16, December 6, 2013, Beijing.

THE GOOD MULTINATIONAL

Para. 2, IF-14, December 4, 2013; IF-11, November 29, 2013.

THE ASSOCIATION LEADER . . .

Para. 1, IF-12, December 2, 2013. In this and the other two portraits sketched here, I have altered a few minor details to protect the identity of my informants. To make their comments more readable, I have also omitted ellipses where other, less relevant words were spoken.

THE HEALTH EDUCATOR . . .

Para. 1, IF-18, December 11, 2013, Beijing.

THE UNIVERSITY SCIENTIST . . .

Para. 1, IF-17, December 10, 2013, Beijing.

Food Industry Documents (UCSLF)

THE GOOD MULTINATIONAL

Para. 4, Email with responses, Alex Malaspina to Barbara Bowman, cc Suzanne Harris, "Re: Daily European News Flash," June 25, 2015. Includes email from Alex Malaspina to Clyde Tuggle, Ed Hays, Suzanne Harris, Clyde Ferguson, cc James Hill, James Behnke, John C. Peters, John Lupien, Junshi Chen, and Huaying Zhang, "Subject: Daily European News," June 25, 2015.

Coda Two

1. Though ILSI-China's influence on Chinese policy may have dissipated, the branch continued to promote energy balance science through at least 2017, inviting James Hill to speak in 2017 and hosting a major international conference on Chinese obesity, with the regulation of energy balance as its theme, in 2017. See Hill, Curriculum Vitae; ILSI-China, "2017 Conference on Obesity."

2. ILSI-China contributed to this first NCD plan by developing a set of concrete tools, or "appropriate technologies," for chronic disease prevention (oil- and salt-measuring dispensers, BMI calculator, waist circumference measuring tape) that the health ministry procured in bulk and distributed countrywide. ILSI-affiliated experts also helped create the concrete obesity targets for the plan (IF-3, November 4, 2013, Beijing; IF-8, November 20, 2013, Beijing).

3. A knowledgeable and readily available overview of these developments is WHO Health Promotion Department, "Healthy China." For some key details, see Xiaodong Tan, Shibo Kong, and Haiyan Shao, "New Strategies."

4. The three are Healthy China 2030 ("Central Committee of the CCP and State Council Issue the 'Outline of the Healthy China 2030 Plan'"), the National Nutrition Plan (2017–30) (General Office of the State Council, "General Office of the State Council Issued"), and the China Medium- and Long-Term Plan for the Prevention and Treatment [i.e., Control] of Chronic Diseases in China (2017–25) (General Office of the State Council, "General Office of the State Council's Notice").

5. Chinese Nutrition Society, *Dietary Guidelines for Chinese Residents (2016)*. This is the second of six recommendations. About once a decade, the society, a semipublic nonprofit organization, prepares these guidelines for issuance by the government. The 2016 guidelines were the fourth set. I briefly introduced the nutrition society in chapter 5, where I mentioned that some members of the Popkin group worked on the 1997 guidelines.

6. Calorie counts come from Fatsecret, "Foods."

7. General Office of the State Council, "The General Office of the State Council's Notice."

8. Ministry of Health and Other 15 Departments, "Notice from the Ministry of Health and Other 15 Departments on the Issuance of the 'China Chronic Disease Prevention and Treatment Work Plan (2012–15).'"

9. General Office of the State Council, "General Office of the State Council Issued the Notice." The quotation is from the section "Basic Principles."

10. Healthy China Promotion Committee, "Healthy China Action"; General Office of the State Council, "General Office of the State Council Issued the Notice."

11. General Office of the State Council, "General Office of the State Council's Notice on the Prevention and Treatment of Chronic Diseases in China." The quotation is in the section "Basic Principles." See also Healthy China Promotion Committee, "Healthy China Action (2019–30)."

12. Xiong-fei et al., "Epidemiology and Determinants of Obesity in China"; Zeng et al., "Clinical Management"; Wang Youfa et al., "Health Policy," 446.

13. Swinburn et al., "Global Syndemic of Obesity, Undernutrition, and Climate Change."

14. Wang Youfa et al., "Health Policy," 449, 446.

15. Xiong-fei et al., "Epidemiology and Determinants of Obesity in China," 382, 381.

16. Wang Youfa et al., "Health Policy," 451, 454.

17. Chan, "Obesity and Diabetes."

18. Xiong-fei et al., "Epidemiology and Determinants of Obesity in China," 383.

19. Dickson, *Party and the People*, 1–10; Saich, *From Rebel to Ruler*, 456–61.

20. On those bureaucratic pathologies, see esp. Fewsmith, *Rethinking Chinese Politics*; Dickson, *Party and the People*, 65–98.

21. The case of chronic disease policy is far from unique. In the field of environmental health, Yanzhong Huang has shown that scientific policy making too often relies on weak or no evidence and inaccurate or fake data, undermining the party's ability to effectively manage environmental degradation and its effects on human health. See Huang, *Toxic Politics*, 85–116. In her in-depth study of environmental policy, Elizabeth Lord found that the gathering of data has been so hampered by political restrictions at the village level that the resulting science has been incapable of answering the policy questions it was created to answer. See Lord, "China's Eco-dream."

UNMARKED CITATIONS
Interviews
STILL NO SODA TAX

Para. 2, IF-32, September 26, 2016, Boston; IF-36, November 22, 2016, Boston.

Conclusion

1. Brenan, "What Percentage of Americans Consider Themselves Overweight?" US CDC statistics can be found in Martin et al., "Attempts to Lose Weight."

2. McKenzie, *Getting Physical*, 55.

3. These changing agendas around exercise are mapped out in McKenzie, as well as in Petrzela, *Fit Nation*. The journalist Greg Critser makes similar observations about the fitness boom of the 1970s. See Critser, *Fat Land*, 67–68.

4. Surgeon General of the United States, *Surgeon General's Call to Action to Prevent and Decrease Overweight and Obesity*.

5. US Department of Health and Human Services, *2008 Physical Activity Guidelines*.

6. Hill and Wyatt, with Aschwanden, *State of Slim*.

7. Swift, Johannsen, et al., "Role of Exercise," esp. 47.

8. Bassett et al., "Step Counting."

9. Chin, Kahathuduwa, and Binks, "Physical Activity and Obesity," esp. 1234–37.

10. Swift, McGee, et al., "Effects of Exercise and Physical Activity," esp. 210.

11. Pontzer, *Burn*.

12. Petrzela, *Fit Nation*, 344.

13. Greenhalgh, "Making China Safe for Coke"; Greenhalgh, "Soda Industry Influence on Obesity Science and Policy in China."

14. Greenhalgh, "Inside ILSI."

15. Perlroth, "Spyware's Odd Targets."

16. Coca-Cola Company, "Coca-Cola Statement," provided to the author by the *BMJ*.

17. Pulley, "Bloomberg: Coca-Cola Severs Long-Term Ties."

18. ILSI, "Setting the Record Straight."

19. Vidry, Letter. I am deeply grateful to Robert (Rob) Dilworth, the journals director at Duke University Press, for sharing this with me and permitting me to use excerpts here (email from Robert Dilworth, May 9, 2022).

20. Steele et al., "Are Industry-Funded Charities Promoting 'Advocacy-Led Studies'?"

21. ILSI-Global, "Common Misconceptions, Dr. Alex Malaspina."

22. ILSI-Global, "ILSI Federation Announces Collective, Global Rebranding."

23. This statement is based on examination of the branch posters displayed at the 2014–16 annual meetings (thus covering activities undertaken between 2013 and 2015). ILSI, "Posters from ILSI Entities around the World."

24. Van Tulleken, *Ultra-processed People*, 59.

25. Popkin et al., "Towards Unified and Impactful Policies."

26. Gómez, *Junk Food Politics*.

27. Gómez, 311.

28. See Popkin et al., "Towards Unified and Impactful Policies"; Mialon, Jaramillo, et al., "Involvement of the Food Industry"; Mialon, Corvalan, et al., "Food Industry Political Practices in Chile."

29. In Colombia, for example, not only does cooperation between industry, government, and the media remain largely unquestioned, but public health experts advocating stronger regulations feel unsafe criticizing the food industry. See Mialon, Charry, et al., "'Architecture of the State."

30. Popkin et al., "Towards Unified and Impactful Policies," 466.

Appendix Two

1. By human subjects approval I mean exemption from requirements to protect human subjects.

2. ILSI North America, "Bylaws, Conformed as of June 17, 2015."

Works Cited

Adams, Vincanne. "Metrics of the Global Sovereign: Numbers and Stories in Global Health." In *Metrics: What Counts in Global Health*, edited by Vincanne Adams, 19–54. Durham, NC: Duke University Press, 2016.

Allison, David B. Curriculum Vitae. October 2, 2023. School of Public Health, University of Indiana, Bloomington.

America on the Move Foundation, Inc. "America on the Move and More Than 1,500 YMCAs Kick Off Steptember." *US Newswire*, September 4, 2007; *Gale OneFile: Health and Medicine*. link.gale.com/apps/doc/A168325016/HRCA?u=anon ~669e40b9&sid=sitemap&xid=4676fb87.

———. IRS Form 990, 2008, attachment. Available through GuideStar Pro.

American Society for Nutrition. "American Society for Nutrition and ASN Foundation Awards Recipients." June 17, 2020. https://nutrition.org/wp-content/uploads/2020/06/American-Society-for-Nutrition-Awards_Prior-Recipients-06-17-20.pdf.

———. "ASN Fellows." https://nutrition.org/asn-foundation/fellows/.

Anderson, Benedict. *Imagined Communities: Reflections on the Origins and Spread of Nationalism*. London: Verso, 1983.

Anschutz Health and Wellness Center, University of Colorado. "Energy Balance Experts from Six Continents Join Forces to Reduce Obesity." Press release, March 31, 2015. https://anschutzwellness.com/energy-balance-experts-six-continents-join-forces-reduce-obesity/. Link no longer active.

Applebaum, Rhona S. "Balancing the Debate." PowerPoint slides for talk delivered at the Food Industry: Trends and Opportunities, 29th International Sweetener Symposium, Coeur d'Alene, ID, August 7, 2012. https://www.phaionline.org/wp-content/uploads/2015/08/Rhona-Applebaum.pdf.

———. Curriculum Vitae. Circa 2004–12. Downloaded from unknown site September 22, 2016. Hard copy in the author's possession.

———. "Promoting Active Healthy Lifestyles: Making a Difference Together." In *2013 Conference on Obesity Control and Prevention in China, Theme: Appropriate Technology and Tools for Weight Control, December 12–13, 2013, Beijing* (in English and Chinese), edited by ILSI-China, 62–76. Beijing: ILSI-China, 2013.

Archer, Edward. "Guidelines on 'Added' Sugars Are Unscientific and Unnecessary." *Nature Reviews Cardiology* 19 (October 11, 2022): 847. https://doi.org/10.1038/s41569-022-00792-9.

Arnold School of Public Health, University of South Carolina. "Blair Connects Energy Balance Experts World-Wide with New Global Energy Balance Network (GEBN)."

Press release, December 5, 2014. http://www.asph.sc.edu/news/GEBN_launch
.html.

Atkins, Robert C. *Dr. Atkins' Diet Revolution: The High Calorie Way to Stay Thin Forever*.
Philadelphia: David McKay, 1972.

Barlow, C. E., H. W. Kohl III, L. W. Gibbons, and S. N. Blair. "Physical Fitness, Mortality and Obesity." *International Journal of Obesity and Related Metabolic Disorders*
19, supplement 4 (October 1995): S41–S44. https://pubmed.ncbi.nlm.nih.gov/
8581093/.

Bassett, David R., Jr., Lindsay P. Toth, Samuel R. LaMunion, and Scott E. Crouter.
"Step Counting: A Review of Measurement Considerations and Health-Related Applications." *Sports Medicine* 46, no. 7 (July 2017): 1303–15. https://doi.org/10.1007/
s40279-016-0663-1.

BBC News. "China's Latest Online Skinny Fad Sparks Concern." BBC News, March 17,
2021. https://www.bbc.com/news/world-asia-china-56343081.

Behnke, James R. "Profile of James Behnke." Compiled by Walker's Research. http://
www.walkersresearch.com/Profilepages/Show_Executive_Title/Executiveprofile/
J/James_R__Behnke_400143544.html. Accessed October 22, 2023.

Bian, Yanjie. "The Prevalence and the Increasing Significance of Guanxi." *China Quarterly* 235 (September 2018): 597–621. https://doi.org/10.1017/S0305741018000541.

Blair, Steven N. Curriculum Vitae. 2022. https://sc.edu/study/colleges_schools/public
_health/documents/cvs/cv_blair.pdf.

———. "Dr. Steven Blair of Coca-Cola and ACSM's Global Energy Balance Network."
Dated September 10, 2015. Posted by CrossFit. https://www.youtube.com/watch?v
=9xBV_Enlh1A.

———. "Physical Inactivity: A Major Public Health Problem." *Nutrition Bulletin* of the
British Nutrition Foundation 32, no. 2 (June 2007): 113–17. https://doi.org/10.1111/j
.1467-3010.2007.00632.x.

———. "Physical Inactivity: The Biggest Public Health Problem." *British Journal of
Sports Medicine* 43, no. 1 (2009): 1–2.

———. "Statement from Dr. Steven N. Blair, P.E.D., Vice President, Global Energy
Balance Network." August 19, 2015. https://gebn.org/news/home/item/69-steven
-blair-statement.

———. "Studies of Energy Balance: Crucial for Understanding and Managing the
Obesity Epidemic." In *2013 Conference on Obesity Control and Prevention in China,
Theme: Appropriate Technology and Tools for Weight Control, December 12–13, 2013,
Beijing* (in English and Chinese), edited by ILSI-China, 19–36. Beijing: ILSI-China,
2013.

Blair, Steven N., Gregory A. Hand, and James O. Hill. "Energy Balance: A Crucial Issue
for Exercise and Sports Medicine." *British Journal of Sports Medicine* 49, no. 15 (August 2015): 970–71. https://doi.org/10.1136/bjsports-2015-094592.

Blair, S. N., H. W. Kohl III, R. S. Paffenbarger, D. G. Clark, K. H. Cooper, et al. "Physical Fitness and All-Cause Mortality: A Prospective Study of Healthy Men and
Women." *JAMA* 262, no. 17 (November 3, 1989): 2395–401. https://doi.org/10.1001/
jama.262.17.2395.

Blair, Steven N., and Kenneth E. Powell. "The Physical Activity Movement Comes of
Age: The Evolution of the Physical Activity Field." *Journal of Physical Education,
Recreation and Dance* 85, no. 7 (September 2014): 9–12. https://doi.org/10.1080/
07303084.2014.937174.

Bloomberg News. "Coca-Cola's Market Share Is Said to Shrink." *New York Times*, February 19, 2000. https://www.nytimes.com/2000/02/19/business/coca-cola-s-market-share-is-said-to-shrink.html.

Boseley, Sarah. "WHO 'Infiltrated by Food Industry.'" *Guardian*, January 8, 2003. https://www.theguardian.com/uk/2003/jan/09/foodanddrink.

Brandt, Allan M. *The Cigarette Century: The Rise, Fall, and Deadly Persistence of the Product That Defined America*. New York: Basic, 2007.

Bray, George, Samara Joy Nielsen, and Barry M. Popkin. "Consumption of High-Fructose Corn Syrup in Beverage May Play a Role in the Epidemic of Obesity." *American Journal of Clinical Nutrition* 79, no. 4 (April 2004): 537–43. https://doi.org/10.1093/ajcn/79.4.537.

Brenan, Megan. "What Percentage of Americans Consider Themselves Overweight?" Gallup, *The Short Answer*, January 3, 2022. https://news.gallup.com/poll/388460/percentage-americans-consider-themselves-o.erweight.aspx.

Brownell, Kelly D. "Get Slim with Higher Taxes." *New York Times*, December 15, 1994. https://www.nytimes.com/1994/12/15/opinion/get-slim-with-higher-taxes.html.

Buyckx, Maxime. LinkedIn profile. https://www.linkedin.com/in/maxime-buyckx-bb58814/. Accessed January 7, 2023.

Callahan, Alice. "Q&A—Katherine Flegal: She Saw the Obesity Epidemic Coming; Then an Unexpected Finding Mired Her in Controversy." *Knowable Magazine*, September 8, 2022. https://knowablemagazine.org/article/society/2022/obesity-research-controversy-woman-scientist.

Cao, Cong. *GMO China: How Global Debates Transformed China's Agricultural Biotechnology Policies*. New York: Columbia University Press, 2018.

Carolina Population Center, University of North Carolina, "China Health and Nutrition Survey." https://www.cpc.unc.edu/projects/china. Accessed November 2023.

Cendrowski, Scott. "Opening Happiness: An Oral History of Coca-Cola in China." *Fortune*, September 11, 2014. https://fortune.com/2014/09/11/opening-happiness-an-oral-history-of-coca-cola-in-china/.

Center for Health Protection, Department of Health, Government of the Hong Kong Special Administrative Region. "Prof. Chen Junshi." December 14, 2018. https://www.chp.gov.hk/en/static/101174.html#:~:text=Dr.,of%20Preventive%20Medicine)%2C%20Beijing.

"Central Committee of the CCP and State Council Issue the 'Outline of the Healthy China 2030 Plan'" ("Zhonggong Zhongyang Guowuyuan yinfa 'Jiankang Zhongguo 2030' guihua gangyao"). October 25, 2016. https://www.fao.org/faolex/results/details/en/c/LEX-FAOC175038/. For the Chinese version, see https://faolex.fao.org/docs/pdf/chn175038.pdf.

Chan, Margaret. "Obesity and Diabetes: The Slow-Motion Disaster. Keynote address at the 47th Meeting of the [US] National Academy of Medicine, October 17, 2016." https://www.who.int/director-general/speeches/detail/obesity-and-diabetes-the-slow-motion-disaster-keynote-address-at-the-47th-meeting-of-the-national-academy-of-medicine.

Chang, Kuang-chi. "A Path to Understanding *Guanxi* in China's Transitional Economy: Variations on Network Behavior." *Sociological Theory* 29, no. 4 (December 2011): 315–39. https://doi.org/10.1111/j.1467-9558.2011.01401.x.

Chen, Chunming, and Frank C. Lu, eds. "Guidelines for Prevention and Control of

Overweight and Obesity in Chinese Adults." *Biomedical and Environmental Sciences* 17, supplement (2004): 1–36.

Chen, Nancy N. *Breathing Spaces: Qigong, Psychiatry, and Healing in China.* New York: Columbia University Press, 2003.

Chin, S.-H., C. N. Kahathuduwa, and M. Binks. "Physical Activity and Obesity: What We Know and What We Need to Know." *Obesity Reviews* 17, no. 12 (2016): 1226–44. https://doi.org/10.1111/obr.12460.

Chinese Nutrition Society. *Dietary Guidelines for Chinese Residents (2016) (Zhongguo jumin shanshi zhinan, 2016).* Beijing: Chinese Nutrition Society. http://en.cnsoc.org/dGuideline/.

Choi, Candice. "Emails Reveal Coke's Role in Anti-obesity Group." APNews-Break, November 24, 2015. https://apnews.com/article/obesity-archive-1fd235360ac94dcf893a87e3074a03a5.

CNN. "US Scientist: Fat Can Be Healthy." *CNN.com,* July 18, 2011. http://www.cnn.com/2001/HEALTH/diet.fitness/07/18/fat.fit/index.html.

The Coca-Cola Company. *2006 Corporate Responsibility Review.* Atlanta, GA: Coca-Cola, 2007. https://www.yumpu.com/en/document/view/36263383/2006-corporate-responsibility-review-the-coca-cola-company.

———. *2007/2008 Sustainability Review: Act. Inspire. Make a Difference; A Dialogue of Progress and Possibility.* Atlanta, GA: Coca-Cola, 2008.

———. *2010/11 Sustainability Report: Reasons to Believe in a Better World.* Atlanta, GA: Coca-Cola, 2011.

———. "Coca-Cola Announces Global Commitments to Help Fight Obesity." Press release, May 8, 2013. https://investors.coca-colacompany.com/news-events/press-releases/detail/229/coca-cola-announces-global-commitments-to-help-fight-obesity.

———. Coca-Cola Form 10-K to the Security and Exchange Commission, Annual Report for the Fiscal Year Ended December 31, 2000, Filed March 7, 2001. https://investors.coca-colacompany.com/filings-reports/annual-filings-10-k?page=3.

———. "Coca-Cola Statement: Statement on British Medical Journal Report—Obesity Policies." January 9, 2019. Provided to the author by *BMJ.*

———. Form 10-K, Annual Report to the US Security and Exchange Commission, for the Period Ending 12/31/03. February 27, 2004. Atlanta, GA: Coca-Cola. https://investors.coca-colacompany.com/filings-reports/annual-filings-10-k?page=3.

———. "Frequently Asked Questions." https://www.coca-colacompany.com/about-us/faq/how-much-sugar-is-in-coca-cola#:~:text=There%20are%2039%20grams%20of,availability%20varies%20based%20on%20geography).&text=Do%20you%20have%20any%20drinks%20with%20fewer%20calories%3F.

———. "Our Commitment to Transparency." *Coca-Cola Journey.* http://transparency.coca-colacompany.com/our-commitment-transparency. Accessed September 30, 2015.

———. "Scientific Research Guiding Principles." January 31, 2016. https://www.coca-colacompany.com/policies-and-practices/scientific-research-guiding-principles. Link no longer active, but a similar article appears at https://www.coca-colacompany.com/media-center/well-being-scientific-research-and-third-party-engagement. Accessed November 12, 2023.

Coca-Cola Journey Australia Staff. "Getting a Grip on Calories: On the Road to Poor Health, Coke Leader Decides to Change. Does He Ever." *Coca-Cola Journey, New Zealand,* June 5, 2013. https://www.coca-colajourney.co.nz/stories/getting-a-grip

-on-calories-on-the-road-to-poor-health-coke-leader-decides-to-change-does-he
-ever. Accessed March 23, 2022. Link no longer active.

Coca-Cola Journey Staff. "Coca-Cola Honors 10 Young Scientists from around the
World." *Coca-Cola Journey*, February 4, 2015. http://www.coca-colacompany
.com/coca-cola-unbottled/coca-cola-honors-10-young-scientists-from-around-the
-world. Accessed July 11, 2017. Link no longer active.

COFCO. "COFCO Coca-Cola." China Oil and Foodstuffs Corporation. https://www
.cofco.com/en/BrandProduct/COFCOCocaCola/. Accessed November 12, 2023.

Cook, Morgan. "UCSD Hires Coke-Funded Health Researcher." *San Diego Union-
Tribune*, September 21, 2016. https://www.sandiegouniontribune.com/news/
watchdog/sd-me-watchdog-ucsd-20160929-story.html.

Crawford, Robert. "Health as a Meaningful Social Practice." *Health: An Interdisciplinary
Journal for the Social Study of Health, Illness and Medicine* 10, no. 4 (October 2006):
401–20. https://doi.org/10.1177/1363459306067310.

———. "Healthism and the Medicalization of Everyday Life." *International Journal of
Health Services* 10, no. 3 (1980): 365–88. https://doi.org/10.2190/3H2H-3XJN-3KAY
-G9NY.

Critser, Greg. *Fat Land: How Americans Became the Fattest People in the World*. Boston:
Houghton Mifflin, 2003.

Dews, P. B. "Foreword." In *Caffeine: Perspectives from Recent Research*, edited by P. B.
Dews, v–vi. Berlin: Springer-Verlag, 1984.

Dickson, Bruce J. *The Party and the People: Chinese Politics in the 21st Century*. Princeton, NJ: Princeton University Press, 2021.

Dietz, William H. "The Response of the US Centers for Disease Control and Prevention to the Obesity Epidemic." *Annual Review of Public Health* 36 (2015): 575–96.
https://doi.org/10.1146/annurev-publhealth-031914-122415.

Dimassa, Carra Mia, and Erika Hayasaki. "L.A. Schools Set to Can Soda Sales." *Los
Angeles Times*, August 25, 2002. https://www.latimes.com/archives/la-xpm-2002
-aug-25-me-soda25-story.html.

Division of NCD Control and Community Health, China CDC. "National Action on
Healthy Lifestyle for All Launched in Beijing." *ILSI Focal Point in China Newsletter*
27, no. 2 (July–December 2007): 10–11.

Douglas, Sandy. "Letter from Sandy Douglas Dated September 22, 2015." http://
transparency.coca-colacompany.com/letter-from-sandy-douglas-dated-22
-september-2015. Accessed December 15, 2016. Link no longer active.

Duckett, Jane. *The Chinese State's Retreat from Health: Policy and the Politics of Retrench-
ment*. Abingdon, Oxon, UK: Routledge, 2011.

Dumit, Joseph. *Drugs for Life: How Pharmaceutical Companies Define Our Health*. Durham, NC: Duke University Press, 2012.

Editor. "Gregory Hand Forced Out as Dean of West Virginia School of Public Health
amid Controversy over Coca-Cola Funding." *Corporate Crime Reporter*, August 2,
2016. https://www.corporatecrimereporter.com/news/200/gregory-hand-forced
-out-as-dean-of-west-virginia-school-of-public-health-amid-controversy-over-coca
-cola-funding/#:~:text=Gregory%20Hand%20has%20been%20forced,conducted
%20for%20a%20new%20dean.

Erikson, Susan L. "Metrics and Market Logics of Global Health." In *Metrics: What
Counts in Global Health*, edited by Vincanne Adams, 147–62. Durham, NC: Duke
University Press, 2016.

Fabricant, M. Chris. *Junk Science and the American Criminal Justice System*. Brooklyn, NY: Akashic Books, 2022.

Farquhar, Judith, and Qicheng Zhang. *Ten Thousand Things: Nurturing Life in Contemporary Beijing*. New York: Zone, 2012.

Farrell, Amy Erdman. *Fat Shame: Stigma and the Fat Body in American Culture*. New York: New York University Press, 2011.

Farrell, Emily T., Michael D. Wirth, Alexander C. McLain, Thomas G. Hurley, Robin P. Shook, Gregory A. Hand, James R. Hebert, et al. "Associations between the Dietary Inflammatory Index and Sleep Metrics in the Energy Balance Study (EBS)." *Nutrients* 15, no. 2 (January 13, 2023): 419. https://doi: 10.3390/nu15020419.

Fatsecret. "Foods." https://www.fatsecret.com/calories-nutrition/. Accessed November 2023.

Fewsmith, Joseph. *Rethinking Chinese Politics*. Cambridge: Cambridge University Press, 2021.

Flegal, Katherine M. "How Body Size Became a Disease: A History of the Body Mass Index and Its Rise to Clinical Importance." In *Routledge Handbook of Critical Obesity Studies*, edited by Michael Gard, Darren Powell, and José Tenorio, 23–39. London: Routledge, 2022.

Foster, Robert T. *Coca-Globalization: Following Soft Drinks from New York to New Guinea*. New York: Palgrave Macmillan, 2008.

Freudenberg, Nicholas. *At What Cost: Modern Capitalism and the Future of Health*. New York: Oxford University Press, 2021.

———. *Lethal but Legal: Corporations, Consumption, and Protecting Public Health*. New York: Oxford University Press, 2014.

Fryar, Cheryl D., Margaret D. Carroll, and Joseph Afful. "Prevalence of Overweight, Obesity, and Severe Obesity among Adults Aged 20 and Over: United States, 1960–62 through 2017–18." National Center for Health Statistics, Health E-Stats, December 2020, rev. January 2021. https://www.cdc.gov/nchs/data/hestat/obesity-adult-17-18/obesity-adult.htm.

Gallup News Service. "Healthcare System." https://news.gallup.com/poll/4708/healthcare-system.aspx. Accessed November 12, 2023.

General Office of the State Council. "The General Office of the State Council's Notice on the Prevention and Treatment of Chronic Diseases in China, Notice of the Medium and Long-Term Planning (2017–2025), State Council General Office Document (2017) No. 12" ("Guowuyuan bangong ting guanyu yinfa zhongguo fangzhi manxingbing zhongchangqi guihua [2017–2025] de tongzhi, guobanfa [2017] 12 hao"). January 22, 2017. https://www.gov.cn/zhengce/content/2017-02/14/content_5167886.htm.

———. "The General Office of the State Council Issued the Notice of the National Nutrition Plan (2017–2030), State Council General Office [2017], Document No. 60" ("Guowuyuan bangongting guanyu yinfa guomin yingyang jihua de tongzhi [2017–2030], guobanfa 60 hao"). June 30, 2017. https://www.gov.cn/zhengce/content/2017-07/13/content_5210134.htm.

Gerstle, Gary. *The Rise and Fall of the Neoliberal Order: America and the World in the Free Market Era*. Oxford: Oxford University Press, 2022.

Gieryn, Thomas F. *Cultural Boundaries of Science: Credibility on the Line*. Chicago: University of Chicago Press, 1999.

Gillam, Carey. "What Is Going On at the CDC? Health Agency Ethics Need Scrutiny."

The Hill, August 27, 2016. https://thehill.com/blogs/pundits-blog/healthcare/293482-what-is-going-on-at-the-cdc-health-agency-ethics-need-scrutiny.

Gómez, Eduardo J. *Junk Food Politics: How Beverage and Fast Food Industries Are Reshaping Emerging Economies*. Baltimore: Johns Hopkins University Press, 2023.

Greene, Jeremy A. *Prescribing by Numbers: Drugs and the Definition of Disease*. Baltimore: Johns Hopkins University Press, 2007.

Greene, Russ. "The Coca-Cola/ACSM 'Exercise Is Medicine' Scheme." *Keep Fitness Legal* (Crossfit.com blog), January 12, 2016. https://keepfitnesslegal.crossfit.com/2016/01/12/the-coca-cola-acsm-exercise-is-medicine-scheme/. Accessed September 9, 2021. Link no longer active.

Greenhalgh, Susan. 2015. *Fat Talk Nation: The Human Costs of America's War on Fat*. Ithaca, NY: Cornell University Press, 2015.

———. "Inside ILSI: How Coca-Cola, Working through Its Scientific Nonprofit, Created a Global Science of Exercise for Obesity and Got It Embedded in Chinese Policy (1995–2015)." *Journal of Health Politics, Policy and Law* 46, no. 2 (April 2021): 235–76. https://doi.org/10.1215/03616878-8802174.

———. "Introduction: Governing through Science; The Anthropology of Science and Technology in Contemporary China." In *Can Science and Technology Save China?*, edited by Susan Greenhalgh and Li Zhang, 1–24. Ithaca, NY: Cornell University Press, 2020.

———. "Making China Safe for Coke: How Coca-Cola Shaped Obesity Science and Policy in China." *BMJ* 364 (January 9, 2019). https://doi.org/10.1136/bmj.k5050.

———. "Soda Industry Influence on Obesity Science and Policy in China." *Journal of Public Health Policy*, 40, no. 1 (March 6, 2019; online January 9, 2019): 5–16. https://doi.org/10.1057/s41271-018-00158-x.

Greenhalgh, Susan, and Li Zhang, eds. *Can Science and Technology Save China?* Ithaca, NY: Cornell University Press, 2020.

Hacking, Ian. *The Social Construction of What?* Cambridge, MA: Harvard University Press, 1999.

Hall, Kevin D. "Did the Food Environment Cause the Obesity Epidemic?" *Obesity* 26, no. 1 (January 2018): 11–13. https://doi.org/10.1002/oby.22073.

———. "Diet versus Exercise in 'The Biggest Loser' Weight Loss Competition." *Obesity* 21, no. 5 (May 2013): 957–59. https://doi.org/10.1002/oby.20065.

———. "Energy Compensation and Metabolic Adaptation: 'The Biggest Loser' Study Reinterpreted." *Obesity* 30, no. 1 (January 2022): 11–13. https://doi.org/10.1002/oby.23308.

Hall, Kevin D., Alexis Ayuketah, Robert Brychta, Hongyi Cai, Thomas Cassimatis, Kong Y. Chen, Stephanie T. Chung, et al. "Ultra-processed Diets Cause Excess Calorie Intake and Weight Gain: An Inpatient Randomized Controlled Trial of Ad Libitum Food Intake." *Cell Metabolism* 30, no. 1 (July 2, 2019): 66–77. https://doi.org/10.1016/j.cmet.2019.05.008.

Hall, Kevin D., Steven B. Heymsfield, Joseph W. Kemnitz, Samuel Klein, Dale A. Schoeller, and John R. Speakman. "Energy Balance and Its Components: Implications for Body Weight Regulation." *American Journal of Clinical Nutrition* 95, no. 4 (April 2012): 989–94. https://doi.org/10.3945/ajcn.112.036350.

Halpern, Nina. "Scientific Decision Making: The Organization of Expert Advice in Post-Mao China." In *Science and Technology in Post-Mao China*, edited by Denis

Fred Simon and Merle Goldman, 157–74. Cambridge, MA: Harvard University, Council on East Asian Studies, 1989.

Han, Xueying, and Richard P. Appelbaum. "China's Science, Technology, Engineering, and Mathematics (STEM) Research Environment: A Snapshot." *PloS ONE* 13, no. 4 (April 2018). https://doi.org/10.1371/journal.pone.0195347.

Hand, Gregory. "Energy Balance: Year 1 of a Longitudinal Study of Weight Change in Young Adults." PowerPoint slides for talk, circa 2014. Source no longer available. For comparable presentation with source, see Hand, "Energy Balance Framework for Weight Management: A Longitudinal Study of Weight Change in Young Adults," PowerPoint slides, cited under Unmarked Citations, Food Industry Documents (UCSFL), chap. 4, para. 3.

Hand, Gregory A., and Steven N. Blair. "Energy Flux and Its Role in Obesity and Metabolic Disease." *European Endocrinology* 10, no. 2 (August 2014): 131–35. https://doi.org/10.17925/EE.2014.10.02.131.

Hand, Gregory A., Robin P. Shook, Amanda E. Paluch, Meghan Baruth, E. Patrick Crowley, Jason R. Jaggers, Vivek K. Prasad, et al. "The Energy Balance Study: The Design and Baseline Results for a Longitudinal Study of Energy Balance." *Research Quarterly for Exercise and Sport* 84, no. 3 (August 22, 2013): 275–86. https://doi.org/10.1080/02701367.2013.816224.

Harvard T. H. Chan School of Public Health. "Sugary Drinks." *The Nutrition Source*, https://www.hsph.harvard.edu/nutritionsource/healthy-drinks/sugary-drinks/. Accessed January 8, 2023.

Healthy China Promotion Committee. "Healthy China Action (2019–30)" ("Jiankang Zhongguo [2019–30]"). July 9, 2019. https://www.gov.cn/xinwen/2019-07/15/content_5409694.htm.

Heber, David. "David Heber, MD, PhD, FACP, FASN: Chairman, Herbalife Nutrition Institute." https://iamherbalifenutrition.com/contributors/david-heber/.

———. "Integrative Approach to Nutrition and Obesity: Colorful Diet and Weight Management." In *2013 Conference on Obesity Control and Prevention in China, Theme: Appropriate Technology and Tools for Weight Control, December 12–13, 2013, Beijing* (in English and Chinese), edited by ILSI-China, 101–8. Beijing: ILSI-China, 2013.

Hill, James O. "Child Obesity: Challenges and Opportunities for Prevention." In *Childhood Obesity: Partnerships for Research and Prevention*, edited by Frederick L. Trowbridge and Debra L. Kibbe, 1–10. Washington, DC: ILSI Press, 2002.

———. Curriculum Vitae. Circa 2018. https://scholars.uab.edu/6698-james-hill . Accessed November 12, 2023.

———. "Reducing Obesity." In *2013 Conference on Obesity Control and Prevention in China, Theme: Appropriate Technology and Tools for Weight Control, December 12–13, 2013, Beijing* (in English and Chinese), edited by ILSI-China, 1–18. Beijing: ILSI-China, 2013.

Hill, James O., Jeanne P. Goldberg, Russell R. Pate, and John C. Peters. "Acknowledgement." *Nutrition Reviews* 59, no. 3 (March 2001): S1. https://doi.org/10.1111/j.1753-4887.2001.tb06979.x.

———. "Discussion." Nutrition Reviews 59, no. 3 (March 2001): S57–62. https://doi.org/10.1111/j.1753-4887.2001.tb06986.x.

———. "Introduction." *Nutrition Reviews* 59, no. 3 (March 2001): S4–S6. https://doi.org/10.1111/j.1753-4887.2001.tb06984.x.

———. "Participants." *Nutrition Reviews* 59, no. 3 (March 2001): S66–74. https://doi
.org/10.1111/j.1753-4887.2001.tb06988.x.

Hill, James O., and John C. Peters. "Environmental Contributions to the Obesity
Epidemic." *Science* 280, no. 5368 (May 29, 1998): 1371–74. https://doi.org/10.1126/
science.280.5368.1371.

Hill, James O., John C. Peters, and Steven N. Blair. "Reducing Obesity Will Require Involvement of All Sectors of Society." *Obesity* 23, no. 2 (February 2015): 255. https://
doi.org/10.1002/oby.20965.

Hill, James O., and John C. Peters, with Bonnie T. Jortberg. *The Step Diet Book: Count
Steps, Not Calories, to Lose Weight and Keep It Off Forever.* New York: Workman,
2004.

Hill, James O., John C. Peters, and Holly R. Wyatt. "Using the Energy Gap to Address
Obesity: A Commentary." *Journal of the American Dietetic Association* 109, no. 11
(November 2009): 1848–53. https://doi.org/10.1016/j.jada.2009.08.007.

Hill, James O., and Frederick L. Trowbridge. "Childhood Obesity: Future Directions
and Research Priorities." *Pediatrics* 101, no. 3, pt. 2 (1998): 570–74. https://doi.org/
10.1542/peds.101.3.570.

Hill, James O., and Holly R. Wyatt, with Christie Aschwanden. *State of Slim: Fix Your
Metabolism and Drop 20 Pounds in 8 Weeks on the Colorado Diet.* New York: Rodale,
2013.

Hill, James O., Holly R. Wyatt, and John C. Peters. "Energy Balance and Obesity." *Circulation* 126, no. 1 (July 3, 2012): 12–32. https://doi.org/10.1161/
CIRCULATIONAHA.111.087213.

———. "The Importance of Energy Balance." *European Endocrinology* 9, no. 2 (August
2013): 111–15. https://doi.org/10.17925/EE.2013.09.02.111.

Hill, James O., Holly Wyatt, Suzanne Phelan, and Rena Wing. "The National Weight
Control Registry: Is It Useful in Helping Deal with Our Obesity Epidemic?" *Journal
of Nutrition Education and Behavior* 37, no. 4 (July–August 2005): 206–10. https://
doi.org/10.1016/s1499-4046(06)60248-0.

Hill, James O., Holly R. Wyatt, George W. Reed, and John C. Peters. "Obesity and the
Environment: Where Do We Go from Here?" *Science* 299, no. 5608 (February 7,
2003): 853–55. https://doi.org/10.1126/science.1079857.

Huang, Yanzhong. *Governing Health in Contemporary China.* London: Routledge, 2013.

———. *Toxic Politics: China's Environmental Health Crisis and Its Challenge to the Chinese State.* Cambridge: Cambridge University Press, 2020.

Ikeda, Joanne, Nancy K. Amy, Paul Ernsberger, Glenn A. Gaesser, Francie M. Berg,
Claudia A. Clark, Ellen S. Parham, and Paula Peters. "The National Weight Control
Registry: A Critique." *Journal of Nutrition Education and Behavior* 37, no. 4 (July–
August 2005): 203–5. https://doi.org/10.1016/S1499-4046(06)60247-9.

ILSI. "2019 ILSI Annual Meeting and Scientific Symposium, Clearwater, Florida."
http://ilsi.org/event/2019-ilsi-annual-meeting/. Accessed February 10, 2019.

———. "Bylaws, Conformed as of January 17, 2015." https://ilsi.org/about/mission/.
Accessed November 15, 2018. Link no longer active.

———. "Chen Chunming, the First President of the Chinese Academy of Preventive
Medicine, Dies at the Age of 93 [May 5, 2018]." *ILSI News* 36, no. 1 (July 18). https://
ilsi.org/ilsi-news-july-2018/. Link no longer active.

———. "Code of Ethics and Organizational Standards of Conduct." March 2009.
https://ilsi.org/wp-content/uploads/2016/05/ILSICodeofEthicsSofC2009.pdf.

———. "An Early Commitment." www.ilsi.org/pages/obesity.aspx. Accessed July 21, 2015. Link no longer active.

———. "Global Network." https://ilsi.org/about/global-network/#:~:text=From%20inception%2C%20ILSI's%20outlook%20has%20been%20international.&text=We%20also%20realize%20health%20decisions,in%20national%20or%20regional%20context. Accessed November 12, 2023.

———. *Global Partnerships for a Healthier World: 2016 ILSI Annual Report*. Washington, DC: ILSI, 2016.

———. "ILSI Response to Globalization and Health." June 6, 2019, edited November 9, 2019. https://ilsi.org/ilsi-response-to-globalization-and-health/.

———. IRS Form 990, 2012, 2013. Available through GuideStar Pro.

———. "Media FAQs." ILSI.org. Accessed June 25, 2015. Link no longer active.

———. "Mission and Operating Principles." https://ilsi.org/about/mission/. Accessed March 11, 2022.

———. "Posters from ILSI Entities around the World" or "Poster Session." https://ilsi.org/event/2014-ilsi-annual-meeting/, https://ilsi.org/event/2015-ilsi-annual-meeting/, and https://ilsi.org/event/2016-ilsi-annual-meeting/. Accessed November 12, 2023.

———. "Setting the Record Straight: ILSI Focal Point in China Response to *BMJ* and *Journal of Public Health Policy*." February 11, 2019. https://ilsi.org/setting-the-record-straight-ilsi-focal-point-in-china-response-to-bmj-and-journal-of-public-health-policy/.

———. "The Value of Annual Meeting Is in the Chance to Connect." *ILSI News*, 31, no. 1 (April 2013). https://ilsi.org/wp-content/uploads/2016/05/April_2013_vfinal.pdf.

ILSI-China. "The 1st Workshop on Restaurant Food and Balanced Diet Convened in Beijing." *ILSI Focal Point in China Newsletter* 25, no. 2 (July–December 2006): 13–15.

———, ed. *2013 Conference on Obesity Control and Prevention in China, Theme: Appropriate Technology and Tools for Weight Control, December 12–13, 2013, Beijing* (in English and Chinese). Beijing: ILSI-China, 2013. (This is a bound volume of conference presentations; editor and publication details are not available; for convenience, I use ILSI-China as the editor.)

———. "2014 Conference on Physical Activity and Health." *ILSI Focal Point in China Newsletter* 41, no. 2 (July–December 2014): 1–4.

———. "2015 Conference on Obesity Control and Prevention in China." *ILSI Focal Point in China Newsletter* 45, no. 2 (July–December 2015): 4–6.

———. "2017 Conference on Obesity Control and Prevention in China." *ILSI Focal Point in China Newsletter* 46 (January–December 2017): 1–4.

———. "Active Public Education on Obesity in China Promoted." *ILSI Focal Point in China Newsletter* 16 (August 2002): 4.

———. "'Beijing Nutrition Week in 2006' Successfully Organized by ILSI FP-China, Carrefour China Foundation for Food Safety, ThinkTank and Food Fortification Office of China CDC." *ILSI Focal Point in China Newsletter* 24, no. 1 (January–June 2006): 1–4.

———. "China Ministry of Health Promulgated the First 'Guidelines for Obesity Prevention and Control of Chinese Adults.'" *ILSI Focal Point in China Newsletter* 18 (July 2003): 1–2.

———. "'[2006] Conference on Obesity and Related Diseases Control in China' Was Successfully Convened in Beijing." *ILSI Focal Point in China Newsletter* 25, no. 2 (July–December 2006): 1–6.

———. "[2011] Conference on Obesity Control and Prevention in China." *ILSI Focal Point in China Newsletter* 35, no. 2 (July–December 2011): 3–4.

———. "'Happy 10' Project Are [*sic*] Expanding in China." *ILSI Focal Point in China Newsletter* 24, no. 1 (January–June 2006): 9–10.

———. "'Happy Ten Minutes' Well Recognized by Students, Parents, School and Government Officials." *ILSI Focal Point in China Newsletter* 22, no. 1 (January–June 2005): 11–12.

———. "Hydration and the Health Symposium." *ILSI Focal Point in China Newsletter* 29, no. 2 (July–December 2008): 2–3.

———. "ILSI Focal Point in China Conducts Supporting Company Meetings." *ILSI Focal Point in China Newsletter* 14 (July 2001): 6.

———. "ILSI Focal Point in China Supports and Participates in the Activities of China Health Forum 2013." *ILSI Focal Point in China Newsletter* 39, no. 2 (July–December 2013): 17–18.

———. "Preface." In *ILSI Focal Point in China: Two Decades of Achievements and Looking to the Future*, 2–3 Beijing: ILSI Focal Point in China, 2013.

———. "ILSI FP-China-Coca-Cola Food Safety and Physical Activity Scholarship Winner Meeting." *ILSI Focal Point in China Newsletter* 38, no. 1 (January–June 2013): 14–15.

———. "Industry Round Table Meeting on 'Strategy on Diet, Physical Activity, and Health' Jointly Organized by MOH and China CDC." *ILSI Focal Point in China Newsletter* 23, no. 2 (July–December 2005): 6–7.

———. "A Joint Meeting of the 'Working Group on Obesity in China' and the 'Working Group on Physical Activity in China' Held in Beijing." *ILSI Focal Point in China Newsletter* 24, no. 1 (January–June 2006): 8–9.

———. "Media Forum on Obesity and Related Diseases Control in China." *ILSI Focal Point in China Newsletter* 25, no. 2 (July–December 2006): 17–18.

———. "People to People International Delegation Visited ILSI Focal Point in China." *ILSI Focal Point in China Newsletter* 25, no. 2 (July–December 2006): 15–17.

———. "Round-Table Meeting on Strategy for NCD Control Convened by ILSI FP-China." *ILSI Focal Point in China Newsletter* 22, no. 1 (January–June 2005): 1–3.

———. "Scientific Advisory Committee of ILSI Focal Point in China Established." *ILSI Focal Point in China Newsletter* 31, no. 2 (July–December 2009): 3–4.

———. "Summary Report—Symposium on the Health Impact of Dietary Fat." *ILSI Focal Point in China Newsletter* 20 (June 2004): 4–5.

———. "Symposium on the Importance of Water as a Nutrient." *ILSI Focal Point in China Newsletter* 24, no. 1 (January–June 2006): 6–8.

———. "Third Asian Conference on Food Safety and Nutrition Held in Beijing." *ISLI Focal Point in China Newsletter* 13 (December 2000): 1–4.

———. "Top Experts Address Seminar on Sweetness." *ISLI Focal Point in China Newsletter* 10 (October 1999): 1–2.

———. "Workshop on Managing Sweetness Held in Beijing." *ILSI Focal Point in China Newsletter* 28, no. 1 (January–June 2008): 1–4.

ILSI-CHP. IRS Form 990, 1999–2004. Available through GuideStar Pro.

ILSI Global. "2013 ILSI Annual Meeting." https://ilsi.org/event/2013-ilsi-annual
-meeting/.

———. "Common Misconceptions, Dr. Alex Malaspina." https://ilsi.org/about/
frequently-asked-questions/. Accessed November 12, 2023.

———. "ILSI Federation Announces Collective, Global Rebranding." May 23,
2022. https://ilsi.org/about/news/ilsi-federation-announces-collective-global
-rebranding/.

———. "Mission and Operating Principles." https://ilsi.org/about/mission/. Accessed
March 10, 2022.

———. "One ILSI." ilsi.org/one/. Accessed March 23, 2022.

ILSI North America. "Bylaws, Conformed as of June 17, 2015." www.ilsi.org/
NorthAmerica. Accessed February 5, 2019. Link no longer active.

———. "Energy Balance 2015 Fact Sheet." www.ilsi.org/NorthAmerica/Pages/EBALC
.aspx. Accessed September 9, 2015.

———. *ILSI North America 2005 Annual Report*. Washington, DC: ILSI North
America, 2006.

———. *ILSI North America 2010–15 Annual Reports*. Washington, DC: ILSI North
America, 2011–16.

———. *ILSI North America 2014–18 Annual Reports*. Washington, DC: ILSI North
America, 2015–19.

———. *ILSI North America 2015 Annual Report*. Washington, DC: ILSI North America,
2016.

Insight. "81 Weight Loss Statistics and Health Benefits in 2023." http://www
.myshortlister.com/insights/weight-loss-statistics. Accessed October 31, 2023.

Institute for the Advancement of Food and Nutrition Sciences (IAFNS, formerly ILSI
North America). "Past Events." https://iafns.org/events/past-events/. Accessed
November 12, 2023.

Internal Revenue Service. "Business Leagues." Updated February 17, 2022. https://
www.irs.gov/charities-non-profits/other-non-profits/business-leagues.

———. "Exemption Requirements—501(c)(3) Organizations." Updated 17, 2022.
https://www.irs.gov/charities-non-profits/charitable-organizations/exemption
-requirements-501c3-organizations.

Internal Revenue Service, District Director. "1985 IRS Tax Code Determination." Letter
to ILSI, May 31, 1985. https://ilsi.org/wp-content/uploads/2016/05/ILSI-IRS-Ltrs
-May-1985.pdf.

Jacobson, Michael F. *Liquid Candy: How Soft Drinks Are Harming Americans*. Washington, DC: Center for Science in the Public Interest, 1998; 2nd ed., 2005. https://
www.cspinet.org/sites/default/files/attachment/liquid_candy_final_w_new
_supplement.pdf.

Jacobson, Michael F., and Kelly D. Brownell. "Small Taxes on Soft Drinks and Snack
Foods to Promote Health." *American Journal of Public Health* 90, no. 6 (June 2000):
854–57. https://doi.org/10.2105/ajph.90.6.854.

Jacobson, Michael F., and Walter Willett. "Letter to the Editor." *New York Times*,
August 13, 2015. https://www.nytimes.com/2015/08/13/opinion/cokes-skewed
-message-on-obesity-drink-coke-exercise-more.html.

Jakicic, John M. Curriculum Vitae. November 2, 2020. https://app.education.pitt
.edu/people/cv/22b53e2e-df85-4bf8-b95c-1da84f4ec1cd.pdf. Accessed March 12,
2022.

James, Jack E. "Caffeine, Health, and Commercial Interests." *Addiction* 89, no. 12 (December 1994): 1595–99. https://doi.org/10.1111/j.1360-0443.1994.tb03760.x.

———. "'Third-Party' Threats to Research Integrity in Public-Private Partnerships." *Addiction* 97, no. 10 (October 2002): 1251–55. https://doi.org/10.1046/j.1360-0443 .2002.00146.x.

Kearns, Cristin E., Laura A. Schmidt, and Stanton A. Glantz. "Sugar Industry and Coronary Heart Disease Research: A Historical Analysis of Internal Industry Documents." *JAMA Internal Medicine* 176, no. 11 (November 2016): 1680–85. https://doi .org/10.1001/jamainternmed.2016.5394.

Kent, Muhtar. "Coca-Cola: We'll Do Better." *Wall Street Journal*, August 19, 2015. https://www.wsj.com/articles/coca-cola-well-do-better-1440024365.

Kibbe, Debra L., Jacqueline Hackett, Melissa Hurley, Allen McFarland, Kathryn Godburn Schubert, Amy Schultz, and Suzanne Harris. "Ten Years of TAKE 10!®: Integrating Physical Activity with Academic Concepts in Elementary School Classrooms." *Preventive Medicine* 52, no. 1 (June 2011): S43–S50. https://doi.org/10 .1016/j.ypmed.2011.01.025.

Klem, Mary L., Rena R. Wing, Maureen T. McGuire, H. M. Seagle, and James O. Hill. "A Descriptive Study of Individuals Successful at Long-Term Maintenance of Substantial Weight Loss." *American Journal of Clinical Nutrition* 66, no. 2 (August 1997): 239–46. https://doi.org/10.1093/ajcn/66.2.239.

Knorr-Cetina, Karin D. *The Manufacture of Knowledge: An Essay on the Constructivist and Contextual Nature of Science*. Oxford, UK: Pergamon, 1981.

Koplan, Jeffery P., and William H. Dietz. "Caloric Imbalance and Public Health Policy." *JAMA* 282, no. 16 (October 27, 1999): 1579–81. https://doi.org/10.1001/jama.282.16 .1579.

Kraus, Charles. "More Than Just a Soft Drink: Coca-Cola and China's Early Reform and Opening." *Diplomatic History* 43, no. 1 (January 2019): 107–29. https://doi.org/10 .1093/dh/dhy060.

Krimsky, Sheldon. *Conflicts of Interest in Science: How Corporate-Funded Academic Research Can Threaten Public Health*. New York: Hot Books, 2019.

Kuczmarski, Robert J., Katherine M. Flegal, Stephen M. Campbell, and Clifford L. Johnson. "Increasing Prevalence of Overweight among US Adults: The National Health and Nutrition Examination Surveys, 1960 to 1991." *JAMA* 272, no. 3 (July 20, 1994): 205–11. https://doi.org/10.1001/jama.1994.03520030047027.

Lampton, David M. *A Relationship Restored: Trends in US-China Educational Exchanges, 1978–84*. Washington, DC: National Academies Press, 1986.

Lauber, Kathrin, Harry Rutter, and Anna B. Gilmore. "Big Food and the World Health Organization: A Qualitative Study of Industry Attempts to Influence Global-Level Non-communicable Disease Policy." *BMJ—Global Health* 6, no. 6 (June 2021). http://dx.doi.org/10.1136/bmjgh-2021-005216.

Li, Duo. "Asia Pacific Clinical Nutrition Society Award for 2018: Professor Chunming Chen." *Asia Pacific Journal of Clinical Nutrition* 27, no. 1 (April 2018): i–ii. https:// apjcn.nhri.org.tw/server/APCNS/2018.pdf.

Liang, Xiaofeng. "Assessment on the Five-Year Implementation of [the] National Healthy Lifestyle Action and Future Direction" (in Chinese). In *2013 Conference on Obesity Control and Prevention in China, Theme: Appropriate Technology and Tools for Weight Control, December 12–13, 2013, Beijing* (in English and Chinese), edited by ILSI-China, 37–44. Beijing: ILSI-China, 2013.

Lipworth, Wendy, and Ian Kerridge. "China's Pharma Scandal and the Ethics of the Global Drug Market." *Conversation*, September 18, 2013. https://theconversation.com/chinas-pharma-scandal-and-the-ethics-of-the-global-drug-market-16424. Accessed November 12, 2023.

Litman, Ethan A., Steven L. Gortmaker, Cara B. Ebbeling, and David S. Ludwig. "Source of Bias in Sugar-Sweetened Beverage Research: A Systematic Review." *Public Health Nutrition* 21, no. 12 (August 2018): 2345–50. https://doi.org/10.1017/S1368980018000575.

Lobosco, Katie. "Med School Returns $1 Million Donation to Coke." *CNN Money*, November 9, 2015. https://money.cnn.com/2015/11/09/news/companies/coca-cola-donation-return/index.html.

Lord, Elizabeth. "China's Eco-dream and the Making of Invisibilities in Rural-Environmental Research." In *Can Science and Technology Save China?*, edited by Susan Greenhalgh and Li Zhang, 115–38. Ithaca, NY: Cornell University Press, 2020.

Ludwig, David S., Karen E. Peterson, and Steven L. Gortmaker. "Relationship between Consumption of Sugar-Sweetened Drinks and Childhood Obesity: A Prospective, Observational Analysis." *Lancet* 357, no. 9255 (February 17, 2001): 505–8. https://doi.org/10.1016/S0140-6736(00)04041-1.

Lynch, Michael, and Steve Woolgar, eds. *Representation in Scientific Practice*. Cambridge, MA: MIT Press, 1990.

Malaspina, Alex. "Celebrating Twenty Years with ILSI Focal Point in China." *ILSI News* 31, no. 2 (August 2013): 1–2. https://ilsi.org/wp-content/uploads/2016/05/ILSI_News_August_vFinal.pdf.

———. "Global Partnerships for a Safer, Healthier World—The Work of the International Life Sciences Institute." *Health Promotion International* 13, no. 3 (September 1998): 187–89. https://doi.org/10.1093/heapro/13.3.187.

———. "Welcome Address: Third Asian Conference on Food Safety and Nutrition, October 3, 2000, Beijing, China." *Biomedical and Environmental Sciences* 13 (2001): vii–xi.

Martin, Crescent B., Kirsten A. Herrick, Neda Sarafrazi, and Cynthia L. Ogden. "Attempts to Lose Weight Among Adults in the United States, 2013–2016." NCHS Data Brief no. 313 (July 2018). https://www.cdc.gov/nchs/data/databriefs/db313.pdf.

Mason, Katherine A. "Divergent Trust and Dissonant Truths in Public Health Science." In *Can Science and Technology Save China?* edited by Susan Greenhalgh and Li Zhang, 95–114. Ithaca, NY: Cornell University Press, 2020.

———. *Infectious Change: Reinventing Chinese Public Health after An Epidemic*. Stanford, CA: Stanford University Press, 2016.

McKenzie, Shelly. *Getting Physical: The Rise of Fitness Culture in America*. Lawrence: University Press of Kansas, 2013.

Mertha, Andrew C. *China's Water Warriors: Citizen Action and Policy Change*. Ithaca, NY: Cornell University Press, 2008.

Mialon, Mélissa, Diego Alejandro Gaitan Charry, Gustavo Cediel, Eric Crosbie, Fernanda Baeza Scagliusi, and Eliana Maria Pérez Tamayo. "'The Architecture of the State Was Transformed in Favor of the Interests of Companies': Corporate Political Activity of the Food Industry in Colombia." *Globalization and Health* 16, no. 1 (October 12, 2020): article 97. https://doi.org/10.1186/s12992-020-00631-x.

Mialon, Mélissa, Camila Corvalan, Gustavo Cediel, Fernanda Baeza Scagliusi, and Marcela Reyes. "Food Industry Political Practices in Chile: 'The Economy Has Always

Been the Main Concern.'" *Globalization and Health* 16, no. 1 (October 27, 2020): article 107. https://doi.org/10.1186/s12992-020-00638-4.

Mialon, Mélissa, Ángela Jaramillo, Patricia Caro, Mauricio Flores, Laura González, Yareni Gutierrez-Gómez, Lina Lay, et al. "Involvement of the Food Industry in Nutrition Conferences in Latin America and the Caribbean." *Public Health Nutrition* 24, no. 6 (April 2021): 1559–65. https://doi.org/10.1017/S1368980020003870.

Michaels, David. *The Triumph of Doubt: Dark Money and the Science of Deception.* Oxford: Oxford University Press, 2020.

Ministry of Health of the People's Republic of China. *Zhongguo chengren chaozhong he feipangzheng yufang kongzhi zhinan* (Guidelines for prevention and control of overweight and obesity in Chinese adults). Beijing: Renmin Weisheng Chubanshe, 2006.

Ministry of Health and Other 15 Departments. "Notice from the Ministry of Health and Other 15 Departments on the Issuance of the 'China Chronic Disease Prevention and Treatment Work Plan (2012–15)'" ("Weishengbu deng 15 bumen guanyu yinfa 'Zhongguo manxingbing fangzhi gongzuo guihua (2012–15) de tongzhi'"). National Health and Disease Control and Prevention Issue [i.e., Document] 34. May 8, 2012. http://www.nhc.gov.cn/cms-search/xxgk/getManuscriptXxgk.htm?id=54755.

Mokdad, A. H., M. K. Serdula, W. H. Dietz, B. A. Bowman, J. S. Marks, and J. P. Koplan, "The Spread of the Obesity Epidemic in the United States, 1991–1998." *JAMA* 282, no. 16 (October 27, 1999): 1519–22, http://doi.org/10.1001/jama.282.16.1519.

Monteiro, Carlos A., Geoffrey Cannon, Renata B. Levy, Jean-Claude Moubarac, Maria L. C. Louzada, Fernanda Rauber, Neha Khandpur, et al. "Ultra-processed Foods: What They Are and How to Identify Them." *Public Health Nutrition* 22, no. 5 (April 2019): 936–41. https://doi.org/10.1017/S1368980018003762, 939.

Moss, Michael. *Hooked: How We Became Addicted to Processed Food.* London: Penguin Random House, 2021.

———. *Salt, Sugar, Fat: How the Food Giants Hooked Us.* New York: Random House, 2013.

Mudd, Michael, and James Hill. "Remarks for ILSI CEO Dinner, Minneapolis, MN, April 8, 1999: Draft April 2, 1999." University of California, San Francisco, Truth Tobacco Industry Documents (1999). https://www.industrydocuments.ucsf.edu/tobacco/docs/#id=gsbf0074.

National Action Office for National Healthy Lifestyle Action. "Quanmin jiankang shenghuo fangshi xingdong jianjie" (Introduction to the National Healthy Lifestyle for All initiative). http://www.jiankang121.cn/ffp.aspx?code=introduce.

Nestle, Marion. *Food Politics: How the Food Industry Influences Nutrition and Health.* 2003. 2nd ed. Berkeley: University of California Press, 2007.

———. *Soda Politics: Taking On Big Soda (and Winning).* Oxford: Oxford University Press, 2015.

———. *Unsavory Truth: How Food Companies Skew the Science of What We Eat.* New York: Basic, 2018.

"News: Coca-Cola Researching Chinese Medicine." October 22, 2007. *Prepared Foods* (newsletter). https://www.preparedfoods.com/articles/106079-news-coca-cola-researching-chinese-medicine.

O'Connor, Anahad. "Coca-Cola Funds Scientists Who Shift Blame for Obesity away from Bad Diets." *New York Times,* August 9, 2015. https://archive.nytimes.com/well

.blogs.nytimes.com/2015/08/09/coca-cola-funds-scientists-who-shift-blame-for
-obesity-away-from-bad-diets/.

———. "Coke's Chief Scientist, Who Orchestrated Obesity Research, Is Leaving." *New York Times*, November 24, 2015. https://archive.nytimes.com/well
.blogs.nytimes.com/2015/11/24/cokes-chief-scientist-who-orchestrated-obesity
-research-is-leaving/#:~:text=Rhona%20S.,the%20Global%20Energy%20Balance
%20Network.

Olcott, Eleanor, Clive Cookson, and Alan Smith. "China's Fake Science Industry: How
'Paper Mills' Threaten Progress." *Financial Times*, March 27, 2023. https://www.ft
.com/content/32440f74-7804-4637-a662-6cdc8f3fba86.

Olinger, David. "CU Expert Who Took Coca-Cola Money Steps Down." *Denver Post*,
March 23, 2016, updated June 6, 2016. https://www.denverpost.com/2016/03/23/
cu-nutrition-expert-who-took-coca-cola-money-steps-down/.

Oreskes, Naomi, and Erik M. Conway. *Merchants of Doubt: How a Handful of Scientists
Obscured the Truth on Issues from Tobacco Smoke to Global Warming*. New York:
Bloomsbury, 2010.

Pan, An, and Frank B. Hu. "Effects of Carbohydrates on Satiety: Differences between
Liquid and Solid Food." *Current Opinion in Clinical Nutrition and Metabolic Care* 14,
no. 4 (July 2011): 385–90. https://doi.org/10.1097/MCO.0b013e328346df36.

Partnership to Promote Healthy Eating and Active Living. IRS Form 990, 2001. Available through GuideStar Pro.

———. "The Partnership to Promote Healthy Eating and Active Living, Inc." Power-
Point presentation, 21 slides. Circa 2004. http://slideplayer.com/slide/4712383/.
Accessed March 16, 2022.

Pate, R. R., M. Pratt, S. N. Blair, W. L. Haskell, C. A. Macera, C. Bouchard, D. Buchner,
W. Ettinger, G. W. Heath, A. C. King, et al. "Physical Activity and Public Health:
A Recommendation from the Centers for Disease Control and Prevention and the
American College of Sports Medicine." *JAMA* 273, no. 5 (February 1, 1995): 402–7.
https://doi.org/10.1001/jama.273.5.402.

Peeke, Pamela M. "Foreword." In *The Step Diet Book: Count Steps, Not Calories, to Lose
Weight and Keep It Off Forever*, by James O. Hill. and John C. Peters, ix–xi. New
York: Workman, 2004.

Pendergrast, Mark. *For God, Country and Coca-Cola: The Definitive History of the Great
American Soft Drink and the Company That Makes It*. 3rd ed. New York: Basic, 2013.

Peregrin, Tony. "Take 10! Classroom-Based Program Fights Obesity by Getting Kids
out of Their Seats." *Journal of the American Dietetic Association* 101, no. 120 (December 2001):1049. https://doi.org/10.1016/S0002-8223(01)00338-8.

Perlroth, Nicole. "Spyware's Odd Targets: Backers of Mexico's Soda Tax." *New York
Times*, February 11, 2017. https://www.nytimes.com/2017/02/11/technology/hack
-mexico-soda-tax-advocates.html.

Peters, John C., and James O. Hill. "Conspiracy or Good Education? Response to Paul
Thacker, 'Coca-Cola's Secret Influence on Medical and Science Journalists.'" *BMJ*
357 (April 13, 2017): j1638. https://www.bmj.com/content/357/bmj.j1638/rr-2.

Peters, John C., Rachel C. Lindstrom, and James O. Hill. "Stepping Up across America:
The Small Changes Approach." *Childhood Obesity* 8, no. 1 (February 2012): 76–78.
https://doi.org/10.1089/chi.2011.0108.

Petryna, Adriana. *When Experiments Travel: Clinical Trials and the Global Search for
Human Subjects*. Princeton, NJ: Princeton University Press, 2009.

Petrzela, Natalia Mehlman. *Fit Nation: The Gains and Pains of America's Exercise Obsession.* Chicago: University of Chicago Press, 2022.

Pinch, Trevor J., and Wiebe E. Bijker. "The Social Construction of Facts and Artifacts." In *The Social Construction of Technological Systems,* edited by Wiebe E. Bijker, Thomas P. Hughes, and Trevor J. Pinch, 17–50. Cambridge, MA: MIT Press, 1987.

Pi-Sunyer, F. Xavier. "The Fattening of America." *JAMA* 272, no. 3 (July 20, 1994): 238–39.

Pomeranz, Jennifer L., and Kelly D. Brownell. "Portion Sizes and Beyond— Government's Legal Authority to Regulate Food-Industry Practices." *New England Journal of Medicine* 367, no. 15 (October 11, 2012): 1383–85. https://doi.org/10.1056/ NEJMp1208167.

Pontzer, Herman. *Burn.* New York: Avery, 2021.

Popkin, Barry M. "The Nutrition Transition in Low-Income Countries: An Emerging Crisis." *Nutrition Reviews* 52, no. 9 (September 1994): 285–98. https://doi.org/10 .1111/j.1753-4887.1994.tb01460.x.

Popkin, Barry M., Simon Barquera, Camila Corvalan, Karen J. Hofman, Carlos Monteiro, Shu Wen Ng, Elizabeth C. Swart, and Lindsey Smith Taillie. "Towards Unified and Impactful Policies to Reduce Ultra-processed Food Consumption and Promote Healthier Eating." *Lancet Diabetes and Endocrinology* 9, no. 7 (July 2021): 462–70. https://doi.org/10.1016/S2213-8587(21)00078-4.

Popkin, Barry M., and Penny Gordon-Larsen. "The Nutrition Transition: Worldwide Obesity Dynamics and Their Determinants." *International Journal of Obesity and Related Metabolic Disorders* 28 (November 2004): S2–S9. https://doi.org/10.1038/ sj.ijo.0802804.

Powys, Betsan. "The Trouble with Sugar." Transcript. *BBC Panorama,* October 10, 2004. http://news.bbc.co.uk/nol/shared/spl/hi/programmes/panorama/ transcripts/thetroublewithsugar.txt.

Proctor, Robert N. *Golden Holocaust: Origins of the Cigarette Catastrophe and the Case for Abolition.* Berkeley: University of California Press, 2011.

Pulley, Brett. "Bloomberg: Coca-Cola Severs Long-Term Ties with Pro–Sugar Industry Group." Bloomberg.com, January 13, 2021. https://www.corporateaccountability .org/media/bloomberg-coca-cola-severs-longtime-ties-with-pro-sugar -industry-group/#:~:text=Food-,Bloomberg%3A%20Coca%2DCola%20Severs %20Longtime%20Ties,With%20Pro%2DSugar%20Industry%20Group&text =Coca%2DCola%20Co.%20has%20ended,pro%2Dsugar%20research%20and %20policies.

Qin, Amy. "Fraud Scandals Sap China's Dream of Becoming a Science Superpower." *New York Times,* October 13, 2017. https://www.nytimes.com/2017/10/13/world/ asia/china-science-fraud-scandals.html.

Qiu, Jane. "Scientists Caught in China's Anti-corruption Sweep." *Nature,* October 16, 2014. https://doi.org/10.1038/nature.2014.16152.

Rodearmel, Susan J., Holly R. Wyatt, Nanette Stroebele, Sheila M. Smith, Lorraine G. Ogden, and James O. Hill. "Small Changes in Dietary Sugar and Physical Activity as an Approach to Preventing Excessive Weight Gain: The America on the Move Family Study." *Pediatrics* 120, no. 4 (October 2007): e869–e879. https://doi.org/10 .1542/peds.2006-2927.

Rosenberg, Irwin H., and Michael R. Taylor. "Foreword." *Nutrition Reviews* 56, no. 5 (May 1998): S1–S2. https://doi.org/10.1111/j.1753-4887.1998.tb01727.x.

Ross, Robert, Amy E. Latimer-Cheung, Andrew G. Day, Andrea M. Brennan, and James O. Hill. "A Small Change Approach to Prevent Long-Term Weight Gain in Adults with Overweight and Obesity: A Randomized Controlled Trial." *Canadian Medical Association Journal (CMAJ)* 194, no. 9 (March 7, 2022): E324–31. https://doi.org/10.1503/cmaj.211041.

Rowe, Sylvia, Nick Alexander, Fergus M. Clydesdale, Rhona S. Applebaum, Stephanie Atkinson, Richard M. Black, Johanna T. Dwyer, et al. "Funding Food Science and Nutrition Research: Financial Conflicts and Scientific Integrity." *American Journal of Clinical Nutrition* 89, no. 5 (May 2009): 1285–91. https://doi.org/10.3945/ajcn.2009.27604.

Sacks, Gary, Boyd A. Swinburn, Adrian J. Cameron, and Gary Ruskin. "How Food Companies Influence Evidence and Opinion—Straight from the Horse's Mouth." *Critical Public Health* 28, no. 2 (2018): 253–56. https://doi.org/10.1080/09581596.2017.1371844.

Saich, Tony. *From Rebel to Ruler: One Hundred Years of the Chinese Communist Party.* Cambridge, MA: Harvard University Press, 2021.

Sallis, R. E. "Exercise Is Medicine and Physicians Need to Prescribe It!" *British Journal of Sports Medicine* 43, no. 1 (January 2009): 3–4.

Schüll, Natasha D. "The Data-Based Self: Self-Quantification and the Data-Driven (Good) Life." *Social Research International Quarterly* 86, no. 4 (Winter 2019): 909–30.

Serodio, Paulo M., Martin McKee, and David Stuckler. "Coca-Cola—a Model of Transparency in Research Partnerships? A Network Analysis of Coca-Cola's Research Funding (2008–2016)." *Public Health Nutrition* 21, no. 9 (June 2018): 1594–607. https://doi.org/10.1017/S136898001700307X.

Shen, Xiaobai, and Robin Williams. "A Critique of China's Utilitarian View of Science and Technology." *Science, Technology, and Society: An International Journal Devoted to the Developing World* 10, no. 2 (September 2005): 197–223. https://doi.org/10.1177/097172180501000202.

Short, Donald. "When Science Met the Consumer: The Role of Industry." *American Journal of Clinical Nutrition* 82, no. 1 (supplement) (July 2005): 256S–258S. https://doi.org/10.1093/ajcn/82.1.256S.

Shum, Desmond. *Red Roulette: An Insider's Story of Wealth, Power, Corruption, and Vengeance in Today's China.* New York: Scribner, 2021.

Simon, Denis Fred, and Merle Goldman, eds. *Science and Technology in Post-Mao China.* Cambridge, MA: Harvard University Asia Center, 1989.

Simon, Michele. *Appetite for Profit: How the Food Industry Undermines Our Health and How to Fight Back.* New York: Nation Books, 2006.

———. *Best Public Relations Money Can Buy: A Guide to Food Industry Front Groups.* Center for Food Safety, May 2013. https://www.foodsafetynews.com/2013/05/best-public-relations-money-can-buy-a-guide-to-food-industry-front-groups/.

Sismondo, Sergio. "Epistemic Corruption, the Pharmaceutical Industry, and the Body of Medical Science." *Frontiers in Research Metrics and Analytics* 6, article 614013 (March 8, 2021): 1–5. https://doi.org/10.3389/frma.2021.614013.

———. *Ghost-Managed Medicine: Big Pharma's Invisible Hands.* Manchester, UK: Mattering, 2018.

Smith Edge, Marianne. "Understanding the Consumer's Knowledge of Energy Balance: Are We Connecting?" Part of a webinar sponsored by ILSI North America,

Academy of Nutrition and Dietetics, ACSM, and IFIC, "Energy Balance at the Crossroads: Translating the Science into Action," Washington, DC, August 28, 2014. http://ilsina.org/event/energy-balance-at-the-crossroads-translating-the-science -into-action/.

Song, Priscilla. *Biomedical Odysseys: Fetal Cell Experiments from Cyberspace to China.* Princeton, NJ: Princeton University Press, 2017.

Steele, Sarah, Gary Ruskin, Lejla Sarcevic, Martin McKee, and David Stuckler. "Are Industry-Funded Charities Promoting 'Advocacy-Led Studies' or 'Evidence-Based Science'? A Case Study of the International Life Sciences Institute." *Globalization and Health* 15, article 36 (2019). https://doi.org/10.1186/s12992-019-0478-6.

Stuckler, David, Gary Ruskin, and Martin McKee. "Complexity and Conflicts of Interest Statements: A Case-Study of Emails Exchanged between Coca-Cola and the Principal Investigators of the International Study of Childhood Obesity, Lifestyle, and the Environment (ISCOLE)." *Journal of Public Health Policy* 39, no. 1 (February 2018): 49–56. https://doi.org/10.1057/s41271-017-0095-7.

Surgeon General of the United States. *Physical Activity and Health: A Report of the Surgeon General.* Washington, DC: National Center for Chronic Disease Prevention and Health Promotion, 1996.

———. *The Surgeon General's Call to Action to Prevent and Decrease Overweight and Obesity.* Rockville, MD: Office of the Surgeon General, 2001.

Suttmeier, Richard P. *Science, Technology, and China's Drive for Modernization.* Stanford, CA: Hoover Institution Press, 1980.

Swift, Damon L., Neil M. Johannsen, Carl J. Lavie, Conrad P. Earnest, and Timothy S. Church. "The Role of Exercise and Physical Activity in Weight Loss and Maintenance." *Progress in Cardiovascular Diseases* 56, no. 4 (January–February 2014): 441–47. https://doi.org/10.1016/j.pcad.2013.09.012.

Swift, Damon L., Joshua E. McGee, Conrad P. Earnest, Erica Carlisle, Madison Nygard, and Neil Johannsen. "The Effects of Exercise and Physical Activity on Weight Loss and Maintenance." *Progress in Cardiovascular Diseases* 61, no. 2 (July–August 2018): 206–13. https://doi.org/10.1016/j.pcad.2018.07.014.

Swinburn, Boyd A., Vivica I. Kraak, Steven Allender, Vincent J. Atkins, Phillip I. Baker, Jessica R. Bogard, Hannah Brinsden, et al. "The Global Syndemic of Obesity, Undernutrition, and Climate Change: The Lancet Commission Report." *Lancet* 393, no. 10173 (February 23, 2019): 791–846. https://doi.org/10.1016/S0140-6736(18)32822-8.

Tan, Xiaodong, Shibo Kong, and Haiyan Shao. "New Strategies to Improve the Health of Chinese People by 2030." *Australian Journal of Primary Health* 23, no. 4 (2017): 307–8. https://doi.org/10.1071/PY16146.

Tancredi, Doris, and John A. Milner. "Energy Balance: A New Paradigm." April 21, 2012. ilsina.org/event/energy-balance-a-new-paradigm/.

Tanner, Murray Scot. "Changing Windows on a Changing China: The Evolving 'Think Tank' System and the Case of the Public Security Sector." *China Quarterly* 171 (September 2002): 559–74. https://doi.org/10.1017/S0009443902000359.

Tatlow, Didi Kirsten. "A Scientific Ethical Divide between China and the West." *New York Times*, June 29, 2015. https://www.nytimes.com/2015/06/30/science/a -scientific-ethical-divide-between-china-and-west.html.

Taubes, Gary. *The Case against Sugar.* New York: Anchor, 2016.

Taylor, Allyn L., and Michael F. Jacobson. *Carbonating the World: The Marketing and*

Health Impact of Sugar Drinks in Low- and Middle-Income Countries. Washington, DC: Center for Science in the Public Interest, 2016.

Thacker, Paul. "Coca-Cola's Secret Influence on Medical and Science Journalists." *BMJ* 357 (April 5, 2017): j1638. https://doi.org/10.1136/bmj.j1638.

ThinkTank Research Center for Health Development (Xintan jiankang fazhan yanjiu zhongxin). "ThinkTank Research Center for Health Development." Website maintained by private research organization founded by three former health officials and one biotech entrepreneur, and licensed by the Ministries of Health, Science and Technology, and Civil Affairs. https://www.healthtt.org.cn/index.html. Accessed November 2023.

Toxicology Forum. "History of the Toxicology Forum." https://toxforum.site-ym.com/page/ToxForumHistory. Accessed November 12, 2023.

US Department of Agriculture and US Department of Health and Human Services. *Nutrition and Your Health: Dietary Guidelines for Americans.* 4th ed. Home and Garden Bulletin 2321995. Washington, DC: USDA, December 1995. https://www.dietaryguidelines.gov/sites/default/files/2019-05/1995%20Dietary%20Guidelines%20for%20Americans.pdf.

US Department of Health and Human Services. *2008 Physical Activity Guidelines for Americans.* Publication U0036. Washington, DC: Office of Disease Prevention and Health Promotion, October 2008.

Van Tulleken, Chris. *Ultra-processed People: The Science Behind Food That Isn't Food.* New York: Norton, 2023.

Vidry, Stéphane. Letter from ILSI Global Director of Operations to Jonathan Oberlander, Editor, *Journal of Health Politics, Policy, and Law.* Circa early October 2020. Provided to author by Duke University Press.

Vogel, Ezra F. *Deng Xiaoping and the Transformation of China.* Cambridge, MA: Harvard University Press, 2011.

Wang, Yeu-farn. *China's Science and Technology Policy: 1949–1989.* Aldershot, UK: Avebury, 1993.

Wang, Youfa, Li Zhao, Liwang Gao, An Pan, and Hong Xue. "Health Policy and Public Health Implications of Obesity in China. Obesity in China 3." *Lancet Diabetes and Endocrinology* 9, no. 7 (July 2021): 446–61. https://doi.org/10.1016/S2213-8587(21)00118-2.

WHO (World Health Organization). *Fiscal Policies for Diet and Prevention of Noncommunicable Diseases (NCDs),* Technical Meeting Report, May 5–6, 2015. Geneva: WHO, 2015. https://www.who.int/docs/default-source/obesity/fiscal-policies-for-diet-and-the-prevention-of-noncommunicable-diseases-0.pdf?sfvrsn=84ee20c_2.

———. *Global Strategy on Diet, Physical Activity and Health.* Geneva: WHO, 2004. https://www.who.int/publications/i/item/9241592222.

———. "The Global Health Observatory, Noncommunicable Diseases." https://www.who.int/data/gho/data/themes/noncommunicable-diseases.

———. "Obesity and Overweight Fact Sheet." June 9, 2021. https://www.who.int/news-room/fact-sheets/detail/obesity-and-overweight#:~:text=Worldwide%20obesity%20has%20nearly%20tripled,%2C%20and%2013%25%20were%20obese.

———. *Obesity: Preventing and Managing the Global Epidemic; Report of a WHO Consultation.* Geneva: WHO, 2000. https://apps.who.int/iris/handle/10665/42330.

———. *Report of the Commission on Ending Childhood Obesity.* Geneva: WHO, 2016.

https://apps.who.int/iris/bitstream/handle/10665/204176/9789241510066_eng
.pdf.

———. "WHO Calls on Countries to Reduce Sugars Intake Among Adults and Children," March 4, 2015. https://www.who.int/news/item/04-03-2015-who-calls-on
-countries-to-reduce-sugars-intake-among-adults-and-children#:~:text=Reducing
%20free%20sugars%20intake%20to,FAO%20Expert%20Consultation%20in
%202002.

WHO Health Promotion Department, "Healthy China." https://www.who.int/teams/
health-promotion/enhanced-wellbeing/ninth-global-conference/healthy-china.
Accessed November 2023.

Wing, Rena R., and Suzanne Phelan. "Long-Term Weight Loss Maintenance." American Journal of Clinical Nutrition 82, no. 1, supplement (July 2005): 222S–225S.
https://doi.org/10.1093/ajcn/82.1.222S.

Wong, David. "Chinese Ethics." In *The Stanford Encyclopedia of Philosophy* (Summer 2021 ed.), edited by Edward N. Zalta. https://plato.stanford.edu/archives/
sum2021/entries/ethics-chinese/.

Xiong-fei, Pan, Limin Wang, and An Pan. "Epidemiology and Determinants of Obesity
in China. Obesity in China 1." *Lancet Diabetes and Endocrinology* 9, no. 6 (June
2021): 373–92. https://doi.org/10.1016/S2213-8587(21)00045-0.

Xu, Jincheng, and Can Gao. "Physical Activity Guidelines for Chinese Children and
Adolescents: The Next Essential Step." *Journal of Sport and Health Science* 7, no. 1
(January 2018): 120–22. https://doi.org/10.1016/j.jshs.2017.07.001.

Xue, Lan, Xufeng Zhu, and Wanqu Han. "Embracing Scientific Decision Making: The
Rise of Think Tank Policies in China." *Pacific Affairs* 91, no. 1 (March 2018): 49–71.
https://doi.org/10.5509/201891149.

Yan, Jiaqi. "The Nature of Chinese Authoritarianism." In *Decision-Making in Deng's
China*, edited by Carol Lee Hamrin and Suisheng Zhao, 3–14. Armonk, NY: M. E.
Sharpe, 1995.

Yan, Yunxiang. "McDonald's in Beijing: The Localization of Americana." In *Golden
Arches East: McDonald's in East Asia*, edited by James L. Watson, 39–76. Stanford,
CA: Stanford University Press, 1997.

Yang, Mayfair Mei-hui. *Gifts, Favors, and Banquets: The Art of Social Relationships in
China*. Ithaca, NY: Cornell University Press, 1994.

Zeng, Qiang, Naishi Li, Xiong-Fei Pan, Lulu Chen, and An Pan. "Clinical Management and Treatment of Obesity in China. Obesity in China 2" *Lancet Diabetes and
Endocrinology* 9, no. 6 (June 2021): 393–405. https://doi.org/10.1016/S2213-8587
(21)00047-4.

Zhai, Fengying, Dawei Fu, Shufa Du, Keyou Ge, Chunming Chen, and Barry M. Popkin. "What Is China Doing in Policy-Making to Push Back the Negative Aspects of
the Nutrition Transition?" *Public Health Nutrition* 5, no. 1A (February 2002): 269–
73. https://doi.org/10.1079/PHN2001303.

Zhao, Wenhua. "EIM Globe and in China: Progress and Future Direction." In *2013
Conference on Obesity Control and Prevention in China*, edited by ILSI-China, 158–61.
Beijing: ILSI-China, 2103.

Zhu, Ling. "Nutritionist Chen Chunming's Health Concepts and Actions" (in Chinese),
compiled from two emails, "Walking with 88-Year-Old Nutritionist," November 20,
2013, and "Visiting Nutritionist Teacher Chen Chunming," April 13, 2015. https://m
.sohu.com/n/532864126/.

Zhu, Weimo. "If You Are Physically Fit, You Will Live a Longer and Healthier Life: An Interview with Dr. Steven N. Blair." *Journal of Sport and Health Science* 8, no. 6 (November 2019): 524–26. https://doi.org/10.1016/j.jshs.2019.09.006.

Zhu, Xufeng. "Government Advisors or Public Advocates? Roles of Think Tanks in China from the Perspective of Regional Variations." *China Quarterly* 207 (September 2011): 668–86. https://doi.org/10.1017/S0305741011000701.

———. *The Rise of Think Tanks in China*. Abingdon, Oxon, UK: Routledge, 2013.

Index

front groups, 10, 11, 13, 20, 42, 122, 260
funding effect, 33, 225, 292n13

GEBN. *See* Global Energy Balance Network (GEBN)
General Foods (company), 44
General Mills (company), 61, 76, 83, 104
"Get Slim with Higher Taxes" (Brownell), 30
Getting Physical (McKenzie), 255
GlaxoSmithKline (GSK), 222
Global Energy Balance Network (GEBN): and Applebaum, 102, 111–16, 118, 121–22, 130, 133–34, 136; and Blair, 102, 111–12, 114–24, 128, 131, 135–36; dissolution of, 103, 128–29; establishment of, 102–3, 111–17; exercise-first and energy balance stances of, 116–21; and Freedhoff, 123–24, 128; and Hill, 102, 111–12, 114, 116–21, 123–25, 128, 131–35; and ILSI, 127, 135–36; and Malaspina, 124–26, 136; and *New York Times* exposé, 9, 123–26, 128–35; and one-company funding model, 121–26; and product-defense science, 113–16, 121–23, 127–29; scientific disagreements within, 117–21; transparency, bias, and ethics of, 103, 113, 115, 122–23, 130–34
global hierarchy of science, 173–75, 241
globalization, 140, 166, 248
Global South, 13, 139, 141, 173, 241
Global Strategy on Diet, Physical Activity and Health (WHO), 166–67, 179, 310n6
Gómez, Eduardo J., 264–65
Gradient (company), 10
Greene, Jeremy A., 17
Greenhalgh, Susan: companies' responses to, 261–62; investigative methods undertaken by, 14–20, 22, 24–26. See also *Soda Science* research methods
guanxi, 145, 196, 198, 200–201, 204, 208, 228, 238, 252. *See also* Chen Chunming (C. M. Chen)
Guidant (company), 72
Guidelines for the Prevention and Control of Overweight and Obesity among Chinese Adults (2003), 163f, 205–6

guiding principles of ILSI ethics, 83, 97

Hall, Kevin, 6
Hand, Gregory, 89t, 90, 93, 106, 110–11, 114, 116, 118–20, 128, 131, 134–36, 185
Happy 10 Minutes, 176t, 189–90, 207–8
Harris, Suzanne, 41, 125, 142, 154, 161, 197, 285
Hays, Ed, 125
Hazda energy balance study, 257–58
healthism, 61–63
Healthy China 2030 (plan), 25, 242, 244, 245t, 246, 249, 251
healthy lifestyle strategies. *See* energy balance; exercise-first solution to obesity epidemic
Heber, David, 184, 187
Heinz (company), 31, 44, 148
Herbalife (company), 184
Hershey (company), 31, 104
Hill, James O., 32–33, 46, 179; and AOM, 68–69, 72; in China, 170, 172, 175, 177, 179–81, 185–86, 207–8, 211, 214, 246; denials of conflicts of interest, 33–35, 47, 65–66, 69–70, 96–97, 128, 131, 134, 220; and diet tokenism, 52, 185; and EBAL, 102–6, 109; on energy balance and energy gaps, 49–51, 57–60, 64, 104; and exercise-first solution, 38–39, 49, 52–55, 60–64, 106, 211, 246, 256; and fitness culture, 56–57; as health entrepreneur, 56, 59; and ILSI, ILSI-China, ILSI-NA, and ILSI-CHP, 32, 35, 37–38, 46, 49, 68, 78, 103, 134–35, 187; and NWCR, 32, 39, 54, 65–67, 91; and perks and incentives of corporate science, 34, 59, 235; and quantified selves, 60–64, 185; as quasi-corporate scientist, 47, 52–53, 69, 87, 96, 98–99, 112, 185; and rhetoric of science, 51–52, 64–68; and *The Step Diet Book*, 56–60, 63, 256–57. *See also* Energy Balance and Active Lifestyle (EBAL) committee; Global Energy Balance Network (GEBN); soda-defense science (China); soda-defense science (United States); Team Coke

International Life Sciences Institute.
See ILSI (International Life Sciences
Institute)
International Obesity Task Force (IOTF),
280, 284
International Study of Childhood Obe-
sity, Lifestyle and the Environment
(ISCOLE), 81, 89t
Isdell, Neville, 77
"Is Sitting a Lethal Activity?," 100

Jacobson, Michael F., 35, 115, 128, 134
Jakicic, John, 105, 136, 300n7
James, Philip (W. P. T.), 164, 202, 207
Jing Jun, 278
Journal of Health Politics, Policy and Law,
259–60
Journal of Public Health Policy, 259, 261
*Journal of the American Medical Associa-
tion* (*JAMA*), 30, 53, 90, 158
junk food. *See* ultraprocessed foods
Junk Food Politics (Gómez), 264
junk science, 10–12, 18. *See also* corporate
science
*Junk Science and the American Criminal
Justice System* (Fabricant), 18

Katzmarzyk, Peter, 89t, 90
Kellogg (company), 31, 33
Kent, Muhtar, 129–30, 143, 182
KFC, 157, 194–95, 207
Knowles, Michael E., 44
Kraft Foods (company), 12, 31, 33, 37, 41,
44, 52, 61, 76, 83, 104–5
Kraus, Charles, 143
Kumanyika, Shiriki, 284

Lancet, 116, 249–51
Lancet Commission on Obesity, 6
Lavie, Carl (Chip), 89t, 90, 96
leader visits to ILSI branches, 154
Liquid Candy (Jacobson), 35
Li Yanping, 285
Ludwig, David S., 75

Macdonald, Ian, 113–14, 133
Malaspina, Alex, 44; and Applebaum,
78; and Blair, 112; and Chen Chun-

ming, 144, 150, 158–59, 161, 189–90,
197, 202; and childhood obesity,
48–49, 54–55; Coca-Cola, product
defense of, 235; and conclusion-first
science, 48; and exercise-first solution, 39, 48, 54, 161; and GEBN, 124–
26; and Happy 10 Minutes, 189–90,
208; and Hill, 112, 125, 161; and ILSI,
31, 44, 112, 125, 142, 150, 154, 262; and
ILSI-China, 136, 140, 142–44, 145,
147–48, 150, 161, 197, 235; and ILSI
CHP, 48–49, 161, 189; obesity, focus
on, 78, 159, 161; and Peters, 124–25;
and Take 10!, 54–55, 161; and WHO,
235. *See also* Coca-Cola; ILSI (International Life Sciences Institute);
ILSI-China; ILSI-Global; ILSI North
America (ILSI-NA)
Mann, Jim, 152
manufacturing doubt, 11
Mao Zedong, 14, 146–47, 199, 207
marketization, 21, 166, 170, 195, 224–25,
237, 248
Mars Snackfoods US (company), 31, 83,
104, 113, 148, 170, 191, 194–95, 260
Mayo Clinic, 72
McCormick Science Institute Research
Award, 135
McDonald's, 31, 76, 143, 157, 256
McKenzie, Shelly, 255, 258
mercenary science, 11, 13, 34, 260
Michaels, David, 9–10, 18, 20, 23, 134
Ministry of Health (MOH), 146, 153,
166, 171, 176t, 188, 191, 193, 201,
203–8, 214–15, 217–18, 286. *See also*
ILSI-China
Mondelēz International (company), 104
Moss, Michael, 6, 40
Mudd, Michael, 37–38, 40–41

National Academy of Medicine (United
States), 250
National Collegiate Athletic Association
(NCAA), 123
National Healthy Lifestyle for All Action
(China), 206–8, 212f
National Institutes of Health (NIH), 33,
65, 133, 146